CONTEMPORARY MATHEMATICS

Mathematics and General Relativity

Proceedings of a Summer Research Conference held June 22–28, 1986

AMERICAN MATHEMATICAL SOCIETY

VOLUME 71

Titles in This Series

Volume

1 **Markov random fields and their applications**, Ross Kindermann and J. Laurie Snell

2 **Proceedings of the conference on integration, topology, and geometry in linear spaces**, William H. Graves, Editor

3 **The closed graph and P-closed graph properties in general topology**, T. R. Hamlett and L. L. Herrington

4 **Problems of elastic stability and vibrations**, Vadim Komkov, Editor

5 **Rational constructions of modules for simple Lie algebras**, George B. Seligman

6 **Umbral calculus and Hopf algebras**, Robert Morris, Editor

7 **Complex contour integral representation of cardinal spline functions**, Walter Schempp

8 **Ordered fields and real algebraic geometry**, D. W. Dubois and T. Recio, Editors

9 **Papers in algebra, analysis and statistics**, R. Lidl, Editor

10 **Operator algebras and K-theory**, Ronald G. Douglas and Claude Schochet, Editors

11 **Plane ellipticity and related problems**, Robert P. Gilbert, Editor

12 **Symposium on algebraic topology in honor of José Adem**, Samuel Gitler, Editor

13 **Algebraists' homage: Papers in ring theory and related topics**, S. A. Amitsur, D. J. Saltman, and G. B. Seligman, Editors

14 **Lectures on Nielsen fixed point theory**, Boju Jiang

15 **Advanced analytic number theory. Part I: Ramification theoretic methods**, Carlos J. Moreno

16 **Complex representations of GL(2, K) for finite fields K**, Ilya Piatetski-Shapiro

17 **Nonlinear partial differential equations**, Joel A. Smoller, Editor

18 **Fixed points and nonexpansive mappings**, Robert C. Sine, Editor

19 **Proceedings of the Northwestern homotopy theory conference**, Haynes R. Miller and Stewart B. Priddy, Editors

20 **Low dimensional topology**, Samuel J. Lomonaco, Jr., Editor

21 **Topological methods in nonlinear functional analysis**, S. P. Singh, S. Thomeier, and B. Watson, Editors

22 **Factorizations of $b^n \pm 1$, b = 2, 3, 5, 6, 7, 10, 11, 12 up to high powers**, John Brillhart, D. H. Lehmer, J. L. Selfridge, Bryant Tuckerman, and S. S. Wagstaff, Jr.

23 **Chapter 9 of Ramanujan's second notebook—Infinite series identities, transformations, and evaluations**, Bruce C. Berndt and Padmini T. Joshi

24 **Central extensions, Galois groups, and ideal class groups of number fields**, A. Fröhlich

25 **Value distribution theory and its applications**, Chung-Chun Yang, Editor

26 **Conference in modern analysis and probability**, Richard Beals, Anatole Beck, Alexandra Bellow, and Arshag Hajian, Editors

27 **Microlocal analysis**, M. Salah Baouendi, Richard Beals, and Linda Preiss Rothschild, Editors

28 **Fluids and plasmas: geometry and dynamics**, Jerrold E. Marsden, Editor

29 **Automated theorem proving**, W. W. Bledsoe and Donald Loveland, Editors

30 **Mathematical applications of category theory**, J. W. Gray, Editor

31 **Axiomatic set theory**, James E. Baumgartner, Donald A. Martin, and Saharon Shelah, Editors

32 **Proceedings of the conference on Banach algebras and several complex variables**, F. Greenleaf and D. Gulick, Editors

33 **Contributions to group theory**, Kenneth I. Appel, John G. Ratcliffe, and Paul E. Schupp, Editors

34 **Combinatorics and algebra**, Curtis Greene, Editor

35 **Four-manifold theory**, Cameron Gordon and Robion Kirby, Editors

36 **Group actions on manifolds**, Reinhard Schultz, Editor

37 **Conference on algebraic topology in honor of Peter Hilton**, Renzo Piccinini and Denis Sjerve, Editors

Titles in This Series

Volume

CONTEMPORARY MATHEMATICS

Volume 71

Mathematics and General Relativity

Proceedings of the AMS-IMS-SIAM Joint Summer Research Conference held June 22—28, 1986 with support from the National Science Foundation

James A. Isenberg, Editor

AMERICAN MATHEMATICAL SOCIETY

Providence • Rhode Island

The AMS-IMS-SIAM Joint Summer Research Conference in the Mathematical Sciences on Mathematics in General Relativity was held at the University of California, Santa Cruz on June 22–28, 1986, with support from the National Science Foundation, Grant DMS-8415201.

1980 *Mathematics Subject Classification*. (1985 *Revision*). Primary 83C, 35J, 35K, 35L, 53B, 53C, 32-XX.

Library of Congress Cataloging-in-Publication Data

AMS-IMS-SIAM Joint Summer Research Conference in the Mathematical Sciences on Mathematics in General Relativity (1986 : University of California, Santa Cruz)
Mathematics and general relativity : proceedings of the AMS-IMS-SIAM joint summer research conference held June 22-28, 1986 with support from the National Science Foundation / James A. Isenberg, editor.
p. cm. -- (Contemporary mathematics, ISSN 0271-4132 ; v. 71)
Includes bibliographies.
ISBN 0-8218-5079-2
1. General relativity (Physics)--Mathematics--Congresses.
2. Mathematical physics--Congresses. I. Isenberg, James A.
II. Title. III. Series.
QC173.6.A47 1986
530.1'1--dc19 88-9685
CIP

Printed in the United States of America.
This volume was printed directly from author-prepared copy.
The paper used in this book is acid-free and falls within the guidelines established to ensure permanence and durability.

CONTENTS

PREFACE

General relativity is an exquisite marriage joining the mathematics of differential geometry to the physics of gravitation. Since its beginnings over seventy years ago in the work of Einstein, Hilbert, and others, it has been a very successful and prosperous alliance for both partners. From the physical point of view, its success may be measured by the accurate agreement of Einstein's classical theory with all gravitational experimentation and astrophysical observation done thus far, and by the theory's prediction of new and interesting effects for which there is much evidence (e.g., black holes, cosmological expansion, gravitational waves, and gravitational lenses.) Mathematically, general relativity has stimulated the fertile study of pseudo-Riemannian as well as Riemannian geometry, and it has motivated important work in complex geometry, in topology, and in the study of both hyperbolic and elliptic PDE systems. In addition to the direct benefits, very important advances for both mathematics and physics have developed from some of general relativity's legacies: Yang-Mills theory, supersymmetry-supergravity theory, and superstring theory.

Reflecting the very wide range of directions in which general relativity research is currently proceeding, the discussions at our Santa Cruz conference (held in June, 1986) involved a large number of topics. The common feature in most was a concentration on questions with serious mathematical content. The intent of those discussions, as well as of this volume of the proceedings, is to keep the mathematical side of general relativity fresh by acquainting more mathematicians and physicists with such questions and the progress that has been made in answering them.

One of the most important recent advances in general relativity has been the proof of the Positive Mass Theorem by Schoen and Yau, and independently by Witten. This success led to a revival in interest in one of the classic questions in general relativity: Is there an expression which provides a local measure of energy density in a general spacetime? The first three articles in this volume discuss attempts to find such an expression. Penrose gives a status report on his twistorial version – what he calls "Quasi-Local Mass," or more generally "Quasi-Local Angular Momentum-Energy-Momentum." (One can calculate all these quantities using his approach). Christodoulou and Yau (§2) focus on another version, first proposed by Hawking. They also discuss the general features one should look for in the desired quantity, as well as some of the analytical problems which arise in trying to nail down a good definition. One of the necessary features of any good definition of localized mass is that if one calculates the mass enclosed in a sphere in an asymptotically flat spacetime and then lets the radius of the sphere approach infinity, the limit must (exist, and) agree with the (well-defined) total mass of the spacetime. Shaw (in §3) discusses this requirement, as well as what happens for large but finite spheres, in the Penrose version, the Hawking version, and certain other versions of localized mass. Note that although there is one unique total mass for any well-behaved asymptotically flat spacetime, there are a number of alternative formulas for calculating it. Sorkin (in §4) presents a new one here, derived from a variational principle, which has a number of virtues.

Another interesting outgrowth of the work on the Positive Mass Theorem (especially in Witten's formulation) is the "New Variables" Program for canonical quantization recently initiated by Ashtekar. This approach replaces the familiar canonical variables (i.e., Cauchy data) for general relativity – the intrinsic three-metric and the extrinsic curvature of a Cauchy hypersurface – by certain spinorial ones which are closely tied to self-duality conditions on the spacetime curvature. In §5, Ashtekar introduces his program, and discusses its utility for the study of the "half-flat Einstein equations." Then in the following two contributions, Smolin and Jacobson discuss the progress that they and others have made in setting up a canonical quantization of Einstein's theory using Ashtekar's formalism. Though still far from achieving this goal, their work is promising.

Self-duality (and anti self-duality) is the focus of §§8-9. Boyer, in §8, considers the problem of determining which compact orientable four-dimensional manifolds admit metrics with self-dual or anti-self-dual Weyl curvature. Then in §9, Flaherty shifts attention to Yang-Mills theory. He sketches a method for constructing Yang-Mills instantons (i.e., self-dual connections) on certain complex surfaces (quaternionic Siegel spaces) which he describes. In both Boyer's and Flaherty's contributions, complex geometry plays an important role.

In Wald's article, as well as in most of the remaining ones in this volume, one is back to working with real Lorentz signature spacetimes. Wald, after giving general definitions of "gauge theory" and "spin-one" or "spin-two" field theories, looks for examples of spin-one and spin-two gauge theories other than the familiar Yang-Mills and Einstein theories. In the spin-two case, he uncovers interesting new possibilities.

One of the outstanding problems in classical (i.e., nonquantum) general relativity at present is that of determining the relationship between initial data for a given spacetime and the global properties of that spacetime. Mathematically, this is essentially the problem of long time existence and stability of the Einstein partial differential equations. [In the jargon of general relativity, at least part of this issue is referred to as the Strong Cosmic Censorship question.] Klainerman and Christodoulou have recently made considerable progress in studying certain aspects of this problem. They focus on the conjecture that any set of sufficiently small Cauchy data (i.e., sufficiently close to Cauchy data for flat spacetime) leads to a globally complete (singularity free) spacetime solution of Einstein's equations. In §11, Klainerman reports on preliminary results related to their efforts to prove this conjecture.

Any study of the relationship between spacetime initial data and global properties relies on the premise that the Cauchy problem is well-posed. For the vacuum Einstein equations, for the Einstein-Maxwell and Einstein-Yang-Mills theories, and for certain others, well-posedness has long been established. In §12, Knill and Whitman show that Rosen's bimetric theory is also well-posed; and the work of Szczyrba (§13) sets up the analysis of this property for one of the quadratic curvature Lagrangian theories. When one considers fluid fields, however, problems can arise. These are made vivid in the contribution of Hiscock and Lindblom (§14), where it is found that certain seemingly natural field theories for fluids with dissipation can have disastrous instabilities and noncausal behavior. On the other hand, they find that certain other field theories for dissipative fluids may be able to avoid these problems.

One way to better understand what happens in complicated field theories, such as those with fluids, might be to carry out a systematic canonical analysis, complete with reduction by the gauge group of the theory. Marsden's contribution (§15) discusses some aspects of this approach, including what to do if one does not have a complete symplectic formulation of the theory, but rather must work with a looser formulation involving a Poisson manifold.

Einstein's theory, as well as most of the others examined in §§10-15, is based on a PDE system which is roughly hyperbolic in nature. Hamilton, however, looks at a rather different system of PDEs in studying "Ricci flow" on various two, three, and four dimensional manifolds. His system is $(\partial/\partial\tau)g = -2\mathrm{Ric}[g] + 2/n\langle R\rangle g$, where g is a Riemannian metric on a compact n-dimensional manifold, $\mathrm{Ric}[g]$ is its Ricci curvature, and $\langle R\rangle$ is the average of the scalar curvature over the manifold. This system is parabolic, and in studying its longtime existence and convergence properties, one learns much about Riemannian manifolds. Results for three and four dimensional manifolds are discussed elsewhere; here, in §16, the analysis on two-dimensional manifolds is studied.

Although some spacetime solutions of the Einstein equations (e.g. those of "small Cauchy data" as in the Christodoulou-Klainerman conjecture) may avoid singularities, the Hawking-Penrose theorem shows that most do not. Therefore, the study of singularities in spacetimes – what forms they might take and where they will occur – has been an important branch of general relativity at least for the last twenty years. A key issue is whether or not the singularities that form will involve infinite curvature invariants, corresponding physically to infinite tidal forces. The Hawking-Penrose theorems do not say. In §17, Kupeli uses the relatively virgin tool of the "shape function" on Cauchy hypersurfaces to study this problem. In §18, Schoen addresses a more specific question: Does a localized, highly compressed, concentration of matter always generate sufficiently strong gravitational effects to cause the local spacetime to collapse into a black hole? He reports on recent work (with Yau) which gives sufficient (mathematical) conditions for the collapse to occur.

Kossowski also discusses singularities in his contribution (§19), but his emphasis is not on those involving spacetime curvature blowups. Rather, he focuses on examining null hypersurfaces immersed in pseudo-Riemannian manifolds, and looks at incompleteness of geometric structures intrinsic and extrinsic to the immersion. In doing so, he introduces an interesting tensor which serves as a connection of sorts for the intrinsic (degenerate!) metric of the null hypersurface.

While general relativity as a theory of classical physics and as a mathematical system works very well, attempts to make it in some way compatible with the quantum principle have met with little success. Admittedly there are no experimental results to serve as a guide, but still there is much theoretical, mathematical, and aesthetic incentive to find a gravitational theory which is compatible with quantum physics and has Einstein's theory as a classical limit. With confidence in the prospects for a standard (particle theory type) perturbation method of quantization of Einstein's theory largely gone, there is renewed interest in canonical quantization techniques, among others. Ashtekar's spinor variable approach (§§4-6) has heightened this interest. In §20, Kuchar addresses one of the long standing problems of the canonical quantization approach as applied to Einstein's equations: how to factor order the operators so that the constraint algebra is (in an appropriate sense) preserved. He studies some model finite-dimensional systems and is able to find an appropriate factor ordering, which in addition is covariant under all relevant transformations. The relevant transformations for general relativity are parametrized by the diffeomorphism group Diff(M) of the spacetime manifold M and are known to be especially tricky to handle in quantization. Friedman and Witt, in §21, discuss some of the mathematical properties of Diff(M), especially concentrating on the analysis of the fundamental group π_0[Diff(M)] and consider the relevance of these properties for the canonical quantization of general relativity as well as for the rest of physics (e.g., for the existence of fermions). Yasskin's contribution (§22) discusses a different aspect of the mathematics of Diff(M). He considers how the diffeomorphism group of a spacetime manifold M relates to the diffeomorphism groups of the leaves of a spatial hypersurface foliation (codimension one) on the one hand and of the curves of a timelike congruence on the other. This could be useful in understanding the dynamical analysis of general relativity and perhaps ultimately it could be useful in quantization.

Another approach attempting to tie together general relativity and quantum ideas has been initiated by Hartle and Hawking. Conceptually based on Feynman path integration, the Hartle-Hawking program calculates a "ground state wave function for the universe" via action-weighted integration over all metrics on compact four dimensional manifolds bounded by a single three-dimensional hypersurface. Ultimately, one ends up working with solutions to a form of the Wheeler-DeWitt equation. In §23, D'Eath and Halliwell try to add fermionic fields to the game. They can do so if the fermion fields are dealt with as a perturbation on a background geometry.

A much more radical scheme for finding a quantum theory of gravity is to look for it as part of a superstring theory. Currently this is perhaps the most widely supported approach among theoretical physicists, if not among general relativists. Peskin (in §24) discusses recent work in superstring field theory, focussing on how to achieve a gauge invariant and covariant formulation. Then in §25, Shaw describes how twistor space constructions can be useful in finding complete solutions of the constraints of string theory without having to impose the light cone gauge conditions.

While a wide range of topics is discussed in the contributions to this volume, it should be noted that this book does *not* purport to provide a comprehensive look at current research in general relativity. Little is said about astrophysics, cosmology, experimental tests, or numerical modeling. Even in mathematical general relativity, our focus, only a portion of current research is covered. Still, we hope this book provides a good representative look at some of the mathematical issues being studied in general relativity.

We wish to thank the NSF, the AMS, and SIAM for their generous support of our conference. We also thank Ms. Marcia Almeida and her staff for their help in preparing the typed copy of the manuscript in this volume.

James Isenberg
Eugene, Oregon

Contemporary Mathematics
Volume **71**, 1988

ASPECTS OF QUASI-LOCAL ANGULAR MOMENTUM

Roger Penrose

I wish to make some remarks concerning the expression that I suggested a few years ago (Penrose 1982; and as subsequently slightly modified, c.f. Penrose and Rindler 1986), which provides a possible definition for the total energy-momentum and angular momentum surrounded by a spacelike 2-surface $\mathscr{S}$, with S^2 topology, in a general-relativistic (four-dimensional) space-time $\mathcal{M}$. The definition is quasi-local in the sense that it refers only to the geometry of $\mathscr{S}$ and the extrinsic curvature quantities for its embedding in $\mathcal{M}$. The definition does not, as things stand, enable one to pick out the family of 4-momentum components from the complete complex of quantities that are defined. Indeed, in the general case, there is, as yet, no unambiguous choice for the definition of even a scalar rest-mass in terms of this complex of quantities. The difficulty here is that there are sometimes different invariant definitions for this rest-mass, and it could be simply a matter of making a "correct" choice from among the various possibilities. The prospect of separating out a 4-momentum from the angular momentum in an invariant way is obstructed in a more fundamental way, however. It is the main purpose of this note to explain why this must be so.

I should first recall the definition of the complex $A_{\alpha\beta}(=A_{\beta\alpha})$ of these quantities:

$$A_{\alpha\beta}Z^{\alpha}Z^{\beta} = \frac{-1}{16\pi}\oint \eta\, R_{abCC'DD'}\,\omega^{C}\omega^{D}\varepsilon^{C'D'}\; dx^{a}\wedge dx^{b}$$

0271-4132/88 $1.00 + $.25 per page

integrated over a spacelike 2-surface $\mathcal{S}$ ($\cong S^2$), where Z^α stands for the 2-surface twistor defined by the field ω^A on $\mathcal{S}$, subject to the <u>2-surface twistor equation</u>

$$\eth'\omega^0 = \sigma'\omega^1, \qquad \eth\omega^1 = \sigma\omega^0$$

(ω^0 and ω^1 being the components of ω^A referred to a spin-frame o^A, ι^A whose flagpoles are orthogonal to $\mathcal{S}$), and where η is proportional, over the surface $\mathcal{S}$ to the determinant

$$\begin{vmatrix} \underset{1}{\omega}^0 & \underset{2}{\omega}^0 & \underset{3}{\omega}^0 & \underset{4}{\omega}^0 \\ \underset{1}{\omega}^1 & \underset{2}{\omega}^1 & \underset{3}{\omega}^1 & \underset{4}{\omega}^1 \\ \eth\underset{1}{\omega}^0 & \eth\underset{2}{\omega}^0 & \eth\underset{3}{\omega}^0 & \eth\underset{4}{\omega}^0 \\ \eth'\underset{1}{\omega}^1 & \eth'\underset{2}{\omega}^1 & \eth'\underset{3}{\omega}^1 & \eth'\underset{4}{\omega}^1 \end{vmatrix}$$

for independent solutions $\underset{1}{\omega}^A,\ldots,\underset{4}{\omega}^A$ of the 2-surface twistor equation. (See Penrose and Rindler 1984 for the notation used here.) In the weak-field limit of general relativity, or at null or spacelike infinity (when $\mathcal{M}$ is asymptotically flat), there is a Hermiticity condition reducing the ten complex components of $A_{\alpha\beta}$ to ten real ones, namely:

$$A_{\alpha\gamma}\bar{I}^{\beta\gamma} = \bar{A}^{\beta\gamma}I_{\alpha\gamma}$$

(Penrose 1982, Penrose and Rindler 1986, Shaw 1983), where the <u>infinity twistor</u> $I_{\alpha\beta}$, subject to

$$\bar{I}^{\alpha\beta} = \frac{1}{2}\,\varepsilon^{\alpha\beta\gamma\delta} I_{\gamma\delta}, \qquad I_{\alpha\beta} = -I_{\beta\alpha}, \qquad \det(I_{\alpha\beta}) = 0\ ,$$

has a (flat-space) definition

$$I_{\alpha\beta} = \begin{pmatrix} 0 & 0 \\ 0 & \varepsilon^{A'B'} \end{pmatrix}$$

in the description whereby any k-valent twistor has 2^k spinor parts, each of which is a spinor field. These spinor fields are related to one another via certain differential equations. In the case of Z^α, the spinor parts are

$$Z^\alpha = (\omega^A,\ \pi_{A'})$$

where, in Minkowski space $\mathbb{M}$, we have

$$\nabla_{AA'}\omega^B = -i\varepsilon_A{}^B\pi_{A'}, \qquad \nabla_{AA'}\pi_{B'} = 0,$$

and the relations between the spinor parts of a polyvalent twistor are generated from these (Penrose and Rindler 1986). For the Hermiticity condition on $A_{\alpha\beta}$ we require not only an infinity twistor but, in addition, a complex-conjugacy operator

$$Z^\alpha \to \bar{Z}_\alpha$$

which is also a duality operation. This is equivalent to providing each twistor Z^α with its "twistor norm"

$$Z^\alpha\bar{Z}_\alpha = \omega^A\bar{\pi}_A + \pi_{A'}\bar{\omega}^{A'}$$

which is Hermitian, with signature (++--).

In conformally flat space-time we can still use a description of twistors in terms of spinor parts, but we now have

$$\nabla_{AA'}\omega^B = -i\varepsilon_A{}^B\pi_{A'}, \qquad \nabla_{AA'}\pi_{B'} = -iP_{ABA'B'}\omega^B$$

where

$$P_{ab} = \frac{1}{12} R\, g_{ab} - \frac{1}{2} R_{ab}.$$

The expression for the twistor norm is the same as for $\mathbb{M}$, as given above. In deSitter space-time, with cosmological constant $\lambda = 6\Lambda$, we have

$$\nabla_{AA'}\omega^{B} = -i\varepsilon_{A}{}^{B}\pi_{A'}, \qquad \nabla_{AA'}\pi_{B'} = -i\Lambda\ \varepsilon_{A'B'}\ \omega_{A}$$

so ω^A and $\pi_{A'}$ are essentially on an equal footing. The spinor-part representation of the deSitter-space infinity twistor is

$$I_{\alpha\beta} = \begin{pmatrix} \Lambda\varepsilon_{AB} & 0 \\ 0 & \varepsilon^{A'B'} \end{pmatrix}.$$

As in Minkowski space, its complex conjugate is its dual (i.e. "twistor-real"), but now

$$\det(I_{\alpha\beta}) = \Lambda$$

and we have

$$I_{\alpha\gamma} I^{\beta\gamma} = \Lambda\, \delta_{\alpha}^{\beta} \quad .$$

In a general curved space-time $\mathcal{M}$, one cannot provide an integrable concept of twistor in terms of such spinor fields, but restricted to $\mathcal{S}$ we have

$$\eth\omega^{0} = -i\rho'\pi_{1'}, \qquad \eth'\omega^{1} = -i\rho\pi_{0'} \quad .$$

At any one point of $\mathcal{S}$, one can write down the previous expression for the "twistor norm", namely

$$\omega^{0}\bar{\pi}_{0} + \omega^{1}\bar{\pi}_{1} + \pi_{0'}\bar{\omega}^{0'} + \pi_{1'}\bar{\omega}^{1'}$$

but this is generally not constant over $\mathcal{S}$. The condition for it to be constant for all 2-surface twistors Z^{α} is that $\mathcal{S}$ be uncontorted (e.g. cf. Jeffryes 1984), i.e. that $\mathcal{S}$ can be embedded in some conformally flat real space-time in such a way that both its intrinsic geometry (induced metric) and its extrinsic curvature quantities ($\rho, \rho', \sigma, \sigma'$) are unaltered from their values in $\mathcal{M}$. In these uncontorted cases, this provides an unambiguous definition for $Z^{\alpha}\bar{Z}_{\alpha}$ and, therefore, for the complex-conjugation map $Z^{\alpha} \to \bar{Z}_{\alpha}$. When $\mathcal{S}$ is contorted, on the other hand, we seem to need some means of averaging this local twsitor norm expression over $\mathcal{S}$. There appear to be various inequivalent ways of achieving this averaging (cf. Penrose 1982), but none has been suggested which is conformally invariant. It is not altogether clear whether such conformal invariance is desirable since general relativity is not, after all, a conformally invariant theory, but the definition of a 2-surface twistor is a conformally invariant one.

With regard to $I_{\alpha\beta}$, we certainly would not expect conformal invariance. In $\mathbb{M}$, we have the local definition

$$I_{\alpha\beta}Z_1^{\alpha}Z_2^{\beta} = \varepsilon^{A'B'}\,{}_1\pi_{A'}\,{}_2\pi_{B'} = {}_1\pi_{0'}\,{}_2\pi_{1'} - {}_1\pi_{1'}\,{}_2\pi_{0'}$$

so we might expect to average this expression over $\mathcal{S}$ in a general $\mathcal{M}$ (the expression being certainly not constant in general), together with some possible involvement of the η-factor. The resulting "averaged" $I_{\alpha\beta}$ will not generally be of the Minkowski kind (i.e. "simple": $\det(I_{\alpha\beta}) = 0$) but will be a deSitter-type infinity twistor. In the particular situations that arise when $\mathcal{S}$ is taken to either spacelike or null infinity, we do indeed get a Minkowski-type infinity twistor, and the previously mentioned Hermiticity condition for $A_{\alpha\beta}$ indeed holds. However, the general problem of a Hermiticity relation for $A_{\alpha\beta}$ with a general (even uncontorted) $\mathcal{S}$ remains unsolved.

We can see, from various considerations, that if there is such a Hermiticity relation, it certainly has to be generally with a deSitter-type infinity twistor and not a Minkowski-type one. One such consideration arises from an examination of Tod's (1983) results for $\mathcal{S}$ lying in a conformally flat hypersurface of time-symmetry. Here we find that when $\mathcal{S}$

surrounds a source region $\mathcal{R}_i$ (once) we have the same angular momentum twistor ${}_iA_{\alpha\beta}$, irrespective of the precise location of $\mathcal{S}$, and similarly if it surrounds $\mathcal{R}_i \cup \mathcal{R}_j$ (once) we always get

$$ {}_{ij}A_{\alpha\beta} = {}_iA_{\alpha\beta} + {}_jA_{\alpha\beta} \quad . $$

However, defining the corresponding rest-masses via the equation

$$ m^2 = -\frac{1}{2} A_{\alpha\beta}\bar{A}^{\alpha\beta} $$

(which is well-defined since $\mathcal{S}$ is uncontorted), we find

$$ m_{ij} < m_i + m_j \quad . $$

Such an inequality would be impossible if the various angular momentum twistors always had the form

$$ A_{\alpha\beta} = \begin{pmatrix} 0 & p_{A'}{}^{B'} \\ p^{A'}{}_{B'} & 2i\mu^{A'B'} \end{pmatrix} $$

(p_a being the 4-momentum and $\mu^{A'B'}$ describing the angular momentum) which occurs when the Hermiticity relation for $A_{\alpha\beta}$ holds with a standard Minkowski-type infinity twistor. However an entry other than zero can occur in the upper left-hand corner with a deSitter-type Hermiticity relation, and this is indeed what happens here.

An even more elementary illustration occurs with the Schwarzschild space-time, where we consider surfaces $\mathcal{S}$ sharing the spherical symmetry of the space-time and lying in the t = 0 hypersurface $\mathcal{H}$. This hypersurface is one which admits Tod's (1984) 3-surface twistors, which implies that (as in the more general time-symmetric case just considered) the spaces of 2-surface twistors for any $\mathcal{S}$ lying in $\mathcal{H}$ can be canonically identified with one another - and with the space $\mathbb{T}(\mathcal{H})$ of 3-surface twistors for $\mathcal{H}$. However, there is no appropriately unique infinity twistor for

$\mathcal{H}$. If we start with $\mathcal{S}$ out at spacelike infinity i^0, we obtain an infinity twistor which represents that point: i^0. But if we move $\mathcal{S}$ continuously (and spherically symmetrically) inwards from i^0 until it reaches the horizon, and then out again on the opposite Kruskal sheet, we finally reach the "opposite" spacelike infinity point $\hat{\imath}^0$. Now from the point of view of $\mathbb{T}(\mathcal{H})$ (i.e. of the conformal geometry of $\mathcal{H}$) the point $\hat{\imath}^0$ seems like the origin when "viewed" from i^0, whereas i^0 itself seems like the origin when "viewed" from $\hat{\imath}^0$. Thus we get two distinct Minkowski-type (i.e. simple) corresponding infinity twistors, say

$$I_{\alpha\beta} \quad \text{and} \quad \hat{I}_{\alpha\beta} \quad .$$

For an arbitrary (spherically symmetrical) location of $\mathcal{S}$, we must, by symmetry get some linear combination of these two. This combination passes smoothly from $I_{\alpha\beta}$ itself to $\hat{I}_{\alpha\beta}$ itself and, in between, we must (and do) get infinity twistors which are of deSitter type.

In fact the Hermiticity relation for $A_{\alpha\beta}$ does hold with this deSitter-type infinity twistor in this case, but it is a very special situation and we do not know what to expect generally. However, the consideration of deSitter-type infinity twistors seems inescapable. Accordingly there will, in general, be no way of projecting out the 4-momentum from the full complex of quantities in $A_{\alpha\beta}$.

References

Jeffryes, B.P. (1984) Two-surface twistor and conformal embedding in Asymptotic Behaviour of Mass and Spacetime Geometry, Proc. Corvallis Oregon 1983, ed. F.J. Flaherty (Springer-Verlag, Berlin)

Penrose, R. (1982) Quasi-local mass and angular momentum in general relativity, Proc. Roy. Soc. London **A381**, 53-63

Penrose, R. & Rindler, W. (1986) Spinors and Space-Time, Vol. 1, Two-Spinor Calculus and Relativistic Fields (Cambridge University Press, Cambridge)

Penrose, R. & Rindler, W. (1986) Spinors and Space-Time, Vol. 2, Spinor and Twistor Methods in Space-time Geometry (Cambridge University Press, Cambridge)

Shaw, W.T. (1983) Twistor theory and the energy-momentum and angular momentum of the gravitational field at spatial infinity, Proc. Roy. Soc. London A390, 191-215

Tod, K.P. (1983) Some examples of Penrose's quasi-local mass construction, Proc. Roy. Soc. London A388, 457-77

Tod, K.P. (1984) Three-surface twistors and conformal embedding, Gen. Rel. Grav. 16, 435-43

Roger Penrose
Mathematical Institute
24-29 St Giles
Oxford University
Oxford OX1-3LB
United Kingdom

Institute for Theoretical Physics
University of California
Santa Barbara, CA 93106

Contemporary Mathematics
Volume 71, 1988

SOME REMARKS ON THE QUASI-LOCAL MASS

D. Christodoulou and S.-T. Yau

In our joint work with Klainerman, we studied the global nonlinear stability problem for the Einstein equation where we proved that if the initial conditions are close to the trivial data, the spacetime is globally hyperbolic and close to the flat spacetime. It is clearly greatly desirable to study the strong field when the initial data is no longer close to the trivial data. In this case, it is almost certain that we have to understand certain conserved (or quasi-conserved) quantities which can control the field in a more local manner. In other words, we expect some concept of quasi-local mass will be useful.

Several definitions of quasi-local have been introduced by different authors. There are certain properties that quasi-local mass must have and these can be described as follows:

[1] It should be zero for the flat spacetime.

[2] It should be identical with the standard definition if the spacetime is spherically symmetric and the quasi-local mass is evaluated on the spheres. In particular, for the centered spheres in the Schwarzchild spacetime, the quasi-local mass should be the standard mass.

[3] For an asymptotoically flat slice, the quasi-local mass of the coordinate sphere should be asymptotic to the ADM mass.

[4] For an asymptotically null slice, the quasi-local mass of the coordinate sphere with respect to standard radial coordinates should be asymptotic to the Bondi mass.

[5] For an apparent horizon, the quasi-local mass should be the irreducible mass which is $\left(\frac{\text{Area}(A)}{16\pi}\right)^{1/2}$

[6] The quasi-local mass should be positive and be monotone in a suitable sense.

None of the known definitions have satisfied all the above properties. The most recent definition of R. Penrose is perhaps closest to doing so. However, it is difficult to compute Penrose's mass and it is also difficult to verify property [6] of the above requirement. S. Hawking has a different definition of quasi-local mass which verifies all the above properties except [6]. It is defined by the following formula

$$M(S) = \left(\frac{\text{Area}(S)}{16\pi}\right)^{1/2}\left(1 - \frac{1}{4\pi}\int_S \rho\mu\right)$$

where ρ is the convergence of the outer null normal and $-\mu$ is the convergence of the inner null normal.

The Hawking definition has the clear advantage of computability. It is sensitive to the geometry of the two surface. The purpose of this note is to demonstrate that the Hawking mass is indeed good, if the geometry of S is nice. We shall call this surface S "round" if it is a subset of a three dimensional spacelike slice M, and if among two dimensional surfaces in M which enclose the same volume as S does, S has least area.

Let us now study the geometry of such a surface. By computing the second variation of the area of S over M, we can conclude that for any function f such that $\int_S f = 0$, we must have

$$\int_S |\nabla f|^2 - \int_S (\Sigma R_{ij} \nu^i \nu^j + \Sigma h^{ij} h_{ij}) f^2 \geq 0 \tag{1}$$

Here R_{ij} is the Ricci tensor of the three dimensional manifold M, ν^i is the normal of S in M and h_{ij} is the second fundamental form of S in M. By the Gauss equations,

$$\Sigma h_{ij} \nu^i \nu^j = R - K + \det (h_{ij}) \tag{2}$$

where R is the scalar curvature of M, and K is the Gauss curvature of S.

Therefore, combining 1 and 2, we have

$$\int_S |\nabla f|^2 \geq \int \left(R - K + \left(\frac{\mathrm{tr}\, h_{ij}}{2}\right)^2\right) f^2 + \frac{1}{4}\int (\lambda_1 - \lambda_2)^2 f^2 \tag{3}$$

where λ_1 and λ_2 are the principle curvatures of S.

Let Ψ be a conformal map which maps S onto the standard sphere $S^2 \subset R^3$ with degree d_s. By moving Ψ with the conformal group of S^2, one can find Ψ (see [LY]) so that, as a vector valued function,

$\int_S \Psi = 0$. Hence, each component of Ψ satisfies condition 1. In particular, each component Ψ^i of Ψ can be put into equation 3.

Since the Dirichlet integral is a conformal invariant, the left hand side of c can be computed and is equal to $\frac{8\pi d_s}{3}$ for each Ψ^i. As $\Sigma(\Psi^i)^2 = 1$, we conclude from 3 that

$$4\pi(2d_s + 1 - g\,(S)) \geq \int_S R + 3\int_S \left(\frac{\mathrm{tr}\, h_{ij}}{2}\right)^2. \tag{4}$$

When S is topologically a sphere, we can take $d_s = 1$ and we obtain

$$1 - \frac{1}{4\pi}\int_S \left(\frac{\operatorname{tr} h}{2}\right)^2 \geq \frac{1}{12\pi}\int_S R \qquad (5)$$

with equality only if S is umbilical.

In particular, we deduce the following

Theorem. The Hawking mass of a two dimensional surface S is not less than

$$M(s) \geq \frac{1}{12\pi}\left(\frac{\operatorname{Area}(S)}{16\pi}\right)^{1/2}\int_S R_M \qquad (6)$$

when S has genus zero and is "round" in a spacelike slice M. Here R_M is the scalar curvature of M. Equality can hold in 6 only if S is umbilical in M.

Corollary. On a maximal slice M of a spacetime which satisfies the local energy condition, the Hawking mass of any round two sphere is non-negative.

The proof of this theorem shows that, in the definition of Hawking mass, it is perhaps desirable to take into account the topology of S. One can show that d_s is bounded by $\left(\frac{g(S)+1}{2}\right) + 1$.

Finally, let us mention that for any spacelike slice which is asymptotic to the Schwarzchild solution up to the second order, there exists a unique foliation by round spheres at infinity of the slice. This is being studied with G. Huisken. It is of interest to study the evolution of such round surfaces.

References

[LY] P. Li and S.-T. Yau, A new conformal invariant and its application to the Willmore conjecture and the first eigenvalue of compact surfaces, Invent. Math., **69** (1982), 269-291.

D. Christodoulou
Dept. of Mathematics
Syracuse University
Syracuse, New York 13244

S.-T. Yau
Dept. of Mathematics
University of California (San Diego)
La Jolla, California 92109

Contemporary Mathematics
Volume 71, 1988

QUASI-LOCAL MASS FOR "LARGE" SPHERES

William T. Shaw[1]

ABSTRACT. The quasi-local measures of mass introduced by Ludvigsen and Vickers, Hawking, and Penrose are examined for large spheres approaching null infinity along an outgoing null hypersurface N. The mass formulae can be calculated in terms of an asymptotic series in inverse powers of r, and some preliminary results for stationary space-times are described. In such space-times it is found that the Hawking and Ludvigsen-Vickers definitions are sensitive to the choice of N, but the Penrose mass does not contain any such unphysical elements. A measure of mass applicable at large but not infinite radii may be very useful in numerical simulations of general relativity, and it is argued that the Penrose mass is a suitable measure.

1. INTRODUCTION. The purpose of this note is to describe the properties of some quasi-local mass formulae in a context which may be of practical importance: the mass contained within a "large" sphere. A potential application of this is to numerical relativity where it is useful to have a viable measure of "total mass" which does not have to be computed at infinity. Ideally, such a measure should be asymptotic to the Bondi mass m_B[1], when computed on suitable outgoing null surfaces, and possess an asymptotic series (derived from general analytical considerations) of the form

$$m(r) = m_B + A_1 r^{-1} + A_2 r^{-2} + \dots \tag{1}$$

in terms of a suitable coordinate r. Numerical computation of the measure on a series of spheres may, by comparison with the analytical formula (1), lead to good estimates for m_B.

At present, there is probably no quasi-local measure of mass which is both universally applicable and universally accepted. There are, however, several definitions in the literature. A second motivation for considering "large spheres" is to test these definitions in a new area. Usually it is feasible to carry out calculations non-perturbatively in three contexts:

1980 Mathematics Subject Classification (1985 Revision). 83C.
[1]Supported in part by the NSF.

a) At infinity in general asymptotically flat spaces;
b) For finite surfaces in very special spaces, typically with a high degree of symmetry;
c) Numerically.

Perturbation techniques allow one to consider "small" spheres and "large" spheres, considered as perturbations of a point and a sphere at infinity respectively. Here the large sphere behaviour of three mass definitions will be considered: those of Hawking [2], Penrose [3] and Ludvigsen and Vickers [4].

The Hawking mass has been considered by S.T. Yau, also in this volume. Its limit at infinity was described by Eardley [5] and its small sphere behaviour by Horowitz and Schmidt [6]. Some properties of the Penrose mass have been described by R. Penrose, also in this volume. A more detailed account, with an extensive bibliography and a description of a modification to the original formula [3] which deals with some anomalies in the small sphere behaviour is given by Penrose and Rindler [7]. The Penrose formula actually defines ten quantities as an energy-momentum and angular momentum structure. The mass is the norm of this structure. The existence of a unique norm is not always guaranteed, although in many cases there is an obvious candidate. Tod [8] has given a large supply of examples where the norm is well-defined and the mass is physically reasonable. The properties of the formula at infinity are discussed in [7] and by Shaw [9,10], Dray and Streubel [11] and Dray [12]. The case of small surfaces has been considered by Kelly, Tod and Woodhouse [13].

The third quasi-local definition that will be considered here is that of Ludvigsen and Vickers [4], who proposed to use a 2-form (defined in their proof of the positive mass theorem [14]) as giving a measure of mass on finite 2-surfaces S(r) which are cross-sections of an outgoing null surface N reaching null infinity $\mathcal{I}$. The definition involves writing down spinors satisfying certain differential equations on N, with boundary conditions at infinity. This definition is not truly quasi-local, since the quantities involved do not just depend on S(r), but also on a propagation rule for getting from S(r) to $\mathcal{I}$. This approach defines a momentum vector dual to the translations of the B.M.S. group [1]. Here the mass will be defined to be the norm of this vector.

2. COORDINATES. Let M be a space-time which is asymptotically flat and empty at null infinity. The space-time possesses a boundary $\mathcal{I}$ which is a null 3-surface provided the stress tensor falls off sufficiently fast. Penrose and Rindler [7] give a detailed account of the conditions for the existence of $\mathcal{I}$ and its properties. Let S be any 2-surface (sphere) cross-section of $\mathcal{I}$ and let N(S) be the unique outgoing null hypersurface with $N(S) \cap \mathcal{I} = S$. Let r be an affine parameter for the null geodesic generators l^a of N, scaled so that if

$S_r(N)$ is the sphere in N of constant r, then as $r\to\infty$, the area A of $S_r(N)$ satisfies

(2) $$A = 4\pi r^2 + O(1).$$

A quantity of some importance for what follows is the shear of N. The shear is the trace-free part of $\nabla_a l_b$ projected onto $S_r(N)$, and can be represented by a complex function $\sigma \doteq m^a m^b \nabla_a l_b$, where m^a and its complex conjugate are complex null vectors spanning the tangent space to $S_r(N)$. The vectors are normalized against the metric g_{ab} so that $m^a \bar{m}^b g_{ab} = -1$. More generally, the projection of any tensor into $S_r(N)$ can be represented by a set of "spin-weighted" scalars, by contraction of all its indices with m^a and $\bar{m}^a$, where the spin weight s is given by : s = (no. of m's) - (no. of $\bar{m}$'s). Thus σ has spin weight s = 2. Differentiation within $S_r(N)$ can be represented by a pair of spin-raising/lowering operators. Here it is only necessary to define a standard operator $\eth$ ("edth"). This can be defined by its action on scalar functions f with s = 0 by

(3) $$\eth f = \lim_{r\to\infty}\{r m^a \nabla_a f\} .$$

The operator $\eth$ and its complex conjugate extend naturally to spin-weighted objects.

The shear σ has an expansion of the form

(4) $$\sigma = \sigma^0\, r^{-2} + O(r^{-4}).$$

The quantity σ^0 , the asymptotic shear of N, plays an important role. It is highly sensitive to the choice of N. For example in Minkowski space-time, the surfaces N with $\sigma^0 = 0$ are precisely the future halves of the null cones of space-time points. In a general asymptotically flat space-time, the dynamic rate of change of σ^0 is a measure of the outgoing gravitational radiation. In stationary space-times, with no gravitational radiation, it makes sense to identify the asymptotically shear-free hypersurfaces with the points of an imagined flat space-time. The notion of a translation can then be realized as a map from a surface N to a new surface N' such that the asymptotic shear is unchanged. More general changes in the choice of N are called supertranslations, and induce a change in the shear in a stationary space-time. If a quantity is defined with respect to a null hypersurface N, one can then ask about its origin-dependence, i.e., how that quantity transforms under changes in N. For example, the total charge should be origin-independent, but the total dipole moment has a certain transformation law. In simple 3-vector notation, if q is the total charge and $\underline{d}$ is the dipole moment, one expects an asymptotic realization of the simple rule:

$$(5) \qquad \underline{d}(\underline{r}) = \underline{d}(\underline{0}) + q\underline{r}.$$

The difficulty with many asymptotic definitions is that they do not reproduce the correct origin-dependence. (The definition of angular momentum at $\mathcal{I}$ is particularly contentious in this respect.)

The space-times that will be considered here are solutions of the Einstein-Maxwell equations near infinity. That is, asymptotically the only matter fields are purely electromagnetic in nature. The results are at present known in any detail only for stationary space-times. It is useful to describe the space-time in terms of its multipole structure. For a general Einstein-Maxwell space-time the only conserved quantities are the total charge and the Newman-Penrose constants [15]. The Bondi mass m_B is both non-negative and non-increasing. For a stationary space-time m_B is constant and it makes sense to define several higher moments of the fields. There are two origin-independent quantities $\underline{J}$ and $\underline{m}$, given as Cartesian 3-vectors and representing the total orbital angular momentum and magnetic moment respectively. There are two corresponding origin-dependent quantities $\underline{D}$ and $\underline{d}$ representing the total mass dipole moment and the total charge dipole moment respectively. The origin-dependence of $\underline{d}$ is realized as an asymptotic form of (5), with a similar relation for $\underline{D}$ with q replaced by m_B. All these quantities can be defined in terms of integrals over cross-sections of $\mathcal{I}$ (see [16] for details).

3. RESULTS. Firstly, for the Hawking mass m_H one finds that in any asymptotically flat Einstein-Maxwell space-time (AFEM), for any choice of N,

$$(6) \qquad m_H = m_B - Hr^{-1} + O(r^{-2}),$$

where $H \geq 0$ and $H = H_1 + H_2$, with $H_1 \geq 0$ and $H_2 \geq 0$, and H_1 is purely electromagnetic. In a stationary AFEM,

$$(7) \qquad H_1 = q^2/2.$$

In a general AFEM, with $dS = d\theta d\phi \sin(\theta)$,

$$(8) \qquad H_2 = (4\pi)^{-1} \int dS\{1/2|\sigma^0|^2 + |\eth\bar{\sigma}^0|^2\}$$

The form of H_2 indicates that the Hawking mass runs into problems at first order in r^{-1}. The definition is highly sensitive to the value of σ^0. In flat space-time it is easy to arrange $\sigma^0 \neq 0$, so that $H_2 > 0$! Now suppose that one considers only stationary AFEM and restricts attention to those N with $\sigma^0 = 0$. Then it is straightforward to calculate some further terms. One finds

$$(9) \qquad m_H = m_B - 1/2q^2r^{-1} - 1/3\{\underline{m}^2 + \underline{d}^2\}r^{-3} - \{\underline{J}^2 + \underline{D}^2\}r^{-3} + O(r^{-4}).$$

The electromagnetic terms are the same as would appear in flat space-time. In the presence of a monopole term it makes sense to have a third order origin-dependent term quadratic in the charge dipole moment. However, the gravitational contribution has an unacceptable origin-dependent term. For example, in the Schwarzschild space-time, on an asymptotically shear-free N,

$$(10) \qquad m_H = m_B - \underline{D}^2r^{-3} + O(r^{-4}),$$

so that the first term in the expansion is origin-dependent. This is milder than the dependence on the shear, but still unacceptable. In the same space-time, if one considers only those null surfaces on which both σ^0 and $\underline{D}$ vanish, then $m_H = m_B$ to all orders in r, a property shared by the Penrose formula (the Hawking and Penrose formula give the same answer on a surface of spherical symmetry in any spherically symmetric space-time, as shown by Tod [8]).

A positive feature of the asymptotic Hawking formula is its behaviour in the Kerr geometry. On a surface with $\sigma^0 = 0 = \underline{D}$, it takes the form

$$(11) \qquad m_H = M - \underline{J}^2r^{-3} + O(r^{-4}),$$

where M $(=m_B)$ is the mass parameter and $|\underline{J}| = Ma$, where a is the specific angular momentum. For slowly rotating black holes, with a<<M, the black hole radius is at $r \sim 2M$. Substitution of this into the asymptotic Hawking formula (11) gives $M - a^2/(8M)$, which gives precisely the first two terms in an expansion of the irreducible mass of the black hole in an expansion in powers of a.

In the approach of Ludvigsen and Vickers, an energy-momentum 4-vector is defined by $S_r(N)$ as a vector $P^a(r)$ dual to the space of BMS translations [1], with

$$(12) \qquad P^a(\infty) = P^a{}_B ,$$

where $P^a{}_B$ is the Bondi energy-momentum. For any AFEM space-time, and for any choice of N, one finds

$$(13)\qquad P^a(r) = P^a{}_B - H^a{}_1 r^{-1} + O(r^{-3}),$$

where $H^a{}_1$ is purely electromagnetic, given by integrating the asymptotic stress tensor against the asymptotic translations. Note that the first order shear terms present in the Hawking definition are absent, so that the cross-section sensitivity is considerably diminished. When the space-time is stationary, for any N one finds

$$(14)\qquad P^a(r) = t^a(m_B - 1/2q^2r^{-1}) + O(r^{-2}),$$

where $P^a{}_B = m_B t^a$ and $t^a t_a = +1$. However, at third order this formula suffers from inappropriate origin-dependence. When $\sigma^0 = 0$, the mass defined by the Ludvigsen-Vickers momentum is

$$(15)\qquad m_{LV} = m_B - 1/2q^2r^{-1} - 1/3(\underline{m}^2 + \underline{d}^2)r^{-3} - 1/4(\underline{J}^2 + \underline{D}^2)r^{-3} + O(r^{-4}).$$

This suffers from the same problems as the Hawking formula due to the term quadratic in $\underline{D}$.

The Penrose mass is the hardest to calculate but is also the best behaved in terms of cross-section sensitivity. For any N in a stationary AFEM, the Penrose mass m_P satisfies

$$(16)\qquad m_P = m_B - 1/2q^2r^{-1} - \mu r^{-3} + O(r^{-4}).$$

When $\sigma^0 = 0$ it is straightforward to calculate μ, and one finds

$$(17)\qquad \mu = 1/3\ (\underline{m}^2 + \underline{d}^2) + \eta\underline{J}^2.$$

There is no term in $\underline{D}^2$. The value of η depends on a choice of norm structure and also on whether the original [3] or modified [7] definition is used. One way of constructing a norm is to write down the inner product appropriate in flat space-time. If it is constant on the surface this is used, as in [8]. Otherwise one can seek some minimal modifications to make it constant. A second approach, which can lead to a different norm, is construct the norm so that the full angular momentum structure has the same reality properties as in flat space-time. A detailed discussion of these issues is outside the scope of this note. However, all the procedures considered so far lead to values of η

in the range $0 \leq \eta \leq 8$. It is certainly possible to choose the norm so that $\eta = 1$, but clearly some more natural and universally applicable procedure is needed.

A more detailed account of these ideas is to appear elsewhere [16]. The general procedure for calculating these results is to construct an asymptotic solution of the Einstein-Maxwell equations along a given N, together with rules for transforming the solution under changes in N. The definitions are then applied order by order. The full angular momentum structure associated with the Penrose formula is also calculated in [16] for stationary AFEM. The non-stationary case is considerably more complicated and is under investigation. An understanding of the non-stationary case is needed before these results may be useful in numerical simulations, where the dynamical evolution of the solution is the main point of interest. The results from the stationary case lend support to the idea that the Penrose mass may be useful in this context.

BIBLIOGRAPHY

1. Bondi, H., "Gravitational waves in general relativity," Nature, 186 (1960), 535.
Bondi, H., Van der Burg, M.G.J. and Metzner, A.W.K., "Gravitational waves in general relativity. VII. Waves from axi-symmetric isolated systems," Proc. R. Soc. Lond., A269 (1962), 21-52.
Sachs, R.K., "Gravitational waves in general relativity. VIII. Waves in asymptotically flat space-time," Proc. R. Soc. Lond. A270 (1962), 103-126.

2. Hawking, S.W., "Gravitational radiation in an expanding universe," J. Math. Phys., 9 (1968), 598-604.

3. Penrose, R., "Quasi-local mass and angular momentum in general relativity," Proc. R. Soc. Lond., A385 (1982), 53-63.

4. Ludvigsen, M. and Vickers, J.A.G., "Momentum, angular momentum, and their quasi-local null surface extensions," J. Phys. A16 (1983), 1155-1168.

5. Eardley, D.M., In Sources of Gravitational Radiation (L. Smarr, ed.), pp. 127-138. University Press, Cambridge, 1979.

6. Horowitz, G.T. and Schmidt, B., "Note on gravitational energy," Proc. R. Soc. Lond., A381 (1982), 215-224.

7. Penrose, R. and Rindler, W., Spinors and space-time Vol. 2, University Press, Cambridge, 1986.

8. Tod, K.P., "Some examples of Penrose's quasi-local mass construction," Proc. R. Soc. Lond., A388 (1983), 457-477.

9. Shaw, W.T., "Twistor theory and the energy-momentum and angular momentum of the gravitational field at spatial infinity," Proc. R. Soc. Lond., A390 (1983), 191-215.

10. Shaw, W.T., "Symplectic geometry of null infinity and two-surface twistors," Class. Quant. Grav. 1 (1984), L33-L37.

11. Dray, T. and Streubel, M., "Angular momentum at null infinity," Class. Quant. Grav. 1 (1984), 15-26.

12. Dray, T., "Momentum flux at null infinity," Class. Quant. Grav. 2 (1985), L7-10.

13. Kelly, R.M., Tod, K.P. and Woodhouse, N.M.J., "Quasi-local Mass for Small Surfaces," Oxford preprint, 1986.

14. Ludvigsen, M. and Vickers, J.A.G., "A simple proof of the positivity of the Bondi mass," J. Phys. A15 (1982), L67-L70.

15. Newman, E.T. and Penrose, R., "10 exact gravitationally conserved quantities," Phys. Rev. Lett. 15 (1965), 231-233.
Exton, A.R., Newman, E.T. and Penrose, R., "Conserved quantities in the Einstein-Maxwell theory," J. Math. Phys. 10 (1969), 1566-1570.

16. Shaw, W.T., "The Asymptopia of Quasi-local Mass and Momentum: I; General Formalism and Stationary Space-times," M.I.T. preprint, 1986, to appear in Class. Quant. Grav.

MATHEMATICS DEPARTMENT,
MASSACHUSETTS INSTITUTE OF TECHNOLOGY,
CAMBRIDGE, MASSACHUSETTS 02139.

Contemporary Mathematics
Volume **71**, 1988

CONSERVED QUANTITIES AS ACTION VARIATIONS

Rafael D. Sorkin[1]

ABSTRACT. An expression for gravitational conserved quantities is presented and shown to be the variation of a first-order Action defined with respect to a background *connection*. When the field equations hold, the expression takes the form of an integral over a codimension-two surface of infinite radius. Its integrand consists of the Komar tensor plus a term linear in the difference between the background connection and that of the spacetime metric. The formula given applies in spacetimes with Kaluza-Klein boundary conditions, as well as in ones which are asymptotically flat in the ordinary sense.

The expression I have written on the blackboard,

$$\oint_\infty \frac{1}{2} dS_{\mu\nu}[\nabla^\mu \xi^\nu + (\Gamma^\mu_{\alpha\alpha} g^{\alpha\alpha} - \Gamma^\alpha_{\mu\alpha} g^{\mu\mu})\xi^\nu] \tag{1}$$

is intended to be a general formula for the conserved quantity $P(\xi^\lambda; \mathring{\nabla}_\mu)$ corresponding to the asymptotic symmetry generated by the vector field ξ^μ. Before going through its derivation from the gravitational Action, I would like to recommend it to you by mentioning some of its virtues. But first let me explain my notation, especially the symbol $\Gamma^\alpha_{\mu\nu}$.

1980 Mathematics Subject Classification (1985 Revision): 83C40.

[1]Supported in part by NSF Grant PHY-8318350.

The metric $g_{\mu\nu}$ is an asymptotically flat[2] solution of the Einstein equation, $G_{\mu\nu} = 0$, in 3 + 1 spacetime dimensions. (If it were not a solution then a 3-space integral $\int - G^{\mu}_{\lambda}\xi^{\lambda} dS_{\mu}$ would have to be added to $P(\xi;\mathring{\nabla})$.) The symmetric tensor $\Gamma^{\mu}_{\alpha\beta}$ represents the difference between the "true" or "metric" connection $\nabla_{\mu} = \nabla_{\mu}[g_{\alpha\beta}]$ and an unphysical *background* connection $\mathring{\nabla}_{\mu}$ on which $P(\xi^{\lambda};\mathring{\nabla}_{\mu})$ explicitly depends. (The sign of $\Gamma = \nabla - \mathring{\nabla}$ is determined by $\nabla_{\mu}v^{\nu} = \mathring{\nabla}_{\mu}v^{\nu} + \Gamma^{\nu}_{\alpha\mu}v^{\alpha}$. I will assume for simplicity that $\mathring{\nabla}$ is torsionless so that Γ is symmetric.) The vector ξ^{λ} must then be an exact symmetry of the background:

$$[\mathring{\nabla}_{\mu}, \mathcal{L}_{\xi}] = 0;$$

and the surface integral in which it occurs is over a closed 2-surface "at spatial infinity". Thus P depends on $\mathring{\nabla}$ both directly through Γ and indirectly via the way in which $\mathring{\nabla}$ determines ξ; on the other hand, notice that only the value of $\mathring{\nabla}$ in a neighborhood of infinity influences the value of P. The surface element $dS_{\mu\nu}$ occurring in (1) is normalized so that Stokes' Theorem reads

$$\oint \frac{1}{2} dS_{\mu\nu}W^{\mu\nu} = \int dS_{\mu}\nabla_{\nu}W^{\mu\nu},$$

with dS_{μ} reducing to $\sqrt{-g}\ d^3x$ for a t = constant hypersurface. Because of the skewness of $dS_{\mu\nu}$ I have not bothered to antisymmetrize explicitly the expression within the square brackets in (1). Finally, $8\pi G = c = 1$, and the respective expressions $\nabla^{\mu}\xi^{\nu}$, $\Gamma^{\mu}_{\alpha\alpha}g^{\alpha\alpha}$, and $\Gamma^{\alpha}_{\alpha\mu}g^{\mu\mu}$ can be written even more unambiguously as $g^{\mu\alpha}\nabla_{\alpha}\xi^{\nu}$, $\Gamma^{\mu}_{\alpha\beta}g^{\alpha\beta}$ and $\Gamma^{\alpha}_{\alpha\lambda}g^{\lambda\mu}$.

[Virtues:]

Because the expression (1) for $P(\xi;\ \mathring{\nabla})$ is derived from an Action principle certain desirable features result automatically:

---

[2]The degree of falloff needed depends on the conserved quantity being evaluated. I do not know the optimum requirements for all cases, but for energy the usual weak falloff $(g - \eta) \sim r^{-(1/2)-\epsilon}$, $\partial g \sim r^{-(3/2)-\epsilon}$ will suffice (cf. [1]). Further asymptotic conditions will emerge, implicitly or explicitly, during the derivation.

One sees easily why P reduces to a flux integral at spatial infinity, and how its conservation comes about. Moreover, the relation of P to an Action variation δS facilitates the proof of general theorems, notably the result that any two of the following properties imply the third: symmetry, extremality, and satisfaction of the field equations. Here "symmetry" refers to invariance under motion along ξ^{μ}, and "extremality" means that $P(\xi; \overset{\circ}{\nabla})$ is unchanged by small variations of the metric (or other fields if they are present). For example, for ξ^{μ} asymptotically timelike, the theorem asserts in particular that any stationary solution is an extremum (critical point) of the total energy. From this in turn the existence of various gravitational solitons can be excluded (cf. [1])). If there is time[3] I will indicate later how the particular implication, stationary + solution → extremum, is derived.

A second virtue of (1) is that it provides a single unified expression for all ten geometrical conserved quantities. Unlike the Komar integral, it does not need a different normalization for angular momentum than for energy. Hence there is no restriction on which symmetries it can accept. (For example ξ can be *any* linear combination of time translation and spatial rotation generators.)

A third virtue of (1) is that, like the Komar integral, it is what might be called "fully covariant" and "potentially quasi-local". That is, it takes the form of an integral of a fully four-dimensional integrand over an arbitrary two-surface [or in higher-dimensional generalizations on $(n$-$2)$-surface]. In contrast the ADM-type surface integrals are not 4-dimensional in form because they contain expressions which are defined only with respect to the spacelike hypersurface in which the 2-surface of integration is embedded. Because of the potential to integrate over any 2-surface, and not merely ones at infinity, we might in principle define quasi-local expressions for mass and the other conserved quantities. However it is not clear what rule, if any, could restrict the largely arbitrary

---

[3]There wasn't.

extension of $\mathring{\vartheta}_\mu$ (and therefore of ξ^λ) to non-asymptotic regions of spacetime in such a way to render these quasi-local expressions of any interest. An obvious first question along such lines is whether some choice of $\mathring{\vartheta}$ would yield one or more of the other, more genuinely quasi-local masses which have been discussed in this conference.

A fourth feature of (1) is that the divergence of its (anti-symmetrized) integrand typically falls off like ξ/r^4 at large radii. This is a big help in discussing the conservation of P, and also in connection with the crucial, interlinked issues of background-dependence and of "additivity". Appropos of the former issue, notice that (at least when $g_{\mu\nu}$ is a solution) the value of P is insensitive to modifications of the background in the interior of spacetime: it can depend at worst on how $\mathring{\vartheta}$ is chosen near infinity.

The latter issue refers to the desire that conserved quantities be additive over widely separated isolated systems. In the absence of such additivity, conservation laws would lose most of their value. For example it would not then be possible to relate energy changes in an astrophysical system to the gravitational radiation emitted by the system. Nor, to pass from the big to the tiny, would it be possible to argue that a spin-$\frac{1}{2}$ geon [2] must be stable because the gravitons it might omit can carry off only integral angular momentum.

A fifth virtue of our definition of P, or really of the particular way it has been written in (1), is that it relates very simply to the well-known spinorial (and manifestly background independent!) expression for energy-momentum

$$\mathrm{Im} \oint \bar{\psi}\, \gamma^\mu \not{\xi}\, \gamma^\nu\, \psi\, dS_{\mu\nu}\,. \tag{2}$$

On one hand this makes possible a simplified presentation [6] of the positive energy proof [5] based on (2), and on the other hand, it leads directly to a proof of the background independence of $P(\xi;\mathring{\vartheta})$, at least for translational ξ^μ, i.e. for energy and linear momentum. It also leads in 5 dimensions to a simplified proof [6] of the so-called Bogomol'ny inequality [7], (electric charge)^2 + (magnetic charge)^2 $\leqslant$ (mass)$^2/2$.

Indeed the generalization of $P(\xi; \overset{\circ}{\nabla})$ to Kaluza-Klein spacetimes is relatively easy precisely because (1) does not require the introduction of a flat metric (or indeed any metric) near infinity, as for example, the ADM expression did until recently [3]. For the monopole spacetime in 5 dimensions [4] (and even for the vacuum in typical higher-dimensional cases) the topology near infinity precludes the existence of any flat metric at all. Nonetheless, there is a natural choice of background and J. Lee has shown that it yields the correct value of the monopole mass, when used in (1).

As well as generalizing to higher dimensions, it looks as though the derivation of (1) ought to generalize to higher derivatives, i.e. to theories including terms like $(\text{Riemm})^2$ in the Lagrangian. In fact I think such an application might lead to a better understanding of the "zero-mass theorem" of reference [8].

The final virtue I would like to mention is the close relationship (1) bears to many common, or recently proposed expressions for gravitational conserved quantities. In addition to (2), which I have just discussed, these expressions include those of Arnowitt-Deser-Misner [9], Landau-Lifschitz [10], Freud [11] (which includes the so-called Einstein pseudo-tensor), Nester [12] (if $\overset{\circ}{\nabla}_\mu \xi^\nu = 0$), and especially Chrusciel [13]. In fact expression (1) for $P(\xi; \overset{\circ}{\nabla})$ reduces precisely to (3.10) of [13] if a background *metric* $f_{\mu\nu}$ is introduced and $\overset{\circ}{\nabla}_\mu$ taken to be the associated torsion-free connection $\nabla_\mu[f_{\alpha\beta}]$. Conversely (1) may be viewed as a generalization of the Chrusciel expression; and then, perhaps, the flexibility of allowing non-metric backgrounds may be proposed as one "extra virtue" of the expression to whose derivation I now turn.

[The Idea Behind the Derivation:]

Before having to deal with the special complications that accompany general covariance, it may be helpful to recall how geometrical conserved quantities arise in the mechanics of non-relativistic point particles (cf. [14]). Let us contemplate then, a system of N interacting particles whose world lines between times t_i and t_f are described by the N functions, $q_k: [t_i,t_f] \to \mathbb{R}^3$, $k = 1$ to N. The Action associated to this portion of the system's history is

(3)
$$S = \int_{t_i}^{t_f} dt\ L(q(t),\dot{q}(t),t).$$

Now, while leaving untouched the initial endpoints $(q_k(t_i),t_i)$ of the particle worldlines, let us apply to their final endpoints an overall (infinitesimal) spacetime displacement $(\delta x,\ \delta t)$:

(4)
$$\delta q_1 = \delta q_2 = \cdots = \delta q_N = \delta x, \quad \delta t_f = \delta t.$$

Assuming the equations of motion are satisfied by the unvaried worldlines, we obtain from this variation,

$$\delta S = -E\delta t + P\cdot\delta x,$$

or in more "relativistic" notation,

(5)
$$\delta S = P_\mu \delta x^\mu.$$

Here of course, P and E are respectively the total linear momentum and energy of the system of particles, P_μ is the corresponding spacetime (co)vector $(-E,\ P)$ and δx^μ is the spacetime vector with components $(\delta t,\ \delta x^1,\ \delta x^2,\ \delta x^3)$. Had we in addition applied an infinitesimal *rotation* $\delta\omega$ to the final particle configuration, we would have obtained an additional contribution to δS of $J\cdot\delta\omega$, J being the total angular momentum.

From eq. (5), which we may take as *defining* energy and momentum, the conservation of P_μ follows very directly. In the first place, it ensues from (5) that for the more general variation in which final *and* initial configurations are independently displaced,

(6)
$$\delta S = P_\mu(f)\delta x^\mu(f) - P_\mu(i)\delta x^\mu(i).$$

Let us apply (6) to the particular variation for which the worldline portions undergo an *overall* spacetime translation through $(\delta t,\ \delta x)$. On one hand (6) then reduces to

$$\delta S = [P_\mu(f) - P_\mu(i)]\delta x^\mu.$$

On the other hand, the Action S is invariant under translation, assuming that no external forces are acting. Setting $\delta S = 0$ for arbitrary δx^{μ}, we conclude that $P_{\mu}(f) = P_{\mu}(i)$.

Now let us pause to admire two properties of the Action – its finiteness and its "additivity" – which figured crucially, if tacitly, in the above derivation. Had we not restricted the integral in (3) to a finite time-interval, or had the Action been infinite for some other reason, the variation (5), which defined P_{μ}, would have been meaningless. And had the Action not been additive over isolated subsystems or "clusters" of particles, the resulting conserved quantities would have inherited its non-additivity, rendering them (as I argued above) useless. Indeed finiteness itself would seem in practice to presuppose additivity because, unless the system in question comprehends all existing particles, what we call "its" Action is really only a term in the Action of some larger system of particles. But such a term can be meaningfully split off only insofar as the larger Action actually decomposes as a sum of terms, each of which involves only those particles comprising the subsystem to which it corresponds.

[Complications Due to Gravity:]

In attempting to formulate the definition (5) for General Relativity we immediately meet problems with finiteness and additivity, as well as ambiguities stemming from the underlying diffeomorphism-invariance of the theory. Corresponding to the r^{-2} falloff of the gravitational force (or in particle language the masslessness of the graviton) one expects a typical asymptotically flat metric g to approach flatness like $1/r$, with $\partial g \sim 1/r^2$ and $\partial\partial g \sim 1/r^2$. Given this behavior, a simple counting of powers of $1/r$ would lead one to expect the integral defining the second-order action,

(7) $$S = \frac{1}{2}\int R\, d^4\mathrm{V} ,$$

to behave like $\int r^{-3} d^3x \; dt \sim \int (dr/r) \int dt$, which diverges logarithmically even if the region of integration is contained in a bounded interval of time. Now it is true (and perhaps secretly vital) that for solutions of the vacuum Einstein equations S is nevertheless trivially finite because $R = 0$. However (5) asks us to evaluate S for metrics $g + \delta g$ which do not (and in general cannot) satisfy the equations of motion, and for such metrics no fortuitous cancellation of the individual terms in R can be expected to occur. The consequent failure of (7) to converge absolutely means that, even if it turned out to be finite, S would likely fail to be additive because the interference terms between the long-range fields of distinct isolated systems would not go to zero as the systems receded from one another.

Moreover, the very meaning of the variation involved in (5) becomes unclear for gravity because here there is no underlying Minkowski metric with respect to which one could define a spacetime translation or rotation of (the final configuration of) the physical metric, $g_{\mu\nu}$. And even if there were such an underlying metric we would still face the question of what boundary conditions to impose on δg at spatial infinity, a question that did not even arise for point particles.

[Meaning of δ for Gravity:]

A partial answer to the question of how to define δ emerges from the following reflections. To displace a system in practice means to displace it with respect to its physical environment. For an isolated gravitational system described by an asymptotically flat metric, the environment is represented by the $r \to \infty$ region, which is idealized as approaching perfect flatness, even though in reality other systems will be present there, and even though spacetime on the largest scales is almost certainly not flat at all. Thus, if the displacement δ is to affect only the system under consideration, and not its implicit surroundings, then it must vanish as $r \to \infty$. And if it is to effect, for example, a time-translation of the system, it must have the form of Lie derivative with respect to a vector field ξ^μ which goes over asymptotically to the time-translation generator defined by the asymptotic metric. Only in this way, for example, would it cause events occurring within the isolated system to be recorded at the appropriately altered times by fixed clocks located in the asymptotic region.

In light of these reflections we may formulate boundary conditions on the variation δ. To that end let Ω be the region whose Action we wish to evaluate, and which we assume is contained between two (asymptotically spacelike and flat) hypersurfaces $\mathcal{H}_i$ and $\mathcal{H}_f$; and let ξ^μ be the asymptotic Killing vector corresponding to the conserved quantity we wish to define. Then our earlier conditions $\delta q(i) = 0$, $\delta t_i = 0$ become here:

$$(8) \qquad \delta g_{\mu\nu} = 0 \quad \text{near} \quad \mathcal{H}_i\,; \quad \delta\mathcal{H}_i = 0.$$

On the other hand, our earlier choices of the endpoint variations, $\delta q(f)$ and δt_f, generalize respectively to the two parts of the following condition.

$$(9) \qquad \delta g_{\mu\nu} = -\,\mathcal{L}_\xi\, g_{\mu\nu} \quad \text{near} \quad \mathcal{H}_f\,; \quad \delta\mathcal{H}_f = \xi^\mu.$$

(Here the second equality means that Ω should be altered by dragging its final boundary, $\mathcal{H}_f$, through ξ^μ.) Finally, there is the condition that the environment remain unaffected:

$$(10) \qquad \delta g_{\mu\nu} \to 0 \quad \text{as} \quad r \to \infty\,.$$

Notice that this last condition would conflict with (9) if ξ were not Killing at $r = \infty$, i.e. if we did not have,

$$(11) \qquad \mathcal{L}_\xi\, g_{\mu\nu} \to 0 \quad \text{as} \quad r \to \infty\,.$$

[The Role of the Background Connection:]

Although conditions (8) - (11) go a long way toward uniquely specifying δ, they do nothing about the problem that the resulting variation leads, when coupled with (7), to a formally divergent expression for δS, even for the case of energy-momentum, where ξ itself is bounded as $r \to \infty$. Perhaps this divergence is trying to tell us that our division of spacetime into regions before, between and after the hypersurfaces $\mathcal{H}_i$ and $\mathcal{H}_f$ is inappropriate; or perhaps it is a symptom of some even more fundamental error in our conceptions; or perhaps it is merely a harmless technical nuisance. In any case it can be ameliorated to a great degree by the familiar technique of converting S to a first-order expression by adding a suitable total divergence. Unfortunately such a divergence cannot be defined without introducing a (largely arbitrary) background structure into the spacetime. The structure used has variously been a coordinate system, a tetrad field, a metric, or a connection. I will choose the last alternative as the minimal structure needed for the purpose at hand.

With the aid of such a background connection $\mathring{\nabla}_\mu$, we can easily recognize the divergence we want as the first term in the right hand side of the identity,

$$R = \nabla_\mu(\Gamma^\mu_{\alpha\alpha} g^{\alpha\alpha} - \Gamma^\alpha_{\alpha\mu} g^{\mu\mu}) + (\Gamma^\mu_{\nu\alpha}\Gamma^\nu_{\mu\alpha} - \Gamma^\beta_{\beta\mu}\Gamma^\mu_{\alpha\alpha} + \mathring{R}_{\alpha\alpha}) g^{\alpha\alpha}.$$

Here $\Gamma = \nabla - \mathring{\nabla}$ as defined earlier, and $\mathring{R}_{\alpha\beta}$ is[4] the Ricci-tensor of $\mathring{\nabla}$. Removing the corresponding divergence from S yields the "equivalent" first order Action

$$S^{(1)} = \frac{1}{2}\int_\Omega R\, dV - \frac{1}{2}\oint_{\partial\Omega} \omega^\mu\, dS_\mu, \tag{12}$$

$$= \frac{1}{2}\int [\mathring{R}_{\alpha\alpha} + \Gamma^\mu_{\nu\alpha}\Gamma^\nu_{\mu\alpha} - \Gamma^\beta_{\beta\mu}\Gamma^\mu_{\alpha\alpha}] g^{\alpha\alpha} dV,$$

where

$$\omega^\mu = \Gamma^\mu_{\alpha\alpha} g^{\alpha\alpha} - g^{\mu\mu}\Gamma^\alpha_{\mu\alpha} \tag{13}$$

and $dV = \sqrt{-g}\, d^4x$. Notice that the integral over $\partial\Omega$ includes a contribution from spatial infinity, as well as from $\mathcal{H}_i$ and $\mathcal{H}_f$.

Now let us assume that

$$\Gamma^\alpha_{\beta\gamma} = 0(r^{-2}) \text{ and } \mathring{R}_{\alpha\beta} = 0(r^{-4}).$$

Then each term in the second line of (12) is $0(r^{-4})$; whence $S^{(1)}$ will

[4]With respect to these conventions for arbitrary connections ∇: $[\nabla_\mu, \nabla_\nu]v^\alpha = R^\alpha_{\beta\mu\nu} v^\beta$, $R_{\mu\nu} = R^\alpha_{\mu\alpha\nu}$.

be finite and additive as long as the hypersurfaces $\mathcal{H}_i$ and $\mathcal{H}_f$ are *parallel* near infinity. On the other hand if one of these surfaces is boosted with respect to the other, the integral defining $S^{(1)}$ will have the character $\int dt\, d^3x\, r^{-4} \sim \int r^3 dr/r^4 \sim \int dr/r \sim lg\ \infty$. The extra power of r here is contributed by the t-integral, whose limits of integration now grow linearly with spatial radius, r. I think the notorious extra difficulty in defining the Lorentz-generators $J_{\mu\nu}$ compared to the translation generators P_μ traces partly to this residual failure of finiteness.[5]

As a byproduct of our effort to improve the falloff of the Lagrangian at spatial infinity, our Action $S^{(1)}$ is no longer invariant under arbitrary diffeomorphisms. We should thus restrict ourselves for consistency to variations under which $S^{(1)}$ is invariant, i.e. to generators ξ^μ such that $\mathcal{L}_\xi$ commutes with $\mathring{\nabla}_\mu$. This is implicitly a restriction on $\mathring{\nabla}$; and conversely it ties ξ^μ so closely to the background that no independent ambiguity remains beyond that present in $\mathring{\nabla}_\mu$ itself.[6] Finally, let me record the further "condition on the variation δ" which merely reminds us that $\mathring{\nabla}$ is indeed a background field rather than a dynamical variable:

$$\delta\mathring{\nabla}_\mu = 0. \tag{14}$$

[5]One way out of this difficulty would be to live in a spacetime of dimensionality $n > 4$. Indeed the corresponding falloff, $\partial g \sim r^{2-n}$ leads to $\int \Gamma\Gamma\, d^n x \sim \int dr\, r^{3-n}$, even for a region of integration with non-parallel bounding hypersurfaces. Unfortunately it is only the asymptotically flat dimensions which count; extra microscopic ones are of no use in this respect.

[6]At least this is so if $\mathring{\nabla}_\mu$ commutes with ξ^μ globally. For topologies other than $\mathbb{R}^4$ such global compatibility can be impossible to arrange; but it would be hard to believe that the validity of (1) could require anything more than compatibility in some neighborhood of spatial infinity. Nevertheless the particular derivation given below does in fact assume that the background connection is *everywhere* Lie derived by ξ^μ.

REMARK. It turns out that the boundary integrand in eq. (12) is closely related to tr K, the trace of the extrinsic curvature or "second fundamental form" of $\partial\Omega$. Unfortunately the latter does not by itself seem to afford any improvement in the finiteness of the Action; but perhaps its connection to (13) could suggest a way to devise a more nearly background-independent equivalent for $S^{(1)}$.

[Computation of $\delta S^{(1)}$:]

We have now almost finished setting up the computation that will lead to the promised formula (1). In order to have an explicit form of $\delta g_{\mu\nu}$ to work with, I will make the "Noether Ansatz"

$$\delta g_{\mu\nu} = -f \, \mathcal{L}_\xi \, g_{\mu\nu}, \tag{15}$$

where f is a scalar function on spactime. In order to satisfy conditions (8)-(10) we may choose for f a bounded function which reduces to zero in a neighborhood of $\mathcal{H}_i$ and to 1 in a neighborhood of $\mathcal{H}_f$. [Notice that the superficially similar Ansatz, $\delta g_{\mu\nu} = -\mathcal{L}_{f\xi} g_{\mu\nu}$ would *not* satisfy (10)!] A particularly convenient choice is $f = f(t)$ where t is a time-function extending $\mathcal{H}_i$ and $\mathcal{H}_f$ to a foliation of Ω.

We could now just compute the change of $S^{(1)}$ under the variation (15), but doing so would not reveal clearly why the final result is a flux integral at infinity rather than an integral over $\mathcal{H}_f$. Remembering that this simplification comes from general covariance, let us, before evaluating $\delta S^{(1)}$, perform the *overall* diffeomorphism generated by $-f\xi^\mu$. This trick has two effects. First it restores the region of integration Ω (which we take to be compact with a "cylindrical" boundary at very large radius) to its unvaried shape and causes $\delta g_{\mu\nu}$ to vanish on $\mathcal{H}_f$ as well as on $\mathcal{H}_i$. Second it introduces a change in $\mathring{\vartheta}_\mu$, though again not on $\mathcal{H}_f$ thanks to the invariance of $\mathring{\vartheta}$ under ξ. The net result is that δ takes the form

$$\delta\Omega = \xi^\mu - \xi^\mu = 0,$$

$$\begin{aligned}\delta g_{\mu\nu} &= -f\, \mathfrak{L}_\xi g_{\mu\nu} + \mathfrak{L}_{f\xi} g_{\mu\nu} \\ &= (\nabla_\mu f)\xi_\nu + (\nabla_\nu f)\xi_\mu, \\ \delta\mathring{\nabla} &= [\mathfrak{L}_{f\xi}, \mathring{\nabla}] \\ &= \mathring{\nabla}_\alpha \mathring{\nabla}_\beta (f\xi^\mu) - f\mathring{\nabla}_\alpha \mathring{\nabla}_\beta \xi^\mu ,\end{aligned}$$

where the last equality is proved by subtracting $0 = [\mathfrak{L}_\xi, \mathring{\nabla}]$ from $\delta\mathring{\nabla}$ and applying the identity

$$[\mathfrak{L}_\eta, \mathring{\nabla}]^\mu_{\alpha\beta} = \mathring{\nabla}_\beta \mathring{\nabla}_\alpha \eta^\mu - \mathring{R}^\mu_{\alpha\beta\nu}\eta^\nu .$$

Now assume $G_{\mu\nu} = 0$ and vary $S^{(1)}$. The result will be an integral over $\partial\Omega$ involving δg, $\delta\nabla$, and $\delta\Gamma = \delta\nabla - \delta\mathring{\nabla}$. It is easy to verify that (i) The terms in $\delta\nabla$ drop out, (they had to since $S^{(1)}$ is only first order in ∂g.) leaving an effective variation in which $\delta\nabla = 0$, $\delta\Gamma = -\delta\mathring{\nabla}$; (ii) The terms in $\mathring{\nabla}\mathring{\nabla} f$ combine to give a total divergence, which can be integrated away because $\partial\partial\Omega = \emptyset$; and (iii) the first derivative ∇f appears only in the combination $dS_\mu \nabla_\nu f - dS_\nu \nabla_\mu f$, where dS_μ is the surface element of $\partial\Omega$.

This last result means that $\delta S^{(1)}$ reduces to some superposition (depending on the restriction of f to $\partial\Omega$) of flux integrals over 2-surfaces embedded in $\partial\Omega$. In the simplest case, we may choose f as the step function $f = f(t) = (t > t_f - \epsilon)$ where $t = t_f$ labels $\mathcal{H}_f$. Then $\delta S^{(1)}$ reduces to

$$(16) \qquad \frac{1}{2} \oint_\Sigma dS_{\mu\nu}\{\nabla^\mu \xi^\nu + \omega^\mu \xi^\nu\},$$

where Σ is the intersection of $\mathcal{H}_f$ with the "cylindrical" portion of $\partial\Omega$. Sending this cylindrical boundary, and hence the 2-surface Σ, to $r = \infty$, we recover $P(\xi; \mathring{\nabla})$ as defined at the outset.

If we don't assume that $G_{\mu\nu} = 0$ then we find instead of (16),

$$\delta S^{(1)} = \int_\Omega -\frac{1}{2} G^{\mu\nu}\delta g_{\mu\nu}\, dV + P(\xi, \mathcal{H}_f; \mathring{\nabla})$$

where P is now (16) with $\Sigma = \partial\mathcal{H}_f$ (a surface at infinity) supplemented by the hypersurface integral,

$$- \int_{\mathcal{H}_f} G^{\mu}_{\lambda} \xi^{\lambda} \, dS_{\mu} ,$$

and Ω is the region between $\mathcal{H}_i$ and $\mathcal{H}_f$.

ADDED NOTE. The derivation given here connects up with the "Noether Operator" framework of [1] as follows (see also [15]). First introduce a generalized Noether operator $T^{\mu}{}_{\lambda}$ via the identity (for arbitrary g, Ω , ξ and f),

$$\delta S^{(1)} = \int_{\Omega} \frac{1}{2} \frac{\delta L^{(1)}}{\delta g_{\mu\nu}} \delta g_{\mu\nu} dV + \oint_{\partial\Omega} f \, T^{\mu}_{\lambda} \cdot \xi^{\lambda} \, dS_{\mu},$$

where $\delta g = -f \mathfrak{L}_{\xi} g$ with $\delta\Omega = f\xi$. This equation appears to determine $T^{\mu}{}_{\lambda}$ uniquely, resolving the ambiguity present in the definition of [1]. When $G_{\mu\nu} = 0$ the resulting vector $T^{\mu}_{\lambda} \cdot \xi^{\lambda}$ is precisely the divergence of the integrand in (1), i.e. it is the current whose integral over $\mathcal{H}$ gives $P(\xi; \mathring{\nabla})$. When $G_{\mu\nu} \neq 0$, $T \cdot \xi$ is augmented by the term we have just encountered, $-G^{\mu}_{\lambda} \xi^{\lambda}$.

When matter is present, T^{μ}_{λ} can be defined in the same way. Usually it will be the sum of the gravitational Noether operator just defined, with the ordinary matter stress-energy tensor, T^{μ}_{λ}(matter).

REFERENCES

[1] B. F. Schutz and R. Sorkin, "Variational Aspects of Relativistic Field Theories, with Application to Perfect Fluids", Annals of Physics, **107** (1977), 1-43.

[2] R. D. Sorkin, "Introduction to Topological Geons", in P. G. Bergmann and v. de Sabbata (eds.), *Topological Properties and Global Structure of Spacetime* (Plenum, 1986).

[3] R. K. Koul, Thesis, Syracuse University, 1986.

[4] R. D. Sorkin, "A Kaluza-Klein Monopole", Phys. Rev. Lett. **51** (1983), 87-90.

[5] E. Witten, "A New Proof of the Positive Energy Theorem", Comm. Math. Phys. **80**, 381 (1981).

[6] J. Lee and R. D. Sorkin, "A Derivation of the Bogomol'ny Inequality for *U*(1) Kaluza-Klein Theory", Syracuse preprint.

[7] G. W. Gibbons and M. J. Perry, "Soliton Supermultiples and Kaluza-Klein Theory", Nuc. Phys. **B248**, 629 (1984); O. M. Moreschi and G. A. J. Sparling, "On the Formulation of the Positive Energy Theorem in Kaluza-Klein Theories", J. Math. Phys. 27, 2402 (1986), JMP 27(9) 2402 (1986).

[8] D. G. Boulware, G. T. Horowitz and A. Strominger, "Zero Energy Theorem for Scale-Invariant Gravity", Phys. Rev. Lett. **50**, 1726 (1983).

[9] R. Arnowitt, S. Deser, C. Misner, "Coordinate Invariance and Energy Expressions in General Relativity", Phys. Rev. **122**, 997 (1961).

[10] L. D. Landau and E. M. Lifshitz, *The Classical Theory of Fields* (Pergamon, 1962), §100.

[11] Ph. Freud, "Uber die Ausdrücke der Gesamtenergie und des Gesamttimpulses eines Materiellen Systems in der Allgemeinen Relativitätstheorie", Ann. Math. **40**, 417 (1939).

[12] J. M. Nester, "A New Gravitational Energy Expression with a Simple Positivity Proof", Phys. Lett. **83A**, 241 (1981).

[13] P. T. Chrusciel, "On the Relation Between the Einstein and the Komar Expressions for the Energy of the Gravitational Field", Ann. Inst. Henri Poincare **42**, 267 (1985).

[14] J. Schwinger, "The Theory of Quantized Fields. I", Phys. Rev. **82**, 914 (1951).

[15] E. Nahmad-Achar and B. F. Schutz, "Conserved Quantities from Pseudotensors and Extremum Theorems for Angular Momentum", Cardiff preprint (1986).

RAFAEL SORKIN
PHYSICS DEPARTMENT
SYRACUSE UNIVERSITY
SYRACUSE, NEW YORK 13244-1130

Contemporary Mathematics
Volume **71**, 1988

A 3+1 FORMULATION OF EINSTEIN SELF-DUALITY

Abhay Ashtekar[1]

ABSTRACT. The complex Einstein equation is considered on a real 4-manifold $M = \Sigma \times R$ where Σ is an arbitrary 3-manifold. A 3+1 decomposition of Einstein's equation is first carried out with emphasis on points where the discussion differs from the real case. New variables, adapted to the (anti-) self-duality condition, are then introduced and the 3+1 equations are recast in terms of them. In the (anti-) self-dual case, these equations simplify considerably; they reduce to just two equations on triads (of vector fields) on Σ which have a simple geometrical interpretation. The resulting characterization of half-flat solutions offers a new tool to investigate the mathematical structure - particularly the issue of exact integrability - of the half-flat Einstein system.

1. INTRODUCTION

Over the past decade, considerable work has been on half-flat solutions to Einstein equation both in the (real) Euclidean and the (complex) Lorentzian regime. (See, e.g. [1].) It came as a surprise that the half-flat condition simplifies the structure of Einstein's equation enormously. In the Lorentzian regime, for example, Newman, Penrose and Plebanski have developed three different approaches to obtain the "general" half-flat solution to Einstein's equation. The analogous problem in the general (non half-flat)

[1]Supported in part by the NSF grants PHY 86-12424 and PHY 17853 supplemented by funds from NASA.

case, on the other hand, remains hopelessly difficult in spite of the fact that it has drawn attention for over fifty years. Another striking feature of these solutions is the discovery by Newman and collaborators that, in spite of non-linearities, the classical S-matrix in the Lorentzian case is trivial [2]. Very few non-linear systems have such striking properties. Indeed, it is tempting to conjecture that the half-flat system may be exactly integrable. If this were to be the case, one could look for the associated system of infinitely many conservation laws and go on to construct an exactly soluble 4-dimensional quantum field theory. Such a model would be of considerable interest to mathematical physics in general since exactly soluble 4-dimensional systems are hard to come by, and to quantum gravity in particular since it may provide considerable insight into the role of non-linear self-dual solutions in quantum gravity. Unfortunately, however, the methods that have been generally used to construct and investigate half-flat solutions are not well-suited to analyse such issues. The H-space techniques of Newman et al. as well as the twistor constructions employed by Penrose in this context, for example, primarily use notions which are non-local in space-time and therefore difficult to use in the traditional setting of Hamiltonian methods.

The purpose of this talk is to present a new formulation of half-flat solutions to Einstein's equation which is better adapted to answer the questions of exact integrability and traditional quantization. The final result is remarkably simple. Fix a real, 3-manifold Σ and a volume element thereon. Consider a triad $V^a{}_{\underset{\sim}{a}}$ of vector fields, $\underset{\sim}{a}$ 1,2,3, satisfying constraint equations:

$$\text{(Div) } V^a{}_{\underset{\sim}{a}} = 0, \tag{1}$$

where Divergence refers to the fixed volume element, and "evolve" it via:

$$\dot{V}^a{}_{\underset{\sim}{a}} = \epsilon_{\underset{\sim}{abc}}[V_{\underset{\sim}{b}}, V_{\underset{\sim}{c}}]^a. \tag{2}$$

Then, by purely algebraic manipulations, one can construct from the solutions to these equations a half-flat solution to Einstein's equation on the real 4-manifold $M = \Sigma \times R$. Conversely, every half-flat solution on M arises from a triad satisfying the above system of equations. Thus, the entire content of the half-flat Einstein equation is contained in equations (1) and (2). If the triad is real, the metric is Euclidean. In general, it is complex.

The plan of this presentation is as follows. Section 2 recalls the standard 3 + 1 decomposition of Einstein's equation for a complex 4-metric on a real 4-manifold $M = \Sigma \times R$. Section 3 introduces the new variables and recasts the various projections of Einstein's equation in terms of them. The new variables consist of a pair $(\tilde{\sigma}^a{}_A{}^B, {}^+A_{aA}{}^B)$ [or, $(\tilde{\sigma}^a{}_A{}^B, {}^-A_{aA}{}^B)$] consisting of a (densitized) soldering-form $\tilde{\sigma}^a{}_A{}^B$ for SL(2,C) (more precisely, complexified SU(2)) spinors on Σ, and a SL(2,C) connection ${}^{\pm}A_{aA}{}^B$ on Σ. [Throughout, $\pm$ will stand for + *or* −.] It turns out that, when the initial value constraints are satisfied, ${}^+A_a$ is a potential for the anti self-dual part of the Weyl tensor and ${}^-A_a$, for the self-dual. Consequently, self-duality can be imposed by setting ${}^+A_a = 0$ [and anti self-duality, by setting ${}^-A_a = 0$.] When this is done, the constraint as well as the evolution equations simplify enormously. Section 4 shows that the simplified equations can be recast in terms of triads of vector fields and become (1) and (2) above. Section 5 compares and contrasts the present, 3 + 1, approach to the half-flat Einstein system with other approaches.

Most of the work presented here was done jointly with Ted Jacobson and Lee Smolin. Since the details of this work will appear elsewhere, here I will omit technicalities and concentrate only on the main thread of reasoning and delicate conceptual issues.

2. PRELIMINARIES

Fix a real 3-manifold Σ and let M be the 4-manifold $\Sigma \times R$. Thus, M is endowed with a natural foliation. Let the leaves of the foliation be given by constant values of a real coordinate t. Consider on M, a complex metric g_{ab}. [Throughout, we shall use

Penrose's abstract index notation. See [3].] Thus, g_{ab} is a complex, second rank, symmetric, non-degenerate tensor field on M. [If g_{ab} happens to be real, we assume that it has signature ++++. All our subsequent conventions will be geared to this signature. In the general, complex case, of course, g_{ab} does not have an unambiguous signature.] Using the standard analytic argument, one can show that there is a unique torsion-free derivative operator ∇ compatible with the metric g_{ab}, i.e., satisfying $\nabla_a g_{bc} = 0$. We define the curvature tensors of g_{ab} via:

$$(3)\qquad \begin{aligned} {}^4R_{abc}{}^dK_d &:= 2\nabla_{[a}\nabla_{b]}K_c\ ; &\quad {}^4R_{ac} &:= {}^4R_{abc}{}^b \\ {}^4R &:= g^{ac}\,{}^4R_{ac} &; \quad {}^4G_{ac} &= ({}^4R_{ac} - \tfrac{1}{2}\,{}^4Rg_{ac}) \end{aligned}$$

where ${}^4G_{ac}$ is the Einstein tensor. The metric g_{ab} also selects an alternating tensor - i.e. a 4-form - ϵ_{abcd} up to sign via:

$$(4)\qquad \epsilon_{abcd}\epsilon_{mnrs}g^{am}g^{bn}g^{cr}g^{ds} = +4!$$

The Riemann tensor of g_{ab} will be said to be half-flat if:

$$(5)\qquad {}^4R_{abcd} = \pm\tfrac{1}{2}\,\epsilon_{ab}{}^{mn}\,{}^4R_{mncd}$$

Note that (5) implies that the metric g_{ab} is also Ricci-flat. If we restrict ourselves to real matrices, we can classify the half-flat metrics into two categories: self-dual and anti self-dual. To do this, fix, once and for all, a real 4-form $\overset{\circ}{\epsilon}_{abcd}$ on M. Then given any metric g_{ab}, choose as its unique alternating tensor the 4-form which satisfies (4) and which has the same orientation as the fixed $\overset{\circ}{\epsilon}_{abcd}$, i.e., which satisfies $\epsilon_{abcd} = f\overset{\circ}{\epsilon}_{abcd}$ for a positive function f. [Note that a nowhere vanishing f exists because all 4-forms on M are proportional and it is either everywhere positive or everywhere negative because the fiducial $\overset{\circ}{\epsilon}_{abcd}$ as well as the ϵ_{abcd} satisfying (4) are all real.] Then, place g_{ab} in the self-dual class if its Riemann tensor satisfies (5) with − sign and in the anti self-dual class if it satisfies (5) with + sign. The labels "self-dual" and "anti self-dual" are convention dependent. But the fact that an unambiguous classification for metrics exists is not. If the metric is complex, on the other hand, in general, the unambiguous classification does not exist. For, in the complex case, the function f relating the alternating tensors of the metric to the fiducial $\mathring{\epsilon}_{abcd}$ is in general complex and we cannot pick out a preferred alternating tensor by requiring that f be positive; it is meaningless, in general, to ask for the alternating tensor of g_{ab} which has the same orientation as a fiducial $\mathring{\epsilon}_{abcd}$. Thus, in the complex [or, Lorentzian] case, one can only ask for half-flat metrics; in general, it is impossible to fix conventions and classify the half-flat metrics into two classes. This simple fact seems to have gone unnoticed in the literature. If we restrict ourselves to a *sub-class* of half-flat metrics [such as Newman's H-spaces] we may have additional structure available [acquired from null infinity, where the "data" for H-spaces is specified] and this may enable us to select a preferred alternating tensor for each metric in the sub-class. In this case, an unambiguous classification would be possible. [Also, as we shall see, if we have spinors, or, more precisely, soldering forms, rather than just metrics at our disposal, we can fix the orientation unambiguously.]

We can now consider the 3 + 1 decomposition of Einstein's equation. Let ζ^a be a smooth vector field on M such that $\zeta_a = g_{ab}\zeta^b$ is everywhere proportional to the co-vector field $\nabla_a t$ and such that $g_{ab}\zeta^a\zeta^b = 1$. [Thus, ζ^a is unique up to sign.] Define the "intrinsic metric" q_{ab} on each t = const. slice via

(6.A) $$q_{ab} = g_{ab} - \zeta_a\zeta_b$$

and the "extrinsic curvature" of this slice via

(6.B) $$K_{ab} = q_a{}^m q_b{}^n \nabla_m \zeta_n.$$

Then, a straightforward calculation enables us to express certain components of the Einstein tensor of g_{ab} in terms of q_{ab} and K_{ab}:

(7.A) $$-2\,{}^4G_{ab}\zeta^a\zeta^b = R + K_{ab}K^{ab} - K^2; \quad \text{and}$$

(7.B) $${}^4G_{mn}q^{ma}\zeta^n = D_m(K^{ma} - Kq^{ma}).$$

Furthermore, if we introduce a vector field t^a, parallel to ζ^a, $t^a = N\zeta^a$, such that, $t^a\nabla_a t = 1$, we can express the "time-evolution" of q_{ab} and K_{ab} in terms of the remaining components of the Einstein tensor:

$$\text{(8.A)} \quad \dot{q}_{ab} \equiv \mathfrak{L}_t q_{ab} = 2N\, K_{ab}; \text{ and,}$$

$$\text{(8.B)} \quad \dot{K}_{ab} \equiv \mathfrak{L}_t K_{ab} = -D_a D_b N + 2N\, K_a{}^m K_{bm} - N\, K\, K_{ab} + N\, R_{ab} - N\, q_a{}^m q_b{}^n\, {}^4R_{mn}$$

Here D, R_{ab} and R are, respectively, the derivative operator, the Ricci tensor and the scalar curvature of the metric q_{ab} and K is the trace, $q^{ab}\, K_{ab}$, of the extrinsic curvature, w.r.t. q_{ab}. Thus, the 4-metric g_{ab} satisfies Einstein's vacuum equation (without cosmological constant) if and only if the pair of fields $q_{ab}(t)$, $K_{ab}(t)$ on M satisfy (7) and (8) with $G_{ab} = {}^4R_{ab} = 0$. This result is completely analogous to the familiar one in real general relativity. [Note, however, that, in the real, Lorentzian case, the initial value problem is also well-posed: Given a pair (q_{ab}, K_{ab}) satisfying the constraint equations, we are assured the existence of the solution g_{ab} for some finite time-interval. Presumably, this result fails to hold in general in the present case, except if we restrict ourselves, e.g., to analytic fields.]

3. NEW VARIABLES

The variables q_{ab} and K_{ab} have direct geometric interpretation. However, both constraint and evolution equations for them are complicated, because, e.g., they involve non-polynomial functionals of q_{ab}. Furthermore, these variables are not adapted to the self-duality condition; none of the equations simplify if we assume that g_{ab} is half-flat. This is the primary reason why the simplicity of self-dual solutions has remained opaque in the usual Hamiltonian formulation. In this section we shall introduce new variables which are free of these problems.

The variables we want to introduce are spinorial. Therefore, we need to extend the general mathematical setting in which we have been working. In addition to tensor-fields $T^{a...b}{}_{c...d}$, let us consider on the 3-manifold Σ, fields $\lambda^{A...B}{}_{M...N}$ with SL(2,C) "internal indices" $A,B, ..., M,N$. Thus, a field λ^A is a cross-section of a vector bundle over Σ whose fibre is a 2-dimensional complex vector space equipped with a volume element ϵ^{AB}. We shall raise and lower the internal indices with the field ϵ^{AB} and its inverse ϵ_{AB}:

$$(9) \qquad \lambda^A = \epsilon^{AB}\lambda_B; \qquad \mu_B = \mu^A\epsilon_{AB}$$

[Note that, the fields ϵ^{AB} and ϵ_{AB} are fixed once and for all, independently of the choice of a metric on Σ. Thus, we are not following the usual relativity convention (see, e.g. [3]) in which the ϵ_{AB} carries information about the metric. That convention is adapted to the case when one deals only with a fixed conformal class of metrics.] To convert the internal indices to spinorial ones, we need a soldering form, $\sigma^a{}_A{}^B$ which "solders" the internal indices to the tangent space of Σ. More precisely, we need an isomorphism between (the 3-complex dimensional) vector-space of trace-free second rank tensors $\lambda_A{}^B$ with internal indices and the complexified tangent space of Σ: $\lambda^a = -\sigma^a{}_A{}^B\lambda_B{}^A$. [Note that SL(2,$C$) refers to the fact that we are considering complexified tangent space to the 3-manifold Σ; it is not the double covering of the Lorentz group associated with a 4-metric on $M = \Sigma \times R$. Thus, the SL(2,C) we have should be thought of as complexified SU(2).] Given a soldering-form $\sigma^a{}_A{}^B$, we can define a complex metric q^{ab} on Σ via:

$$(10) \qquad q^{ab} := -\operatorname{Tr}\sigma^a\sigma^b \equiv -\sigma^a{}_A{}^B\sigma^b{}_B{}^A$$

Then, $\sigma_{aA}{}^B := q_{ab}\sigma^b{}_A{}^B$ is the "inverse" of $\sigma^a{}_A{}^B$. This $\sigma_{aA}{}^B$ may be thought of as a "square-root" of the metric q_{ab}.

We are now ready to introduce the new variables. The first of these is a densitized soldering form $\tilde{\sigma}^a{}_A{}^B$; i.e., an isomorphism between trace-free $\lambda_A{}^B$ and vector-densities $\tilde{\lambda}^a$ of weight one, $\tilde{\lambda}^a = \tilde{\sigma}^a{}_A{}^B\lambda_B{}^A$. Given a $\tilde{\sigma}^a{}_A{}^B$, we can define a metric q^{ab} uniquely via $\operatorname{Tr}\tilde{\sigma}^a\tilde{\sigma}^b = -(\det q)q^{ab}$, and an unweighted soldering-form $\sigma^a{}_A{}^B$ via

$\sigma^a{}_A{}^B = (\det q)^{-1/2}\tilde{\sigma}^a{}_A{}^B$. We will regard $\tilde{\sigma}^a{}_A{}^B$ as primary and $\sigma^a{}_A{}^B$ and q^{ab} as derived objects. Our second variable is going to be a SL(2,*C*)-valued connection ${}^{\pm}\mathcal{D}$ which acts on both tensor and internal indices. [Recall that ± stands for + *or* −; we will need only one of these two connections.] Given a pair $(\tilde{\sigma}^a, {}^{\pm}\mathcal{D})$, we are to recover the usual variables q_{ab} and K_{ab} as follows: q_{ab} is given by (10) and K_{ab} is defined via

$$K_{ab} = -\operatorname{Tr} K_{(a}\sigma_{b)} \equiv -K_{(a|A|}{}^B\sigma_{b)B}{}^A \quad \text{where } K_{aA}{}^B \text{ is given by}$$
$$K_{aA}{}^B\lambda_B = \pm\sqrt{2}\,({}^{\pm}\mathcal{D}_a - D_a)\lambda_A. \tag{11}$$

Here D is the unique torsion-free derivative operator which annihilates $\sigma^a{}_A{}^B$. It is often convenient to use connection 1-forms ${}^{\pm}A_{aA}{}^B$ in place of derivative operators ${}^{\pm}\mathcal{D}$. Fix, once and for all, a flat connection ∂ on Σ and set

$$ {}^{\pm}\mathcal{D}_a\lambda_B = \partial_a\lambda_B + {}^{\pm}A_{aB}{}^A\lambda_A \tag{12}$$

Thus, ${}^{\pm}A_{aA}{}^B$ are related to the spin-coefficients $\Gamma_{aA}{}^B$ of $\sigma^a{}_A{}^B$ via:

$$ {}^{\pm}A_{aA}{}^B = \Gamma_{aA}{}^B \pm \frac{1}{\sqrt{2}}K_{aA}{}^B. \tag{13}$$

Given q_{ab}, K_{ab} and a soldering form $\sigma_{aA}{}^B$ for q_{ab}, we can reverse the above procedure and simply define ${}^{\pm}A_{aA}{}^B$ via (13) or ${}^{\pm}\mathcal{D}$ via (11). Thus, roughly, the first new variable, $\tilde{\sigma}^a$, can be thought of as a square-root of the metric q^{ab}, while the second, ${}^{\pm}A_a$, as a spin-connection that knows both about the intrinsic spin connection Γ_a and the extrinsic curvature K_{ab}. These variables first arose in the context of a Hamiltonian formulation of general relativity [4]. For brevity, here I shall omit both the original motivation and the systematic procedure by which Einstein's equation can be recast in terms of these variables and simply report the final result.

In the passage from q_{ab}, K_{ab} to $\tilde{\sigma}^a$, ${}^{\pm}A_a$, we have introduced three (complex) new degrees of freedom: The SL(2,*C*) transformations on the spinorial indices do not affect the tensorial quantities q_{ab} and K_{ab}. [See equations (10) and (11).] Consequently, we will now have three new constraint equations. These turn out to be simply:

(14) $$0 = {}^{\pm}\mathcal{D}_a \tilde{\sigma}^a{}_A{}^B$$

[Note that this equation is very analogous to the Gauss law constraint in Yang-Mills theory, with $\tilde{\sigma}^a$ playing a role of the Yang-Mills electric field. In the Hamiltonian formulation, it generates precisely the SL(2,C) "gauge rotations" on the internal indices.] Modulo (14) the familiar constraints [equations (7) with $G_{ab} = 0$] can be now written as:

(7'.A) $$\mathrm{Tr}\ \tilde{\sigma}^a\ {}^{\pm}F_{ab} = 0, \text{ and}$$

(7'.B) $$\mathrm{Tr}\ \tilde{\sigma}^a \tilde{\sigma}^b\ {}^{\pm}F_{ab} = 0,$$

where ${}^{\pm}F_{abA}{}^B$ is the curvature of ${}^{\pm}A_{aA}{}^B$:

(15) $${}^{\pm}F_{abA}{}^B = 2\partial_{[a}\,{}^{\pm}A_{b]A}{}^B + {}^{\pm}A_{aA}{}^M\,{}^{\pm}A_{bM}{}^B - {}^{\pm}A_{bA}{}^M\,{}^{\pm}A_{aM}{}^B.$$

These are the constraint equations. The "configuration variable" $\tilde{\sigma}^a$ has 9 components per space-point. Equations (14) and (7') provide us with 3 + 3 + 1 = 7 constraints, leaving us with 2 true degrees of freedom per space-point. The evolution equations for $\tilde{\sigma}^a$ and ${}^{\pm}A_a$ turn out to be:

(8'.A) $$\dot{\tilde{\sigma}}^a = \pm\, 2\sqrt{2}\ {}^{\pm}\mathcal{D}_b(\underset{\sim}{N}\ \tilde{\sigma}^{[a}\tilde{\sigma}^{b]}), \text{ and,}$$

(8'.B) $${}^{\pm}\dot{A}_a = \pm\, \sqrt{2}\ \underset{\sim}{N}\ [\tilde{\sigma}^b, {}^{\pm}F_{ab}].$$

Here, $\underset{\sim}{N}$, a density of weight −1, is the new lapse, related to the more familiar lapse function N [of equation (8)] via $N = (\det\ q)^{1/2}\underset{\sim}{N}$. Using the definitions (10) and (11) of q_{ab} and K_{ab}, one can verify that (8') implies that q_{ab} and K_{ab} evolve via (8). The constraint equations (7') and (14) are preserved under evolution. Thus, given a 1-parameter family of fields $\tilde{\sigma}^a(t)$ and ${}^{\pm}A_a(t)$ satisfying (14), (7') and (8'), on $M = \Sigma \times R$, we obtain a vacuum solution of Einstein's equation, $g^{ab} = (\det\ q)^{-1}\{-\mathrm{Tr}\ \tilde{\sigma}^a\tilde{\sigma}^b + (\underset{\sim}{N})^{-2}\, t^a\, t^b\} = q^{ab} + \zeta^a\zeta^b$.

REMARKS. (i) In the real, Euclidean case, the space of spinors is equipped with a Hermitian conjugation operation, $\lambda_A \to \lambda_A^+$, and the soldering form $\sigma^a{}_A{}^B$ maps Hermitian, trace-free $\lambda_A{}^B$ to real vectors λ^a. In this case, the fields q_{ab}, K_{ab} are real and $\tilde{\sigma}^a{}_A{}^B$, $K_{aA}{}^B$ and ${}^{\pm}A_{aA}{}^B$ are all Hermitian.

(ii) In the general complex case, $(\det q)^{1/2}$ is unique only up to sign. Hence, given $\tilde{\sigma}^a$ and ${}^{\pm}A_a$, we can recover $\sigma^a{}_A{}^B$ and K_{ab} only up to sign. Given a *particular* $\sigma^a{}_A{}^B$, we can select a particular orientation ϵ^{abc} on Σ by fixing a convention:

$$\epsilon^{abc} = -\sqrt{2}\,\sigma^a{}_A{}^B\sigma^b{}_B{}^D\sigma^c{}_D{}^A. \tag{16}$$

[Recall that an unambiguous convention cannot in general be fixed if we are given only the complex metric q_{ab}.] In the present case, since σ^a has a sign ambiguity, so does ϵ^{abc} given by (16). Given a pair $(\tilde{\sigma}^a(t), {}^{\pm}A_a(t))$ satisfying (14), (7') and (8'), however, we can unambiguously select a 4-orientation ϵ^{abcd} as follows. Compute $q_{ab}(t)$ and $K_{ab}(t)$ and choose that unit normal ζ^a to $\Sigma(t)$ for which $K_{ab} = +2\mathcal{L}_\zeta q_{ab}$. Now, σ^a, ϵ^{abc}, K_{ab} and ζ^a are all unique only up to sign. However, $\epsilon^{abcd} := \zeta^{[a}\epsilon^{bcd]}$ is free of these ambiguities. If the 4-metric g_{ab} is given to be half-flat, we can use this ϵ^{abcd} to decide if its Weyl curvature is self-dual or anti self-dual. That is, the pair $(\tilde{\sigma}^a(t), {}^{\pm}A_a(t))$ has more information than the metric g_{ab} itself.

(iii) Given a pair $(\tilde{\sigma}^a, {}^{\pm}A_a)$, the field $K_{aA}{}^B$ is completely determined by (13). From this, we can construct $\underset{\sim}{K}_{ab} := -\mathrm{Tr}\, K_a\sigma_b$ where σ_b is the unique inverse of $\tilde{\sigma}^a$. However, this $\underset{\sim}{K}_{ab}$ will not in general be symmetric in a and b. Its symmetric part gives us, in essence, the extrinsic curvature [equation (11)]. Constraint (14) requires precisely that the anti-symmetric part of $\underset{\sim}{K}_{ab}$ be zero.

(iv) It is only for simplicity that we have required the fiducial connection ∂ to be flat. It is easy to extend the framework to the case when it is not. The form of constraints and evolution equations remain the same; only the expression of ${}^{\pm}F_{ab}$ in terms of ${}^{\pm}A_a$ acquires an extra term, the curvature of ∂.

(v) The fact that $\tilde{\sigma}^a$ is analogous to the electric field in the $SL(2,C)$ Yang-Mills theory suggests that we may try the Yang-Mills half-flat ansatz, $E^{a}{}_{A}{}^{B} = \pm B^{a}{}_{A}{}^{B}$, on the variables $\tilde{\sigma}^{a}$ and A_{a}. However, it turns out that, in the present case, the fields $\tilde{\sigma}^{a}$ and $B^{a}{}_{A}{}^{B} := \epsilon^{abc} F_{bcA}{}^{B}$ have different physical dimensions. [In the above discussion we have suppressed factors of G, Newton's constant. For details, see [4].] Therefore, to impose the Yang-Mills half-flat condition, we need to introduce a new constant, c, of dimensions $[\text{length}]^{-2}$. Then, we can demand that $\tilde{\sigma}^{a}$ and B^{a} be related via $B^{a} = \pm(c/G)(\det q)^{-1/2}\tilde{\sigma}^{a}$. That is, we may just choose any A_{a}, compute its magnetic field B^{a}, and just define $\tilde{\sigma}^{a}$ through the above "Yang-Mills half-flat ansatz". Then, it turns out that the pair $(\tilde{\sigma}^{a}, A_{a})$ *automatically* solves all initial value equations for Einstein's theory with cosmological constant $\Lambda = (3/\sqrt{2})c$! We do not have to impose any constraint equations by hand. Furthermore, in the 4-dimensional solution obtained by the evolution of such "initial data", the Weyl tensor is automatically half-flat. Thus, Yang-Mills half-flat condition does imply Einstein half-flat condition in the case when $\Lambda \neq 0$. [This result was obtained in collaboration with Paul Renteln.] In the case when $\Lambda = 0$, however, the ansatz fails because we cannot recover $\tilde{\sigma}^{a}$ from A_{a}. We shall see in the next section that a new line of attack is needed in this case.

4. THE HALF FLAT CASE

Suppose we are given a pair $(\tilde{\sigma}^{a}(t), {}^{\pm}A_{a}(t))$ satisfying all field equations, (14), (7') and (8'). Then, one can show that the following identity holds:

$$(17) \qquad -\frac{1}{\sqrt{2}}\epsilon^{ab}{}_{m}\, {}^{\pm}F_{abA}{}^{B}\sigma_{nB}{}^{A} = R_{mn} + K_{m}{}^{a}K_{an} - K\, K_{mn} \pm \epsilon_{abm}D^{a}K^{b}{}_{n} = E_{mn} \pm B_{mn}$$

where E_{ab} and B_{ab} are, respectively, the electric and the magnetic parts, $E_{ab} = {}^{4}C_{ambn}\zeta^{m}\zeta^{n}$, $B_{ab} = \frac{1}{2}\epsilon_{am}{}^{pq}\, {}^{4}C_{pqbn}\zeta^{m}\zeta^{n}$, of the Weyl tensor of the 4-metric g_{ab}. Note that, in spite of various sign ambiguities in individual fields that appear in (17), both the right and the left sides of (17) are unambiguous. Thus, in essence, ${}^{\pm}F_{ab}$ carries the information about the anti self-dual and the self-dual parts of the Weyl tensor.

Therefore, self-dual solutions are precisely those for which ${}^{+}F_{ab}$ vanishes. Let us assume, for simplicity, that we can pass to a gauge in which ${}^{+}A_a$ itself vanishes and consider the form (14), (7') and (8') of Einstein's equation with the upper, +, sign. [This corresponds, in the twistor language, to working with the dual twistor space rather than with the twistor space, in the self-dual case. Using the lower, −, sign in the self-dual case gives us a formulation of the "googly problem".] Then, enormous simplifications occur. Equations (7') are satisfied identically, (8'.B) implies that, if ${}^{+}A_a$ starts out being zero on the initial, $t = 0$, hypersurface, it continues to remain so, and the only non-trivial equations, (14) and (8'.A) reduce to:

(18.A) $$\partial_a \tilde{\sigma}^a{}_A{}^B = 0 \ ,$$ and

(18.B) $$\dot{\tilde{\sigma}}^a = 2\sqrt{2}\ \partial_b(\underset{\sim}{N}\tilde{\sigma}^{[a}\tilde{\sigma}^{b]})$$

respectively. These equations can be further simplified by a suitable choice of the lapse-density $\underset{\sim}{N}$. Let $\underset{\sim}{N}$ be chosen to be time-independent and let it satisfy $\partial_a \underset{\sim}{N} = 0$. Let us furthermore introduce a triad $V^a{}_{\underset{\sim}{a}}$ of vector fields and expand out $\underset{\sim}{N}\tilde{\sigma}^a{}_A{}^B$ in terms of this triad and (fixed) Pauli matrices $P^{\underset{\sim}{a}}{}_A{}^B$ satisfying $[P^{\underset{\sim}{a}}, P^{\underset{\sim}{b}}] = (-\sqrt{2})\epsilon^{\underset{\sim}{a}\underset{\sim}{b}\underset{\sim}{c}}P^{\underset{\sim}{c}}{}_A{}^B$, $\underset{\sim}{a} = 1, 2, 3$:

(19) $$\underset{\sim}{N}\,\tilde{\sigma}^a{}_A{}^B = V^a{}_{\underset{\sim}{a}} P^{\underset{\sim}{a}}{}_A{}^B$$

Then, (18) are completely equivalent to

(1) $$\partial_a V^a{}_{\underset{\sim}{a}} = 0,$$ and

(2) $$\dot{V}^a{}_{\underset{\sim}{a}} = \epsilon_{\underset{\sim}{a}\underset{\sim}{b}\underset{\sim}{c}}[V_{\underset{\sim}{b}}, V_{\underset{\sim}{c}}]^a$$

where [,] stands for the Lie bracket between vector fields. Thus, in the self-dual case, the entire content of Einstein's equation is captured in two geometric equations on vector-triads $V^a{}_{\underset{\sim}{a}}$ on Σ .

Finally, we can express the solution g^{ab} directly in terms of the triad $V^a{}_{\underset{\sim}{a}}$. Using the expression $g^{ab} = (\det q)^{-1}\{- \mathrm{Tr}\ \tilde{\sigma}^a\tilde{\sigma}^b + \underset{\sim}{N}^{-2}t^at^b]$ of the 4-metric, we have:

$$g^{ab} = \underset{\sim}{N}\,(\det \hat{q})^{-1/2}(\hat{q}^{ab} + t^at^b) \tag{20}$$

$$\text{where } \hat{q}^{ab} = V^a{}_{\underset{\sim}{a}}\, V^b{}_{\underset{\sim}{b}}\, \delta^{\underset{\sim}{ab}}.$$

This 4-metric is a solution to the Einstein vacuum equation and its Weyl tensor is half-flat. Conversely, given a half-flat 4-metric on M, on any leaf $\Sigma(t)$ of the foliation, the electric part of its Weyl tensor equals the magnetic part (modulo sign) whence the corresponding ${}^+A_a$ can be chosen to be zero. Then, the corresponding $\tilde{\sigma}^a$ automatically satisfies (18) and we can always introduce a triad $V^a{}_{\underset{\sim}{a}}$ satisfying (1) and (2). Thus, (20) is a half-flat metric and every half-flat metric can be cast in the form given by (20).

REMARKS. (i) To obtain anti self-dual solutions, one sets ${}^-A_a = 0$, uses the lower, –, set of equations (14), (7′) and (8′) and obtains a reduced set of equations on $\tilde{\sigma}^a$. The constraint equation is the same as (18.A) but the evolution equation differs from (18.B) by a minus sign. Consequently, given a solution $\tilde{\sigma}^a$ to the self-dual system (18), $-\tilde{\sigma}^a$ is a solution to the anti self-dual system. The change of sign of $\tilde{\sigma}^a$ leaves the 4-metric g^{ab} unaffected but changes the orientation ϵ^{abcd}. [In one case we set ${}^+A = 0$ and in the second, ${}^-A = 0$. Since D remains unaffected under the sign flip of $\tilde{\sigma}^a$, $K_{aA}{}^B$, given by (13), changes sign, whence the 4-orientation is reversed.] Thus, the metric g_{ab} can only be called half-flat. In the real, Euclidean case, we can do better. We can fix a reference orientation, $\mathring{\epsilon}_{abcd}$, and admit only those ($\tilde{\sigma}^a$, ${}^{\pm}A_a$) pairs which yield a ϵ_{abcd} with the same orientation as the reference one. Then, if ($\tilde{\sigma}^a, {}^+A = 0$) is admissible, ($-\tilde{\sigma}^a, {}^-A = 0$) is not, whence a given half-flat metric is either self-dual or anti-self-dual without any ambiguity.

(ii) The evolution equation can be cast in a 4-dimensional form by considering the tetrad $(t^a, V^a{}_{\underset{\sim}{a}})$, and denoting it by $e^a{}_{\underset{\sim}{\alpha}}$ for $\underset{\sim}{\alpha} =$

0,1,2,3. Then, it simply states that:

$$[e_{\underset{\sim}{\alpha}}, e_{\underset{\sim}{\beta}}] = \epsilon_{\underset{\sim}{\alpha}\underset{\sim}{\beta}\underset{\sim}{\gamma}\underset{\sim}{\delta}}[e_{\underset{\sim}{\alpha}}, e_{\underset{\sim}{\delta}}]. \tag{21}$$

Lionel Mason has pointed out that the connection between the present approach and twistor theory may become clearer in terms of (21).

5. DISCUSSION

The reduced form of half-flat Einstein's equation, (1) and (2), is rather simple and geometric; it resembles the Euler equations for a rigid body. The rich literature on generalizations of these equations to infinite dimensional systems suggests that (1) and (2) may be an exactly integrable system. In any case, it is in a form that is well-adapted to a standard Hamiltonian treatment. This, by itself, is a bonus since self-dual systems do not normally accommodate Hamiltonian descriptions; if one considers the phase-space of all complex Yang-Mills fields, for example, the pull-back of the symplectic structure to the (anti) self-dual space is completely degenerate and the restriction of the Hamiltonian to this subspace vanishes identically. The Hamiltonian description and the issue of exact integrability is being investigated in collaboration with Pawel Mazur.

The primary limitation of the present framework arises from the fact that one restricts oneself to manifolds with topology $\Sigma \times R$. Most of the Euclidean instanton solutions, for example, do not accommodate such a topology. Furthermore, our restriction to such manifolds seems to imply that the framework cannot lead to instantons which induce a topology change. On the other hand, the new 3 + 1 form of the field equations is such that one never needs to lower the vector index on $\tilde{\sigma}^a$. Consequently, it is quite possible that solutions to this set of equations exists in which $\tilde{\sigma}^a$ fails to be invertible, say, on a set of measure zero on M. The resulting 4-metric g_{ab} would then be degenerate at such points. [Thus, the set

(14), (7') and (8') may be regarded as a generalization of Einstein's equation, reducing to it in the case when $\tilde{\sigma}^a$ is nondegenerate everywhere.] Recent work of Rafael Sorkin suggests that, in such cases, it may be possible to reinterpret the solution as being regular but representing a topology change.

ACKNOWLEDGEMENT. The main results presented here were obtained jointly with Ted Jacobson and Lee Smolin. I am grateful also to Gary Horowitz, Lionel Mason, Ted Newman and Paul Renteln for numerous discussions.

REFERENCES

1. M. Ko, M. Ludvigsen, E. T. Newman and K. P. Tod, "The Theory of H-Space," Physics Reports, 71 (1981), 51-139.

2. M. Ludvigsen, E. T. Newman and K. P. Tod, "Asymptotically Flat H-Spaces," J. Math. Phys., 22 (1981), 818-823.

3. R. Penrose and W. Rindler, *Space-time & Spinors*, Vol. 1, Cambridge University Press, Cambridge (1984).

4. A. Ashtekar, "A New Phase Space Formulation of General Relativity," to appear in Phys. Rev.; A. Ashtekar, T. Jacobson and L. Smolin, "A New Characterization of Half-Flat Solutions to Einstein's Equation," submitted to J. Math. Phys.

ABHAY ASHTEKAR
PHYSICS DEPARTMENT
SYRACUSE UNIVERSITY
SYRACUSE, NEW YORK 13244-1130

and

INSTITUTE FOR THEORETICAL PHYSICS
UNIVERSITY OF CALIFORNIA
SANTA BARBARA, CALIFORNIA 93106

Contemporary Mathematics
Volume **71**, 1988

QUANTUM GRAVITY IN THE SELF-DUAL REPRESENTATION

Lee Smolin

ABSTRACT. Following a general discussion of some important issues which confront the canonical approach to quantum gravity, recent progress in quantizing general relativity based on Ashtekar's reformulation of canonical general relativity is described. Using these new variables, quantum states of the gravitational field may be constructed which are exactly annihilated by the Hamiltonian constraint of quantum general relativity. These solutions are associated with collections of smooth loops in a three manifold, Σ. For example, it is shown that for each collection of loops in Σ which are smooth and nonintersecting there exists a quantum state of the gravitational field which is annihilated by the Hamiltonian constraint. Solutions involving intersections also exist, and are described briefly.

INTRODUCTION

We know from the fact that the conventional perturbation theory becomes nonsense when the cutoff scale becomes smaller than the Planck length that the solution to the problem of quantum gravity will involve new ideas about the properties of spacetime at very short distances. Broadly speaking, three competing hypotheses as to why the perturbation theory fails at the Planck scale have

---

1980 Mathematics Subject Classification (1985 Revision), 83C45.

Supported in part by NSF grants PHY85-03072 to Yale and PHY82-17853, supplemented by funds from NASA, to the University of California at Santa Barbara.

motivated most attempts to construct a quantum theory of gravity in the last thirty years. The first, and most commonly tried approach, has been based on the presumption that the fault lies in the dynamics of general relativity. This idea has inspired many different attempts to modify the dynamics of the gravitational field in a way which would reduce to Einstein's theory at large distances, but lead to a well defined perturbation theory at all scales. We now know that none of these attempts succeed [1], with the possible exception of string theories [2].

A second approach, more ambitious but less popular, has been to ascribe the fault to the application of the usual formalism of quantum theory to the geometry of spacetime, and to look for radically new ways in which both spacetime structure and quantum theory might emerge from a more fundamental structure [3,4,5]. While it seems quite likely that such an approach is ultimately correct, it has so far proved impossible to take any of the several very interesting proposals which have been made along this line far enough to see if they succeed or fail.

More optimistic than either of these approaches, the third approach asserts that no modifications are required either in the dynamics of general relativity or in the basic formalism of quantum mechanics. The trouble, as Bryce DeWitt, among others, has been arguing for decades, lies merely in the inapplicability of perturbation theory to the study of the short distance behavior of the quantum gravitational field [6-14]. Because of the dimensional nature of the coupling, the quantum fluctuations in the gravitational field are strongly coupled at the Planck scale, so that only with a nonperturbative approach could one hope to get meaningful results.

While both rational and conservative, this last approach has been paid the least attention of the three. The reason is not hard to find. It is simply the lack of any reliable approach by which one might study the dynamics of quantized general relativity at the non-perturbative level. There have been a number of very interesting attempts to get at the short distance behavior of the quantized gravitational field nonperturbatively; these include the summations of infinite classes of diagrams of DeWitt [6] and of Isham, Salam and Strathdee [7], the very interesting observation that the classical self energy is finite in general relativity made by Arnowitt, Deser and Misner [8], work and speculation on spacetime foam by Wheeler [9], Hawking [10] and others [11] and the discussions of the strong couping expansion by Isham [12] and others [13]. All of these attempts yield results which indicate that the short distance behavior could be qualitatively different than that given by perturbation theory. However, all of them are subject to important criticisms of one sort or another, and none of them has ever led to a systematic and controlled calculational procedure which might replace perturbation theory.

However, it often happens in the history of science that an idea that at one time seems worked over and discredited can, in the light of a new development, become once again an exciting and fertile source of ideas and research projects [15]. Certainly the recent history of string theory shows this. However, while most of our attention has been focused on string theory (and for good reason!), a very interesting development has occured which may lead for the first time to a viable nonperturbative approach to the quantization of ordinary general relativity. This new development is Ashtekar's recent construction of what he calls his "new variables" for the Hamiltonian formulation of general relativity [16,17]. In his contribution to this volume [18], Ashtekar has described these new variables, and the enormous simplification that they make possible in certain problems in classical general relativity such as the self dual Einstein equations [19]. It is my purpose in this paper to describe the work which has been done on the application of this new formalism to the quantum problem in the year that we have had it to work with. This work, as the reader will see, is very much "in progress" and much more remains to be done than we have been able to do so far. However, enough has been achieved that I think we can prudently say that we have for the first time a decent chance of resolving the issue of whether general relativity makes sense as a quantum theory.

Roughly speaking, Ashtekar's new variables are a set of coordinates on the phase space of general relativity, the use of which simplifies considerably the Hamiltonian formulation of the theory. Both the constraints and the Hamilton's equations of motion become polynomial in the new variables, whereas in the usual formalism, written in terms of the metric and its conjugate momenta, they involve inverses and square roots of determinants. Furthermore, in terms of these new variables the initial value problem, at both the classical and quantum level, turns out to be very closely related to the initial value problem of a certain Yang-Mills theory. Specifically, the physical phase space of general relativity, which consists of those pairs of coordinates and momenta that satisfy the constraints, is shown to be a submanifold of the physical phase space of this particular Yang-Mills theory [16,17]. This is possible because one of the new variables is a certain connection on three dimensional spinors. This connection plays a kind of a dual role in the formalism in that it functions both as a kind of Yang-Mills connection and as a *complex* coordinate on the phase space of general relativity analogous to the Bargmann, or coherent space, coordinate used to study the quantization of the harmonic oscillator. It is by exploiting both of these aspects of the new variables that much of the power of the new formalism arises.

The use of these new variables has led to several new developments in quantum general relativity. In mentioning them I want to distinguish between two kinds of results: formal results, which involve expressions such as operator products at points or functional differentials and integrals, which are not, strictly speaking defined without the specification of a regularization procedure; and fully regularized results, in which such a regularization procedure has been worked out.

First of all, Ashtekar was able to show last year that his new variables yield a formal solution to the operator ordering problem which has plagued the Hamiltonian approach to quantum gravity for the whole history of the subject [16]. This result is formal, in that a regularization procedure is not employed; on the other hand it is careful, in the sense that the coefficients of terms such as $\delta^3(0)$ and $\partial_a \delta^3(0)$ are kept track of, and are required to cancel.

An important issue, which is currently under investigation, is then to check that this result holds also in the context of a full regularization scheme. One of the ways in which this is being investigated is through a lattice formulation of the theory. This formulation exploits the connection with Yang-Mills theory to construct a regulated formulation of quantum general relativity in the language of a Kogut-Susskind lattice gauge theory [20]. Unfortunately, there is not space to describe this formulation here, but the interested reader may find it described elsewhere [21].

On the lattice one constructs a Hilbert space of states associated with holonomy elements of the connection, taken around closed loops in the lattice. It turns out that the constraints of general relativity, including the notoriously unmanageable Hamiltonian constraint, have a very simple action on such states [21]. This suggested to us that it might be fruitful to work directly in the continuum with states constructed from holonomy elements based on closed loops in space. This led to some surprising developments, which are the main subject of this article.

The basic result, which is described in Section 5, is that it is possible to associate a quantum state of the gravitational field to any collection of closed loops, γ_I in a three manifold, Σ, in such a way that the state is exactly annihilated by the Hamiltonian constraint whenever all of the loops are smooth and nonintersecting [22]. This result is demonstrated both formally and in the context of a consistent regularization procedure. Further, there also exist solutions in the case that there are intersections in the curves [22]. These solutions, which are described in Section 6, involve the superpositions of states based on different ways of routing the holonomy elements through the points of intersection. This leads to an interpretation of the Hamiltonian constraint as an operator that induces interactions between loops of different topologies, an inter- pretation that is also apparent in the lattice formulation. Again, all of these solutions can be demonstrated in the context of a regularization procedure.

These loop states are, however, not yet physical states of the gravitational field because, as they are based on specific choices for the imbeddings of the loops in the three manifolds Σ, they are not invariant under diffeomorphisms of Σ. In order to make them physical states we need to construct linear combinations of them which are left invariant by the action of infinitesimal diffeomorphisms of Σ. This is a difficult problem which is at present rather poorly understood. Some remarks about this are made in reference [23].

The plan for the rest of this paper is as follows. In the next section we go over the basics of the Hamiltonian approach to quantum gravity and indicate four important issues that arise over and over again in this work. In Section 3 the new variables are introduced, and in Section 4 the self dual representation of the quantum theory is described. We are then prepared to discuss the results concerning the action of the constraints on the loop states. The paper then closes in Section 7 with a discussion of the important open problems.

Most of the work that I will describe here was done either by, or in collaboration with, Abhay Ashtekar, Ted Jacobson, and Paul Renteln. There are many technical details which are not discussed in this paper. The reader who is interested in them should consult the original papers [16-19,21-23,30].

2. THE HAMILTONIAN APPROACH TO QUANTUM GRAVITY

From the beginning of the subject, work on quantum gravity has been split into two directions which have come to be called the canonical approach and the covariant approach. The reasons for this division rest on a fundamental tension between the roles time plays in quantum mechanics and in general relativity [24]. The diffeomorphism invariance of general relativity comes into apparent conflict with the need to select a prefered time coordinate for the statement of the quantization procedure. Even if, ultimately, this tension is resolved and a consistent and physically meaningful quantization of general relativity is carried out, the diffeomorphism invariance of general relativity cannot be represented in the quantum theory in the same way that a Yang-Mills gauge symmetry would be [25].

The transition between a classical and a quantum theory is usually made in terms of the Hamiltonian formulation of the theory, in which a time coordinate is singled out as parametrizing the evolution of fields which are assumed to be valued on a space-like surface – a manifold of simultaneous instants of "time". Of course, for a simple system there is an equivalent formulation in terms of the path integral. However, the relationship between the path integral formulation and the original Hamiltonian formulation can only be made for an unconstrained system containing only physical degrees of freedom. When a dynamical system has gauge freedom and hence, in the Hamiltonian formulation, constraints on initial data [26] the connection between the Hamiltonian and the path integral formulations is less direct, and the detailed form of the measure of the path integral is often nct the obvious one [27]. Because of the fact that the gauge symmetry of general relativity involves the time coordinate, which plays a prefered role in the Hamiltonian formulation, this is especially true in quantum gravity.

In this situation, there has been a tendency to make a choice between the two notions of time and hence work either in the canonical theory, which takes the quantum mechanical notion of time more seriously than the general relativistic, or in the covariant theory, where the preference is reversed. This division has been deepened by the fact that almost all of the work done in the covariant formulation has been perturbative, in that it is based on a semiclassical analysis of the functional integral; while almost all of the work in the canonical approach has been, at least in principle, nonperturbative. For this reason the canonical approach has seemed to many people the deeper approach, in that when one works perturbatively one retains a classical background metric, which one uses in the definition of the quantum theory. While some progress is made by doing this, one avoids many of the fundamental issues of quantum gravity which arise when one puts the whole metric in the wavefunction, and therefore forgoes the use of any classical background metric in the definition of the quantum theory.

At the present time, it is clear that the perturbative approach to quantizing general relativity is a dead end, so that if any further progress is to be made we must take the more fundamental non- perturbative approach. This means finding out whether it is possible to do physics in a context in which the whole metric is in the wavefunction. While it is possible to study the covariant functional integral formalism nonperturbatively, two kinds of difficulties arise. First, it has so far proved very difficult to define a reliable numerical algorithm which gives useful information about the short distance structure of the theory. Numerical algorithms such as Monte-Carlo simulation tend to be more successful when the interesting structure happens at large scales rather than accumulating at the cutoff scale, which is what dimensional analysis suggests will be the case in quantum gravity [11]. Second, because one of the constraints is quadratic in the momenta, the computation of the measure of the path integral from the canonical theory has until now only been carried out perturbatively.

In this circumstance it seems that we must fall back on the canonical theory if we are not to give up the project of quantizing general relativity.

As opposed to the covariant perturbation theory, in which roughly the same kinds of problems arise in general relativity and in Yang-Mills theory, one encounters several difficulties when setting up the Hamiltonian quantization for general relativity that are unique to theories of gravity. These follow from the fact that the gauge group of classical general relativity includes the diffeomorphisms of four dimensional spacetime.

One kind of difficulty follows from the fact that by the very act of setting up the Hamiltonian formalism one seems to break the gauge symmetry by singling out a time coordinate and an associated foliation of spacetime by three dimensional surfaces. Of course, the symmetry is not actually broken in the sense that one ultimately shows that, at least in the classical theory, the evolution is independent of how one makes this choice. But there is still a cost in that, as we shall see, the algebra of the constraints which generate the gauge symmetries in the canonical theory is not the same as the algebra of infinitesimal four dimensional diffeomorphisms [25]. We will see that as a result of this the representation of the complete gauge symmetry of the classical theory in the quantum theory has certain problematic aspects.

A second kind of difficulty is associated with the subgroup of the gauge symmetry which is the three dimensional diffeomorphism group of the spacelike surfaces with which spacetime is foliated. The representation of this part of the gauge symmetry is not problematic in the classical theory, but becomes so in the quantum theory when one considers issues associated with regularization and operator ordering. Furthermore, as we shall see, the difficulties associated with the three dimensional diffeomorphisms are much more critical in the self-dual representation because the metric, which is very useful in constructing diffeomorphism invariant objects, is represented by a composite operator which is not well defined in the absence of a regularization procedure. Thus, the implementation of diffeomorphism invariance is a more difficult problem in the self-dual representation than in the usual metric representation.

We will begin by reviewing the basics of the Hamiltonian approach to quantum gravity. Our emphasis here will be on understanding the ways in which, as a result of the fact that the gauge group is the four dimensional diffeomorphism group, Hamiltonian quantum gravity must differ from the Hamiltonian quantization of more conventional theories such as Yang-Mills theory. Our discussion of the Hamiltonian approach to quantum gravity will be organized in terms of four "issues" -- each both the result of the conflict between the roles time and space play in quantum mechanics and general relativity and the origin of an important feature of the Hamiltonian quantization of gravity [28].

The setting for the Hamiltonian formulation of general relativity is a three dimensional surface, Σ, on which live a three metric, q_{ab} and its associated conjugate momentum, p^{ab}. As far as the global features of the problem are concerned there are two interesting contexts that have been studied, the cosmological and the asymptotically flat. In the cosmological context the manifold Σ is assumed to be compact, while in the asymptotically flat context Σ is assumed to be noncompact, and q_{ab} and p^{ab} are required to satisfy certain fall-off conditions as one moves to infinity. These conditions are formulated so that any four manifold which is constructed by evolution from the initial data will be an asymptotically flat four manifold. The details of these fall-off conditions will not concern us here. They are given in Ref. [17] and in the book by Ashtekar cited in references [28].

One might think that the cosmological context would be the preferred context to work in, as it is both more "fundamental" and avoids messy details associated with boundary conditions. However, when one goes to the quantum theory the cosmological problem is fraught with deep and possibly irreconcilable problems of principle [29]. To avoid these we may stick to the asymptotically flat context.

Issue 1: The algebra of the constraints

Because of the four dimensional group of gauge transformations, the initial data on Σ is subject to four constraints. Three of these generate diffeomorphisms on Σ. They take the form,

$$\mathbf{P}_a = D_b p^b{}_a. \tag{2.1}$$

Their algebra (under the natural Poisson brackets associated with the pair (q_{ab}, p^{ab})) is just the algebra of three dimensional diffeomorphisms,

$$\{\mathbf{P}_a(x), \mathbf{P}_b(y)\} = \partial_{[a}\delta^3(x,y)\mathbf{P}_{b]}(x). \tag{2.2}$$

The last constraint is associated with four dimensional diffeomorphisms that involve the time coordinate, and which, hence, generate changes in the choice of the embedding of the three surface Σ in four dimensional spacetime. This constraint is called the Hamiltonian constraint, and takes the form,

$$\mathcal{H} = \frac{G}{\sqrt{q}} (p^{ab}p_{ab} - \tfrac{1}{2} (p_c{}^c)^2) - \frac{1}{G}\sqrt{q}\ R \tag{2.3}$$

It is in the Poisson bracket relations satisfied by this constraint that the first issue emerges. The bracket of $\mathcal{H}$ with the diffeomorphism constraints is not problematic,

$$\{\mathbf{P}_a(x), \mathcal{H}(y)\} = \partial_a \delta^3(x,y)\mathcal{H}(x). \tag{2.4}$$

However, if we take the bracket of $\mathcal{H}$ with itself we find,

$$\{\mathcal{H}(x), \mathcal{H}(y)\} = \partial_a \delta^3(x,y) q^{ab}(x)\mathbf{P}_b(x). \tag{2.5}$$

This is problematic for two reasons. First, it is not what one would expect from the algebra of infinitesimal diffeomorphisms on the four manifold. Second, the structure constant involves a dynamical variable, q^{ab}. This is something that does not occur in situations such as Yang-Mills theory where the gauge group has nothing to do with spacetime. It is in fact directly a reflection of the fact that we have to compromise the gauge symmetry of the theory in setting up the Hamiltonian formalism.

An algebra in which one has structure *functions* involving dynamical variables, rather than simply structure constants, is called an open algebra. An algebra can be open and still be first class, as ours is. But a theory with an open algebra has, at the quantum level, difficulties that more ordinary theories do not have. In particular, when one is careful about both operator ordering and regularization it is not completely clear what we should mean by the condition that the algebra of the constraints closes quantum mechanically. This is discussed in detail in Ref. [22].

Issue 2: The Hamiltonian is made up of the constraints

In the classical theory, the constraint generates a small change in the imbedding of the three surface Σ in the four dimensional spacetime. Thus, if we smear $\mathcal{H}$ with a function, N, (called the lapse) of compact support, the resulting function,

$$\mathcal{H}(N) = \int d^3x \; N(x)\mathcal{H}(x) \tag{2.6}$$

generates a canonical transformation associated with a change of the three surface Σ such as that shown here:

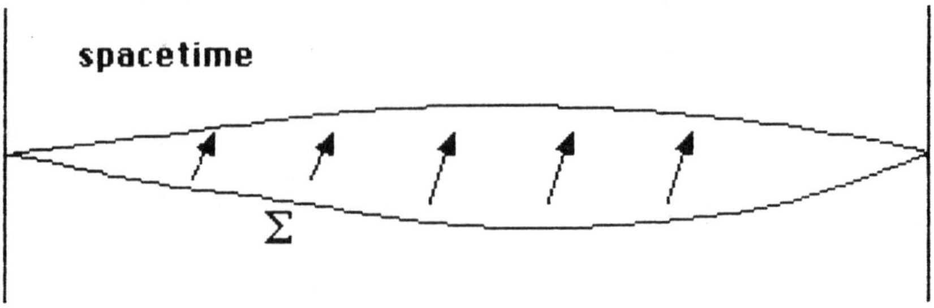

However, note that, locally, this canonical transformation is exactly the same as that associated with the evolution of the fields q_{ab} and p^{ab} under the field equations. As a result of this the local piece of the Hamiltonian, which after all is the object that generates the evolution in the Hamiltonian formalism, must be just a linear combination of the constraints. The only additional piece that the Hamiltonian contains is a boundary term which allows us to distinguish between actual evolution, from the point of view of an observer at infinity, and a change of the choice of three surface Σ that falls off fast enough that it does not result in any evolution of the dynamical quantities seen be the observer at infinity. The difference in these situations is illustrated on the next page.

The Hamiltonian then has the form,

$$H = \int_\Sigma (N\mathcal{H} + N^a \mathbf{P}_a) + \int_{\partial\Sigma} d\delta^b N(\partial_a q_{bc} - \partial_b q_{ac})\delta^{ac} \tag{2.7}$$

In this expression, N and N^a are Lagrange multiplier fields, which are called, respectively, the lapse and the shift. In the classical theory they are required to parameterize the way in which a four dimensional spacetime is generated from the evolution of the dynamical variables on Σ. However, in the quantum theory, a well defined four dimensional spacetime exists to the same extent that trajectories of electrons do, and the only part of the lapse and the shift that are physically meaningful are their asymptotic

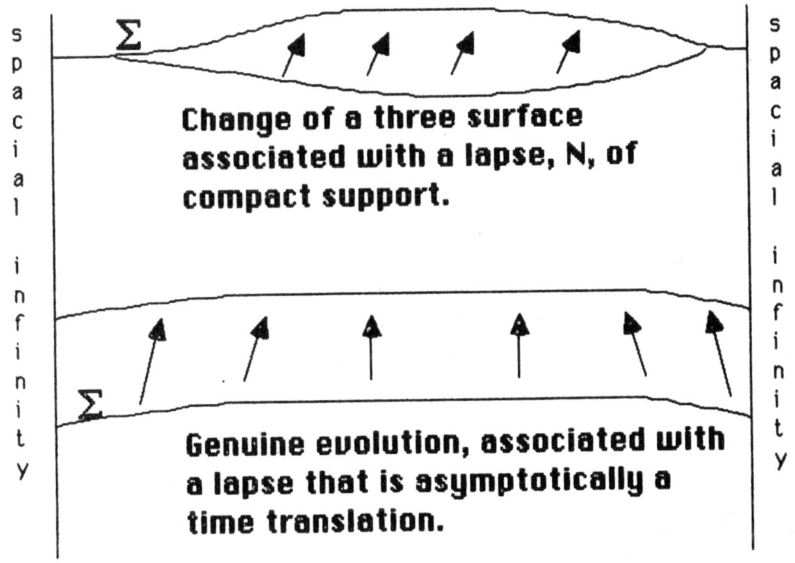

values, which parameterize the evolution of the physical quantities seen by the classical observer at infinity.

The transition from a classical theory to a quantum theory is made, as usual, by constructing a representation space S on which the dynamical variables, q_{ab} and p^{ab} act as operators, in such a way that the canonical commutation relations,

$$[q_{ab}(x),\ p^{cd}(y)] = i\hbar\ \delta^{c}_{(a}\delta^{d}_{b)}\delta^{3}(x,y) \tag{2.8}$$

are satisfied. However, as Dirac taught us [26], when we quantize a theory with first class constraints we must impose the condition that the physical states are contained in that subspace of the representation space which is left invariant by the action of the gauge group of the theory. This means that the states in the physical subspace of the representation space, $S_{\text{phys.}}$ must be annihilated by the constraints of the theory, expressed as operators,

$$\begin{aligned} &\hat{\mathcal{H}}\,|\,\psi\,\rangle = 0 \\ &\hat{\mathrm{P}}_{a}\,|\,\psi\,\rangle = 0, \quad \text{for } |\,\psi\,\rangle \in S_{\text{phys.}} \end{aligned} \tag{2.9}$$

In the case of general relativity, this means that once the physical representation space has been found, the action of the Hamiltonian is just by the boundary term,

$$H| \psi \rangle = \int_{\partial\Sigma} dS^{b} N(\partial_{a} q_{bc} - \partial_{b} q_{ac})\delta^{ac} | \psi \rangle \tag{2.10}$$

Thus, in quantum general relativity, all of the *local* dynamics is expressed by the conditions (2.9) that the constraints annihilate the physical states. This is in contrast to the case of Yang-Mills theory, in which the problem of the restriction to the physical degrees of freedom is distinct from the problem of the imposition of the local dynamics on those degrees of freedom. In general relativity, at both the classical and the quantum level, these two problems are inextricably bound up together in the one condition that the constraints be satisfied.

This essential difference, which is a direct consequence of the fact that the gauge group in general relativity is the spacetime diffeomorphism group, means that the project of constructing quantum general relativity is fundamentally different from the project of constructing quantum Yang-Mills theory.

This difference is reflected strongly in the next issue:

Issue 3: The inner product

As we have just noted, a universal consequence of the presence of a gauge symmetry in a theory is that the physical representation space, $S_{\text{phys.}}$ of the theory is smaller than the original representation space, S on which the canonical commutation relations are defined. This is because the physical states must satisfy the condition that they are annihilated by the operators which generate the gauge transformations.

In order to connect the quantum theory with observations we need to posit an inner product on the representation space -- thus making it into a Hilbert space. In the case of a theory with constraints, there are two possible ways in which we might do this. We might impose an inner product on either the whole representation space, S, or just on the physical representation space, $S_{\text{phys.}}$. A moment's thought suffices to convince one that the latter is the right thing to do. The crucial point is that the physical states will, by virtue of the fact that they live in proper subspace of S, be nonnormalizable under almost any choice of an inner product for the whole representation space. For the same reason, there need not be any simple connection between the physical inner product we impose on $S_{\text{phys.}}$ and some natural inner product we might impose on the whole representation space, S.

One consequence of this observation is that there is no requirement for the constraints to be ordered so that they are Hermitian. As they annihilate all of the states on which the physical inner product is defined, the condition that the constraints be Hermitian under the physical inner product cannot be formulated. Further, any attempt to impose a condition that the constraints be Hermitian with respect to some inner product on the larger space, must flounder on the simple fact that the choice of the latter has absolutely no physical significance.

Now, in a conventional theory, such as Yang-Mills theory, the inner product reflects information about the constraints, and hence about the physical degrees of freedom. But, as the restriction to the physical degrees of freedom tells us nothing, in this case, about the dynamics that the physical degrees of freedom satisfy, the inner product carries no further information about the dynamics of the theory. In particular, in Yang-Mills theory, the inner product cannot know anything about the form of the Hamiltonian (once the constraints are taken out) or about the value of the coupling constant. Two consequences of this is that the inner product for the linearized theory must be identical with the inner product of the fully interacting theory and that the inner product in the perturbation theory must be identical with the inner product of the fully nonperturbative theory.

In general relativity the situation is quite different. The inner product can be imposed only on the subspace of S which is annihilated by all of the constraints. However, the restriction to this subspace is equivalent to the imposition of all of the local dynamics of the theory. Thus, in quantum gravity the inner product cannot be imposed until all of the local dynamics have been solved!

This has the very important consequence that in quantum gravity, as opposed to more familiar theories such as Yang-Mills theory, the inner product can be different for the free and for the interacting theory, for the perturbation theory and for the fully nonperturbative theory. Indeed, to the extent that nonperturbative effects are important for the existence of the theory, we would expect that the inner product contains important information about the nonperturbative dynamics. But, we know on general grounds that quantized general relativity cannot exist unless nonperturbative effects dominate the behavior of the theory at short distances. Thus, in contrast to every case we are familiar with, in quantum gravity we should expect the physical inner product to both be different from that of the free theory, and to reflect short distance nonperturbative structure.

We may note that this is not the only fundamental issue which has been raised concerning the inner product of quantum gravity. Others have been discussed by Kuchar in [24].

Issue 4: Regularization

Like all other quantum field theories, the expressions of the Hamiltonian formulation of quantum gravity involve products of operators at the same point of space, and as a result, it is apparently the case that the theory must be defined through a regularization procedure. However, the regularization of a quantum gravity theory differs in two fundamental respects from the regularization of an ordinary quantum field theory.

The first difference has to do with the relationship of the regularization procedure to the inner product. As we have just discussed, in an ordinary quantum field theory the inner product of the interacting theory is the same as that used in the free theory. This means that the inner product can be defined in the context of the free theory, which can be solved exactly. When one then studies the interacting theory, and faces the problem of regularization, the inner product is already known. One then has at hand a sufficient criterion for a regularization procedure to be adequate -- it must make finite all physical inner products.

In quantum gravity, on the other hand, one has to regularize the constraints in order to set up the problem of finding the physical subspace on which the inner product is to be defined. Thus, one must, to begin with, set up a regularization procedure in the absence of any inner product. Furthermore, the expressions being studied are not observable quantities defined by inner products, but expressions of the form,

$$C(x) \mid \psi \rangle = 0 \tag{2.11}$$

where C is an operator and $\mid \Psi \rangle$ is some state. This difference has an important consequence for the types of regularization that are considered. In an ordinary theory, what is being regularized is an operator product acting on a known Hilbert space which has a fixed inner product. Thus, one regularizes expectation values by modifying the operators and not the states. However, in quantum gravity, both the operator and the space of states are in question when one attempts to interpret the quantum constraint equations of the form (2.11). Thus, there is no a priori reason why the regularization procedure should consist of a modification of the operator rather than the state. In fact, as we shall see, it will turn out that in the self dual representation it is more convenient to regularize the quantum constraint equations through a modification of the states rather than of the operators.

The reader may wonder why these same considerations do not apply to the case of Yang-Mills theory, in which one also has to solve the constraint equations before one sets up the inner product. The difference is that the constraint equations in Yang-Mills theory are linear in the conjugate momenta. Thus, the regularization which is required to define them is not nearly as troublesome as in the case of general relativity, where we have a constraint quadratic in the conjugate momenta.

The second respect in which regularization of a quantum theory of gravity differs from that of an ordinary theory is in the relationship of the regularization procedure to the gauge symmetry. Ideally, one would like to use a regularization procedure which preserves the gauge symmetry of the theory. Two circumstances conspire to make this more difficult to achieve in quantum gravity than in ordinary theories. These are, first that the metric has become an operator so that the wavefunction is defined on a manifold that has no background metric or connection structure; and second, that the gauge symmetry includes the diffeomorphism group of the manifold Σ .

In order to see how these two circumstances interact to make it more difficult to construct an acceptable regularization procedure, let us consider what happens to the usual point splitting procedure in this context. Let us consider how to define the action of an operator product $O(x) = A(x)B(x)$ on a state $|\Psi\rangle$. The point splitting procedure begins by defining the action of a regularized operator,

$$O(x,y) \mid \psi\rangle = A(x)B(y) \mid \psi \rangle. \tag{2.12}$$

We then study the limit of this action as $y \to x$. Right away there are two things that we usually do which we cannot do in this context. First of all, it is usually sufficient, and helpful, to study regularized expectation values rather than regularized operator actions. However, as we mentioned above, because we are not going to be able to define an inner product until after we have found the physical states we are going to have to study the regularization of the action of the constraints directly. Second, we usually parametrize the divergences which arise as $y \to x$ in terms of powers of the physical distance between x and y. However, as there is no background metric on Σ, there is no physical notion of how far apart x and y are. Furthermore, point splitting is usually defined by requiring that the limit $y \to x$ be taken along geodesics. However, since there is also no background connection on Σ, there are no physical geodesics, and any smooth path connecting x and y

is just as good as another.

Of course, one could attempt to remedy these problems by imposing an arbitrary metric and connection in a region containing x and y, and using these to define the regularization procedure. However, the imposition of this additional structure violates diffeomorphism invariance. For example, given any two smooth curves connecting x and y there is a diffeomorphism that takes them into each other, leaving x and y fixed. Thus the imposition of additional structure that picks out a preferred set of curves as "geodesics" breaks the diffeomorphism invariance.

Now, in the familiar metric representation, these problems exist, but they are not as serious as they might be. This is because one is working in a representation based on functionals of the metric in which the metric operator acts by a simple multiplication,

$$\hat{q}_{ab}(x) \mid q\rangle = q_{ab}(x) \mid q\rangle. \tag{2.13}$$

Thus, one can often use the operator for the three-metric in the construction of a regularized operator in a way that allows the regularized operator to be both well defined and diffeomorphism invariant. However, the self-dual representation, which we are about to describe, is based on wavefunctions of a coordinate that does not commute with the three-metric or its Christoffel symbol. Indeed, the metric in this representation is represented itself by a composite operator; it involves a product of functional derivatives at a point. Thus, the metric is not itself well defined without the imposition of a regularization procedure. In particular it cannot be used to provide a notion of distance or of geodesics which could be used in the definition of a regularization procedure. Furthermore, states which satisfy (2.13) no longer exist naively. Whether they exist at all is a delicate question involving details of both the regularization procedure and the short distance behavior of the theory. The results we are going to describe in Sections 5 and 6 suggest that such states do not exist.

As a final point concerning the interaction of diffeomorphism invariance with the problem of regularization, we show that absurd results follow if we attempt to use point splitting to define the action of an operator product on a state which is diffeomorphism invariant. Let $D(\phi)$ denote the operator which generates the action of the diffeomorphism ϕ on states in the theory. Consider a diffeomorphism invariant state $|\Psi\rangle$ for which

$$D(\phi) \mid \psi\rangle = \mid \psi\rangle, \qquad \forall \phi \in \mathrm{Diff}(\Sigma). \tag{2.14}$$

Now, let us try to define the action of the operator product $A(x)B(x)|\Psi\rangle$ by point splitting as in (2.12). Consider the action of a diffeomorphism ϕ, which leaves x fixed, and takes the point y to some other point z, which is distinct from x and y. Because the state is diffeomorphism invariant we have,

$$\begin{aligned} A(x)B(y)| \psi\rangle &= D(\phi)D(\phi^{-1})A(x)B(y)D(\phi)D(\phi^{-1}) \mid \psi\rangle \\ &= D(\phi)A(x)B(z) \mid \psi\rangle. \end{aligned} \tag{2.15}$$

A consequence of this is that we cannot learn anything useful about the local product $A(x)B(x)$ by studying a limit of point split operators $A(x)B(y)$ in which y approaches x. To see this explicitly, let us consider a sequence of points y_i, $i = 1, \infty$ (all distinct from x) such that $\mathrm{Lim}_{i\to\infty} y_i = x$. For each i, define a diffeomorphism ϕ_i which takes y_i to z (as before arbitrary but distinct from x), leaving x fixed. Then it follows that

$$\lim_{i\to\infty} A(x)B(y_i) \mid \psi\rangle = \left(\lim_{i\to\infty} D(\phi_i)\right) A(x)B(z) \mid \psi\rangle. \tag{2.16}$$

The limit may be singular, because there is no smooth diffeomorphism which is the limit of $D(\phi_i)$ as i goes to infinity. But the result can have nothing to do with taking a product of operators at "nearby points", because the point z is fixed and is not in any sense "near x". The point is that when acting on diffeomorphism invariant states, there are only two possibilities for an operator product $A(x)B(y)$. Either x and y are the same point or they are distinct. There is no useful concept of y being near, or approaching, x in the case that the states are diffeomorphism invariant, and there is no physical background metric available with which to measure distances.

3. THE NEW VARIABLES

In this section I will give a brief review of the new variables for canonical general relativity introduced by Ashtekar. These are the basis for the self-dual representation for quantum general relativity which will be described in the following sections. For more details the reader can consult the articles by Ashtekar [18] and Jacobson [30] in this volume, or the original papers [16,17].

In the last section we introduced the phase space $\Gamma = (q_{ab}, p^{ab})$ of general relativity associated with a three dimensional manifold, Σ. Ashtekar's new variables are based not on this phase space, but on an extension of it, in which the space of dynamical variables is enlarged by representing the degrees of freedom of the gravitational field in terms of local frames. This can be done in terms of local frame fields; however, for many purposes it is convenient to express the group of local rotations in terms of certain spinorial variables. We will begin with the latter formalism, as it is the one employed by Ashtekar, and then show the relationship to the more familiar frame field formalism.

We may begin by introducing a set of two component fields, λ^A, on Σ, which parametrize a two dimensional vector bundle over Σ.[a] In addition we may build a dual bundle of fields μ_A. On these bundles we impose two fixed "a priori" structures, which are defined without reference to the metric, or any other fields, on Σ. These are,

1) A symplectic metric, ϵ^{AB}, with inverse ϵ_{AB}, which satisfy,

$$\epsilon^{AB} = -\epsilon^{BA}; \qquad \epsilon^{AB}\epsilon_{AB} = 2$$
$$\lambda_B = \lambda^A\epsilon_{AB}; \qquad \mu^A = \epsilon^{AB}\mu_B \tag{3.1}$$

[a]Upper case latin indices A, B, C, D, ... are two component abstract indices.

2) A Hermitian conjugate operation, $+ : \lambda_A \to \lambda_A^+$, which satisfies the four conditions,

$$
\begin{aligned}
(\alpha\lambda_A + \beta\mu_A)^+ &= \bar{\alpha}\lambda_A^+ + \bar{\beta}\mu_A^+ \\
(\lambda_A^+)^+ &= -\lambda_A \\
\lambda^{+A}\lambda_A &\geqslant 0, \quad \text{equality implying } \lambda_A = 0 \\
(\lambda_A\mu_B)^+ &= \lambda_A^+\mu_B^+
\end{aligned}
\tag{3.2}
$$

There is then a natural symmetry group, which consists of those transformations which leave this fixed structure invariant; it is a local SU(2) gauge group. We should stress that this SU(2) group is defined without any reference to a metric (if one exists) on Σ. Because of this, it is perhaps proper to call the structure defined so far a "prespinor" bundle.

This prespinor bundle can become a genuine spinor bundle if we introduce some structure which relates the SU(2) rotations to the group of SO(3) transformations in the tangent space of Σ which preserve some metric q_{ab}. This is done by introducing a set of soldering fields, $\sigma^a{}_{AB}$. These are defined to be symmetric and hermitian in the spinor indices AB. This guarantees that the $\sigma^a{}_{AB}$ define an isomorphism between the space of symmetric hermitian fields x^{AB} and the tangent space of Σ .

Given a choice of the $\sigma^a{}_{AB}$ we may define a metric q_{ab} by

$$q^{ab} \equiv \sigma^a{}_{AB}\sigma^b{}_{CD}\epsilon^{AC}\epsilon^{BD} = -\operatorname{Tr}\sigma^a\sigma^b. \tag{3.3}$$

Clearly, with this definition, the SU(2) transformations which preserve the fixed structures on the prespinor bundle are mapped onto the SO(3) group of transformations in the tangent space which preserve q_{ab}.

The key idea is then to regard the field $\sigma^a{}_{AB}$ as the fundamental field, and the q_{ab} as a derived field defined through (3.3). As we are working in the Hamiltonian picture we then want to define a conjugate momenta $M_a{}^{AB}$. These together comprise what we will call the extended phase space $\Gamma_E = (\sigma^a{}_{AB}, M_a{}^{AB})$. The symplectic structure on Γ_E is given by,

$$\{f,g\} = \int_\Sigma \left[\frac{\delta f}{\delta M_a{}^{AB}} \frac{\delta g}{\delta \sigma^a{}_{AB}} - \frac{\delta f}{\delta \sigma^a{}_{AB}} \frac{\delta g}{\delta M_a{}^{AB}} \right]. \tag{3.4}$$

The phase space of general relativity, Γ, is then defined to be a submanifold of Γ_E in the following way. The local SU(2) transformations on Γ_E are generated[b] by the function,

$$M_{[ab]} = 0; \qquad M_{ab} = -\,\mathrm{Tr}(M_a \sigma_b). \tag{3.5}$$

As is usual in the Hamiltonian formalism the generator of gauge transformations is to be considered a constraint on initial data. The constraint surface, defined by the condition $M_{[ab]} = 0$, is then identified with the phase space of general relativity, Γ. To do this we may define conjugate momenta for q_{ab} by,

$$p^{ab} \equiv M^{(ab)}. \tag{3.6}$$

These p^{ab} satisfy the Poisson bracket relations,

$$\begin{aligned} \{p^{ab}(x), p^{cd}(y)\} &= \tfrac{1}{2}\, \delta^3(x,y)[M^{[ca]} q^{bd} + \text{symmetrizations}] \\ \{p^{ab}(x), q_{cd}(y)\} &= 2\delta^2_{(c}\delta^b_{d)}\delta^3(x,y). \end{aligned} \tag{3.7}$$

We see that on the constraint surface, the pair (q_{ab}, p^{ab}) satisfy the usual Poisson bracket relations. Thus, the constraint surface may be coordinatized by the pair (q_{ab}, p^{ab}) and identified with Γ. However, it is important to stress that q_{ab} and p^{ab} are functions on the entire extended phase space, but do not satisfy the usual Poisson bracket relations off of the constraint surface.

We may now introduce Ashtekar's new variables. To motivate them, let us recall a few simple facts from the quantization of the harmonic oscillator. That simple system has a phase space which is parameterized by a pair (q,p). As was first pointed out by Bargmann [31], for the quantum theory it is often convenient to introduce a complex coordinate on this (real) phase space, for example,

[b]By Poisson brackets with the symplectic structure (3.4).

$$z = q + ip. \tag{3.8}$$

This is convenient because quantum state functionals can then be represented as holomorphic functions on the phase space (q,p), rather than as ordinary functions on the configuration space (q).

We are about to define a nonlinear analogue of the Bargman coordinate, z, on the phase space of general relativity. For certain technical reasons it turns out to be much more convenient to do this at the level of the extended phase space Γ_E.

The new variable is a certain connection on the spin bundle. In order to define it we must first introduce the ordinary, torsion-free, metric compatible spin connection on Σ. This is defined by the condition,

$$D_a \sigma^b{}_{AB} = 0. \tag{3.9}$$

It then follows that $D_a q_{bc} = 0$. Let us denote the connection coefficients associated to D_a by $\Gamma_{ab}{}^c$ and Γ_{aAB}. We may note that the latter is symmetric and hermitian on the spinor indices, as is required for it to gauge the local SU(2) rotations. We may then define a new connection $\mathcal{D}_a$ acting on spinors,

$$\mathcal{D}_a \lambda_A \equiv \partial_a \lambda_A + A_{aA}{}^B \lambda_B. \tag{3.10}$$

The connection coefficients, A_{aAB} are defined by,

$$A_{aAB} \equiv \Gamma_{aAB} + \frac{i}{\sqrt{2}} \pi_{aAB} \tag{3.11}$$

where the symmetric hermitian spinor π_{aAB} is defined by,

$$\pi_{aAB} \equiv \frac{1}{\sqrt{q}} \left[M_{aAB} - \tfrac{1}{2} \sigma_{aAB} (M_b{}^{CD} \sigma^b{}_{CD}) \right]. \tag{3.12}$$

We see that A_{aAB} is analogous to the Bargmann coordinate z in that its hermitian part Γ_{aAB} contains information only about $\sigma^a{}_{AB}$, while its antihermitian part π_{aAB} contains information about the conjugate momenta M_{aAB}.

As a coordinate on the extended phase space $(\sigma^a{}_{AB}, M_a{}^{AB})$, A_{aAB}

has several remarkable properties. First of all, all of its components commute among themselves,

$$\{A_{aAB}(x), A_{bCD}(y)\} = 0. \tag{3.13}$$

This is a nontrivial property, and at the present time the only proof which is known is by direct, and tedious, calculation [32].*

In the case of the harmonic oscillator, it is often convenient to coordinatize the phase space in terms of z and $\bar{z}$. The analogous coordiates in our case, A_{aAB} and $A^{\dagger}_{aAB}$, are not, however, very convenient because, while $[z,z] = -2i$, the Poisson bracket of A_{aAB} with $A^{\dagger}_{aAB}$ is somewhat complicated. Things are much simpler if we choose as a second coordinate on Γ_E the densitized soldering form,

$$\tilde{\sigma}^{a}{}_{AB} \equiv \sqrt{q}\ \sigma^{a}{}_{AB}. \tag{3.14}$$

We then have,

$$\{A_{a}{}^{AB}(x), \tilde{\sigma}^{b}{}_{MN}(y)\} = \frac{i}{\sqrt{2}}\delta^{(A}{}_{M}\delta_{N}{}^{B)}\delta^{3}(x,y). \tag{3.15}$$

This relation can be seen as analogous to the relation $[q,z] = i$, and indeed the use of the pair $(\tilde{\sigma}^{a}{}_{AB}, A_{aAB})$ to coordinatize Γ_E should be seen as analogous to the use of the pair (q,z) to coordinatize the phase space of the harmonic oscillator. Thus, the condition that the $\sigma^{a}{}_{AB}$ are Hermitian is analogous to the condition that q be real.[c]

When the gauge constraint (3.5) is satisfied, A_{aAB} reduces to the spin connection studied by Sen [33], Witten [34] and others. This connection is, in the classical theory, the projection of the four

[c]In the harmonic oscillator there is another constraint, which is that $p = -i(z - \bar{z})$ be real. It is however, not necessary to explicitly impose both because the symplectic structure $dq\wedge dp$ guarantees that the imaginary part of p has vanishing Poisson brackets with all quantities on the constraint surface $q = \bar{q}$. A similar situation holds for the case of $(\tilde{\sigma}^{a}{}_{AB}, A_{aAB})$.

*Note added in proof: A much simpler derivation of this and other results connected with Ashtekar's formalism is given in Ref. 36.

dimensional left-handed spin connection into the surface Σ.[d] Hence it is closely related to the self-dual piece of the Weyl curvature. This is discussed in more detail in the article by Ashtekar in this volume.

We may now state the constraints of general relativity in terms of the new variables. The SU(2) gauge constraint (3.5) now has the form,

$$G_{AB} = \mathcal{D}_a \tilde{\sigma}^a{}_{AB} = 0. \tag{3.16}$$

In the presence of this constraint, the diffeomorphism and Hamiltonian constraints take the simple forms,

$$C_a = \mathrm{Tr}\, F_{ab} \tilde{\sigma}^b = 0 \tag{3.17}$$

$$C = \mathrm{Tr}\, F_{ab} \tilde{\sigma}^a \tilde{\sigma}^b = 0 \tag{3.18}$$

where $F_{abA}{}^B$ is the curvatuve of the connection A_{aAB},

$$F_{abA}{}^B \equiv 2\, \partial_{[a} A_{b]A}{}^B + [A_a, A_b]_A{}^B. \tag{3.19}$$

It is important to stress that these new forms of the constraints are only equal to the constraints of general relativity on the constraint surface $G_A{}^B = 0$. This is one of the reasons for the enormous simplification in the form of C (3.18) over the form of the usual Hamiltonian constraint (2.3). In essence the latter has been factored into four simpler constraints: the three gauge constraints, $G_A{}^B$ and the new form of the Hamiltonian constraint, (3.18). We see here one reason why the use of the extended phase space is an essential element of this construction.

[d]It is important to note that the components of the Sen connection, to which A_{aAB} is equal on the constraint surface $M_{[ab]} = 0$, do not commute among each other under the usual Poisson bracket structure on Γ. Because of this a representation of the quantum theory which is diagonal in A_{aAB} is possible only if the canonical commutation relations are taken from the Poisson bracket (3.4) on the extended phase space and not from the usual Poisson brackets on the standard phase space Γ.

Before going on to the quantum theory, we will introduce the alternate frame field representation of this formalism, which is sometimes very convenient. We may introduce a *fixed* basis of symmetric Hermitian spinors, $\tau^i{}_{AB}$, normalized so that $\mathrm{Tr}\, \tau^i \tau^j = \delta^{ij}$. Here, $i,j = 1,2,3$ label the three elements of the basis. A familiar choice of the $\tau^i{}_{AB}$ would be the usual Pauli matrices, divided by $-i\sqrt{2}$.

We may then expand our basic variables as,

$$\tilde{\sigma}^a{}_{AB} = \tilde{\sigma}^{ai}\, \tau^i{}_{AB}$$
$$A_{aAB} = A_a{}^i\, \tau^i{}_{AB}\,. \tag{3.20}$$

Now, the $\tilde{\sigma}^{ai}$ are three real vector fields, while the $A_a{}^i$ are the components of a complexified SU(2) connection.

In this notation, the diffeomorphism and Hamiltonian constraints take the forms,

$$G^i = (\mathcal{D}_a \tilde{\sigma}^a)^i = 0 \tag{3.21}$$

$$C_a = F_{ab}{}^i\, \tilde{\sigma}^{bi} = 0 \tag{3.22}$$

$$C = \epsilon^{ijk}\, F_{ab}{}^k\, \tilde{\sigma}^{ai} \tilde{\sigma}^{bj} = 0 \tag{3.23}$$

4. QUANTIZATION IN THE SELF DUAL REPRESENTATION

We now are ready to quantize general relativity using the new variables. We may note that as all of the constraints are of no more than quadratic order in both A_{aAB} and $\tilde{\sigma}^a{}_{AB}$, we could equally well consider a representation diagonal in either variable. This is in contrast to the usual form of the constraints, in which the presence of nonpolynomial functions of q_{ab} makes the choice of a representation diagonal in the momenta p^{ab} practically impossible. Up till this time, only the representation based on functionals of A_{aAB} has been explored in any detail. However, there is no reason to believe that the other choice will not also prove fruitful.

In this and the following sections, we will use the frame field notation introduced at the end of the last section.

As $A_a{}^i$ is a complex coordinate on the phase space Γ_E, a representation based on functionals of $A_a{}^i$ must satisfy some condition of analyticity. By analogy with the Bargmann representation, we will consider a quantization based on the space of holomorphic functionals of $A_a{}^i$. This space of functionals, denoted S, constitutes the arena for the self-dual representation of quantum general relativity.[e]

The Hilbert space of quantum general relativity will be that subspace of S which is annihilated by all of the constraints. In order to represent the constraints G^i, C_a and C as operators on S we need to discuss and resolve three issues: 1) The choice of the operator to represent σ^{ai}, 2) Operator ordering of the constraints, 3) Regularization of the constraints. We will now discuss the first two of these, reserving discussion of regularization for the next section.

By virtue of the Poisson bracket relations, the canonical commutation relations must take the form,

$$[A_a{}^i(x), \tilde{\sigma}^{bj}(y)] = -\sqrt{2}\,\hbar\delta^{ij}\delta_a{}^b\delta^3(x,y). \qquad (4.1)$$

In order to satisfy this we will choose simply,

$$\tilde{\sigma}^{ai}_{(x)} = \frac{\delta}{\delta A_a{}^i(x)}, \qquad (\hbar = 1/\sqrt{2}). \qquad (4.2)$$

We may note that this is analogous to the choice, in the case of the harmonic oscillator of $q = d/dz$. In that case, the choice $\bar{z} = d/dz$ is more popular because it leads to a simpler form for the Hamiltonian. However, they are both equally correct, in the sense that in either case one can choose an inner product such that the operator that represents q is hermitian.

We might then wonder whether there exists an acceptable choice of an inner product such that the choice (4.2) defines a hermitian operator. However, this is not a question that we can ask, because, as we emphasized in Section 2, a physically meaningful inner product can only be defined on the space of

[e]Several important aspects of this representation are discussed in the article by Jacobson in this volume.

states that are annihilated by all of the constraints of the theory. However, as we have not yet even constructed the operators which will be the constraints we do not yet know what form the inner product will take. Furthermore, only operators which commute with all of the constraints will be able to be hermitian with respect to a physical inner product. As $\tilde{\sigma}^{ai}$ does not, by itself, commute with any of the constraints there will never be any context for asking whether it itself is represented by a hermitian operator.

Of course, we want the choice of an operator to represent $\tilde{\sigma}^{ai}$ to lead to a situation in which physical observables are represented on the physical Hilbert space by hermitian operators. However, as far as I know, we have no way at this stage of knowing whether we will make a good choice in this respect. In this situation, we must fall back on other arguments. There are two arguments for the choice (4.2).[f] 1) Simplicity. It is the simplest possible choice, and leads to equations that we can solve. 2) Geometrical quantization. In that language there is a prequantum inner product under which the choice (4.2) is hermitian [36]. However it is not clear whether or not this has any relevance for the physical inner product.

We may now turn to the second issue, which is operator ordering. We choose to make the choice indicated by the equations (3.21 - 3.23). Given this choice, the constraint operators take the following forms,

$$G^i(x)\psi(A) = \mathcal{D}_a \frac{\delta}{\delta A_a{}^i(x)} \psi(A) = 0 \tag{4.3}$$

$$C_a(x)\psi(A) = F_{ab}{}^i(x) \frac{\delta}{\delta A_b{}^i(x)} \psi(A) = 0 \tag{4.4}$$

$$C(x)\psi(A) = \epsilon^{ijk} F_{ab}{}^k(x) \frac{\delta^2}{\delta A_a{}^i(x)\delta A_b{}^j(x)} \psi(A) = 0. \tag{4.5}$$

[f]This question is discussed in more detail in Ref. [22].

The ordering of G^i and C_a is determined by the requirement that when acting on functionals of $A_a^{\ i}$ they generate infinitesimal gauge transformations and diffeomorphisms, respectively. The ordering of C is then motivated by the requirement that the commutation relations of the constraints not lead to inconsistencies. As we discussed in Section 2, this is a rather subtle issue, which is also intertwined with the question of regularization. The reader may consult Ref. [22] for a discussion of this issue.

Finally, we may note that the simplest operator which measures metric information is the inverse metric density (of weight 2),

$$\tilde{q}^{ab}(x) \equiv qq^{ab}(x) = -\sum_i \frac{\delta^2}{\delta A_a^{\ i}(x)\delta A_b^{\ j}(x)} . \tag{4.6}$$

Other operators which measure metric information, such as the metric itself, or the Christoffel symbol, will be still more complicated.

5. SOLUTIONS TO THE HAMILTONIAN AND GAUGE CONSTRAINTS

In this section, I will describe a recently discovered class of wavefunctionals $\Psi(A_a^{\ i})$ which are exactly annihilated by the gauge and the Hamiltonian constraints. Many more details are discussed in [22].

We may begin by noting that there is a state which is, given the orderings (4.3 - 4.5), annihilated by all of the constraints. It is

$$\Psi_0(A) = |\, 0\rangle = 1. \tag{5.1}$$

Although the meaning of this state must be, at this stage, unclear, this is one more exact solution than was ever found in the metric representation. We will now construct a kind of Fock space of states which are exactly annihilated by G^i and C based on this state in the sense that they could not be solutions were $|0\rangle$ not also a solution $|\, 0\rangle$ will thus be referred to (with a certain amount of tongue-in-cheek as the "ground state."

We will do this by looking for solutions to the constraints in the following order: first G^i, then C and finally C_a. We thus begin with G^i. As we remarked in the last section, given the ordering (4.3) this means that $\Psi(A_a{}^i)$ must be gauge invariant, as well as holomorphic.

How do we construct gauge invariant, holomorphic functionals of $A_a{}^i$? We must stress that since in the self dual representation the metric is a second order functional differential operator we do not have the use of a metric or volume element on Σ in the construction of these states. The only way known to us to construct a class of such states is as follows.

To every closed curve γ in Σ, we may associate a state $|\gamma\rangle$ of the form

$$|\gamma\rangle = H(\gamma) = \mathrm{Tr}\, Pe^{\oint ds\, A_a{}^i(\gamma(s))\dot{\gamma}^a(s)T^i} . \tag{5.2}$$

This is a gauge invariant and holomorphic functional of $A_a{}^i$, so that it is of the type required. It is, of course, a rather poor wavefunctional, as it only depends on the value of $A_a{}^i$ on the curve γ. But, we may go on to describe a rather large class of wavefunctionals as follows. Consider any set (finite or infinite) of loops γ_I in Σ. We may associate to this set the state,

$$|\{\gamma_I\}\rangle = \prod_I H(\gamma_I). \tag{5.3}$$

The space containing linear combinations of all such states, together with $|0\rangle$, will be called the Fock space of the loop space on Σ built on $|0\rangle$, and denoted by $\mathcal{F}(\mathfrak{L}_\Sigma)$. It is clearly an important mathematical problem to define this space more rigorously, and then to determine exactly how large a space of gauge invariant holomorphic functionals of $A_a{}^i$ is contained in it.

We now move to the next step, which is consideration of the Hamiltonian constraint, C. We will see that the Hamiltonian constraint acts in a remarkably simple way on states in $\mathcal{F}(\mathfrak{L}_\Sigma)$, and indeed, annihilates a great many of them.

The basic result is the following:

Under a suitable regularization procedure, $C \mid \{\gamma_I\}\rangle = 0$ whenever all of the curves γ_I are smooth and nonintersecting.

There are also solutions in the case that there are intersections. These will be discussed in Section 6.

We will study the action of C on the loop states in two steps. First we consider the action of the naive, unregularized C on $\mid \{\gamma_I\}\rangle$. We will see that this is the product of a divergence and an expression which vanishes due to the algebraic symmetric of (4.5). We then introduce a regularization procedure under which the divergence is removed without affecting the vanishing of the remainder of the expression.

Let us begin with the simplest state, $\mid \gamma\rangle$, which involves a single loop excitation. If we define,

$$U_\gamma(s,t) \equiv Pe^{\int_s^t dv\, A_a{}^i(\gamma(v))\dot{\gamma}^a(v)\tau^i}, \tag{5.4}$$

then it is not hard to see that the first functional derivative of $\mid \gamma\rangle$ is given by,

$$\frac{\delta \mid\gamma\rangle}{\delta A_b{}^j(x)} = \int_0^1 ds\; \delta^3(x,\gamma(s))\dot{\gamma}^b(s)\; \mathrm{Tr}\; U_\gamma(0,s)\tau^j U_\gamma(s,1). \tag{5.5}$$

Notice that this is singular, by virtue of the fact that we are taking a three dimensional functional derivative of a one dimensional object. We may then take the second functional derivative, at the same point:

$$\frac{\delta^2 \mid \gamma\rangle}{\delta A_a{}^i(x)\delta A_b{}^j(x)} = \int_0^1 ds \int_0^1 dt\; \delta^3(x,\gamma(s))\delta^3(x,\gamma(t))$$

$$\times\, \dot{\gamma}^a(s)\dot{\gamma}^b(t)\Big[\theta(s-t)\mathrm{Tr}(U(0,t)\tau^j U(t,s)\tau^i U(s,1)) \tag{5.6}$$

$$+\, \theta(t-s)\mathrm{Tr}(U(0,s)\tau^i U(s,t)\tau^j U(t,1))\Big].$$

Although this expression is singular we may note that it is symmetric in a and b alone. As the only vectorial quantity that could come from the state $\mid \gamma\rangle$ is the tangent vector $\dot{\gamma}^a(s)$, when we take the second derivative we just get the tangent vector twice. We may also note that since, by virtue of the fact that functional derivatives commute like ordinary derivatives, the expression must be, to the extent that it is meaningful at all, symmetric under interchange of the pair (a,i) with the pair (b,j). Thus it will also be symmetric in i and j, at least formally. However, the Hamiltonian constraint (4.5) involves only the piece of the second functional derivative which is antisymmetric separately in the pair ab and the pair ij. Thus, formally, C must annihilate $|\,\gamma\rangle$.

We can see this in detail if we smear C with a lapse N.

$$C(\underset{\sim}{N})\,|\gamma\rangle = \int d^3x\, \underset{\sim}{N}(x)\epsilon^{ijk} F_{ab}{}^{k}(x) \frac{\delta^2\,|\gamma\rangle}{\delta A_a{}^{i}(x)\delta A_b{}^{j}(x)}$$

$$= \left\{-\frac{1}{4}\int_0^1 ds \int_0^1 dt\ \delta^3(\gamma(s),\gamma(t))\underset{\sim}{N}(\gamma(s))F_{ab}{}^{k}(\gamma(s))\dot{\gamma}^a(s)\dot{\gamma}^b(s)\epsilon^{ijk}\delta_{ij}\right\}|\,\gamma\rangle$$

$$= \left\{-\frac{1}{4}\int_0^1 ds\ \frac{\delta^2(0)}{J(s)}\ \underset{\sim}{N}(\gamma(s))F_{ab}{}^{k}(\gamma(s))\dot{\gamma}^a(s)\dot{\gamma}^b(s)\epsilon^{iik}\right\}|\,\gamma\rangle. \tag{5.7}$$

Here J is a Jacobian factor.

We now need to show that this result can be demonstrated in the context of a well defined regularization procedure which eliminates the singular factors in (5.7).

At the end of Section 2 we discussed some of the difficulties which occur when one tries to implement a point splitting regularization in this context. The point splitting regularization of $C\,|\gamma\rangle$ has been studied in detail; the reader who is interested in the details is invited to consult Ref. [22]. However, because of the problems with diffeomorphism invariance which we mentioned above, the result is not completely satisfactory. (It turns out that a subset of the naive solutions are still solutions in the context of the point splitting procedure. However, the selection of that subset is regularization-dependent and breaks diffeomorphism invariance.) Because of these problems we choose to employ a different regularization procedure which is diffeomorphism covariant. This regularization procedure is motivated by the following considerations.

First of all, as we mentioned above, we cannot impose any physically meaningful inner product until after all of the constraints have been solved. The expressions we are interested in regularizing are not expectation values, but are expressions of the form $A(x)B(x)\Psi(A) = 0$. These expressions define a certain space of states, which, with the addition of an inner product defined so that the solutions are normalizable, will comprise the physical Hilbert space. In this context there can be no objection to regularizing expressions of this form by modifying the state functional $\Psi(A)$ rather than the operator.

That it is the definition of the states $|\gamma\rangle$, rather than the operator C which needs to be altered in the regularization procedure is further motivated by the fact that the first functional derivative (5.5) is already singular as a consequence of the one dimensional nature of the state. This suggests that the singular terms in the action of C come from the singular nature of the state, rather than from the fact that C involves second functional derivatives. We will see that it is in fact possible to make the first functional derivatives nonsingular in a way which makes the action of $C(x)$ completely finite.

Consider for simplicity a field theory in which $A_a^{\ i}$ is replaced by a simple one form b_a. Then consider wave functionals of the special form,

$$|M^b\rangle = \exp \int d^3x\, M^b\, b_b \tag{5.8}$$

where M^b is a smooth density of Σ. This functional is infinitely functionally differentiable,

$$\frac{\delta^n |M^b\rangle}{\delta b_{a_1}(x_1)\cdots \delta b_{a_n}(x_n)} = M^{a_1}(x_1)\cdots M^{a_n}(x_n)\,|M^b\rangle. \tag{5.9}$$

This is well defined no matter how many of the points coincide.

We then regularize the states $|\gamma\rangle$ as follows. We imbed the curve $\gamma(s)$ in a two dimensional congruence of smooth nonintersecting curves, $\gamma(s;\sigma)$, parameterized b the pair $\sigma = (\sigma^1,\sigma^2)$. These are chosen such that $\gamma(s;0,0) = \gamma(s)$. The three functions (s,σ^1,σ^2) are then coordinates on an open neighborhood of a segment of γ.

To define the regulated state functionals choose a one parameter family of two dimensional densities, $f_\epsilon(\sigma)$, defined so that,

$$\lim_{\epsilon\to\infty} f_\epsilon(\sigma) = \delta^2(\sigma). \tag{5.10}$$

Now define,

$$B^{f_\epsilon}(s) \equiv \int d^2\sigma f_\epsilon(\sigma) A_a{}^i(\gamma(s,\sigma))\dot{\gamma}^a(s,\sigma)\tau^i. \tag{5.11}$$

Now define the regularized parallel propagator as,

$$U^{f_\epsilon}(s,t) \equiv pe^{\int_s^t du\ B(u)}. \tag{5.12}$$

This is no longer gauge covariant. This is not a serious problem, as we shall see shortly. Now, we may replace the state $|\gamma\rangle$ by the trace of U^{f_ϵ} around a loop,

$$H^{f_\epsilon}[\gamma] \equiv \mathrm{Tr}\ U^{f_\epsilon}(0,1). \tag{5.13}$$

The first functional derivative is,

$$\begin{aligned}\frac{\delta H^{f_\epsilon}[\gamma]}{\delta A_b{}^j(x)} &= \int dt\ \mathrm{Tr}\ U^{f_\epsilon}(0,t)\,\frac{\delta B^{f_\epsilon}(t)}{\delta A_b{}^j(x)}\,U^{f_\epsilon}(t,1)\\ &= \int dt \int d^2\sigma\delta^3(x,\gamma(t,\sigma))\dot{\gamma}^b(t,\sigma)\\ &\quad\times \mathrm{Tr}\ \tau^{\dot{\gamma}} U^{f_\epsilon}(t,\ 1+t).\end{aligned} \tag{5.14}$$

This is not singular. Neither is the action of $C(x)$,[g]

$$\begin{aligned}C(x)H^{f_\epsilon}[\gamma] &= \left\{-\frac{1}{4} f^2(x)F_{ab}{}^k(x)\dot{\gamma}^a(x)\dot{\gamma}^b(x) \times \epsilon^{ijk}\delta_{ij}\right\}H^{f_\epsilon}[\gamma]\\ &= 0.\end{aligned} \tag{5.15}$$

Thus, $H^f(\gamma)$ provides a rigorous exact solution to the constraint C for any smooth $f(\sigma)$ and any smooth nonintersecting γ. We now *define* the action of C on $|\ \gamma\rangle$ as follows. Consider any sequence of smooth $f^\epsilon(\sigma)$ which satisfies (5.11). Clearly we have,

$$\lim_{\epsilon\to\infty} H^{f_\epsilon}[\gamma] = |\ \gamma\rangle. \tag{5.16}$$

We shall then define the action of $C(N)$ on $|\ \gamma\rangle$ by,

$$C(N)\ |\ \gamma\rangle \equiv \lim_{\epsilon\to\infty} C(N)H^{f_\epsilon}[\gamma] = 0. \tag{5.17}$$

---

[g]For simplicity this expression is written in the coordinates $x^a = (s,\sigma^1,\sigma^2)$.

We may note that we are free to do this because $C(N) \mid \gamma\rangle$ is not, to begin with, well defined, being the product of a divergent and a vanishing quantity. However, this prescription will not be well defined unless it is independent of the choice of the set of smearing functions, $f_\epsilon(\sigma)$.[g] However, from (5.15) it is. Therefore, through the definition (5.17) $C(N) \mid \gamma\rangle$ is well defined, and zero.

We must worry about whether gauge invariance is spoiled by this regularization procedure. However, it is easy to see that,

$$\operatorname*{Lim}_{\epsilon\to\infty} G^i H^{f_\epsilon} = G^i \operatorname*{Lim}_{\epsilon\to\infty} H^{f_\epsilon} = G^i \mid \gamma\rangle = 0. \tag{5.18}$$

Because we have not altered the definitions of the operators, it also immediately follows that,

$$\operatorname*{Lim}_{\epsilon\to\infty} [G^i(x),G^j(y)]H^{f_\epsilon} = [G^i(x),G^j(y)] \operatorname*{Lim}_{\epsilon\to\infty} H^{f_\epsilon}. \tag{5.19}$$

Thus, no problems with gauge invariance can result from the use of this regularization procedure.

It is now straightforward to extend this procedure to any state $|\{\gamma_I\}\rangle$ based on a set of smooth, noninteresting loops γ_I. We may, for example, impose the additional condition that the $f_\epsilon(\sigma)$ are of compact support in σ. Given this we can choose a smooth congruence $\gamma(s;\sigma)_I$ surrounding every loop σ_I in such a way that the curves on which the $f_\epsilon(\sigma)$ are nonvanishing are always nonintersecting. Furthermore, once this is done it will remain done even after the action of a smooth diffeomorphism. Hence, all of the steps may be repeated for a smeared state,

$$|\{\gamma_I\}\rangle^{f_\epsilon} \equiv \prod_I H^{f_\epsilon}[\gamma_I]. \tag{5.20}$$

Thus, we can show that, by extending the definition,

$$C(x) \, |\{\gamma_I\}\rangle \equiv \operatorname*{Lim}_{\epsilon\to\infty} C(x) \, |\{\gamma_I\}\rangle^{f_\epsilon} = 0. \tag{5.21}$$

for any set of smooth, nonintersecting curves, γ_I.

[g]It is also independent of the choice of congruence $\gamma(s;\sigma)$, as long as it consists of smooth curves. This may be important when we consider the action of the generators of diffeomorphisms.

6. SOLUTIONS WITH INTERSECTIONS

We come now to our last topic, which is the action of the constraints on states with intersections. This is a complicated and, at present, incompletely understood subject. Here, we will content ourselves with describing an example which shows that solutions to the constraints involving intersections do exist. For details, including those concerning regularization, which will not be addressed here, see [22].

Let us consider a set of states associated with a figure 8 configuration based on two loops γ and δ.

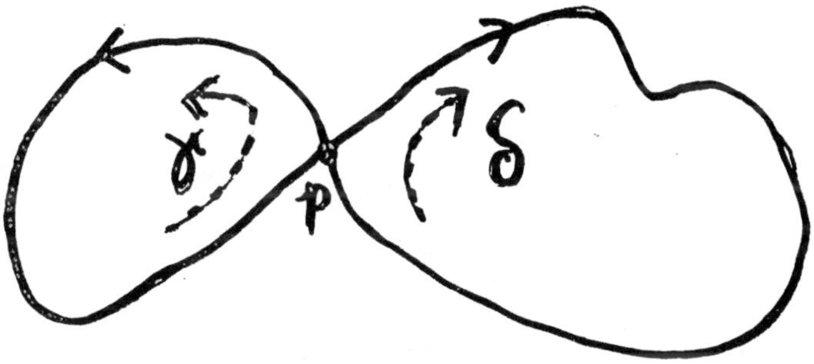

The two loops will be assumed to be smooth, except at their common base point, p. However, we will assume that the figure eight loop $[\gamma\delta]$ is smooth also at p.

Let us consider three states which can be built on these loops,

$$|\circ\circ\rangle \equiv H(\gamma)H(\delta) \tag{6.1}$$

$$|\infty\rangle \equiv H(\gamma\delta) \tag{6.2}$$

$$|\bowtie\rangle \equiv H(\gamma\delta^{-1}) \tag{6.3}$$

Now only two of these states are linearly independent. This is because of the fundamental spinor identity,

$$\delta_A{}^B\delta_C{}^D - \delta_A{}^D\delta_C{}^B - \epsilon_{AC}\epsilon^{BD} = 0. \tag{6.4}$$

As a result we have the identity,

$$|\infty\rangle - |\circ\circ\rangle - |\bowtie\rangle = 0. \tag{6.5}$$

We thus have a two dimensional vector space of states associated with an intersection such as p. We can now apply the Hamiltonian constraint, C, to these states. The result is that there exists only one linear combination of the states which is annihilated by C. Let us, for example, choose as a basis for the intersection states, $|\bigcirc\bigcirc\rangle$ and $|\infty\rangle$. Neither of these by themselves are annihilated by C. The reason is not hard to see, if one recalls that the reason why C annihilated states based on smooth nonintersecting loops was that, at any point, both functional derivatives in C must bring down the same tangent vector $\dot{\gamma}^a(s)$. However, at p there are two linearly independent tangent vectors, $\dot{\gamma}^a(0) = \dot{\delta}^a(1)$ and $\dot{\gamma}^a(1) = \dot{\delta}^a(0)$. Thus, one can get a nonvanishing term proportional to $F_{ab}{}^k \, \dot{\gamma}^a(0)\dot{\delta}^b(0)$.

However, it turns out that $C(p)\,|\bigcirc\bigcirc\rangle$ is proportional to $C(p)|\infty\rangle$ so that one can find a linear combination which vanishes formally. This is,

$$C(N)[|\infty\rangle - 2|\bigcirc\bigcirc\rangle] = 0. \tag{6.6}$$

By using (6.5) this may also be expressed as,

$$C(N)[|\bigcirc\bigcirc\rangle - |\infty\rangle] = 0. \tag{6.7}$$

It is rather complicated to show that this cancellation takes place also when one considers intersections involving regulated states. As one might expect, the regularization of states involving intersections is a good deal more delicate than the case of states without intersections. The regularization of intersections by both point splitting and the state smearing technique we discussed above is carried out in Ref. [22].

We see from (6.6) and (6.7) that the Hamiltonian constraint has the effect of mixing states involving different numbers of closed loops. This is seen also in the lattice theory [21]. This suggests that, when the constraints are satisfied, the Hamiltonian of the theory will contain an interaction by which a state based on two loops that touch at a point, such as $|\bigcirc\bigcirc\rangle$, is transformed into a state involving only one loop such as $|\infty\rangle$.

As some readers may have noticed, this action of the Hamiltonian constraint is in some respects similar to the three string vertex in closed string field theories. Whether there is anything to this similarity, or to the other obvious points of similarity between the formalism which is emerging here and string theory, is an intriguing question for the future.

7. CONCLUSION

In the previous sections we have seen that by using the new self-dual representation of quantum gravity it has been possible to do some things that have not so far been done in the usual metric representation. At present, we can point to three results as evidence that this approach deserves further investigation. These are the formal solution to the operator ordering problem §4 [16], the interpretation of the Hamiltonian constraint as an interaction term between states based on extended objects §5 [21,22] and the existence of a large class of quantum states which are exact solutions to the Hamiltonian constraints §6 [22].

However the canonical road to quantum gravity is both long and steep, and these developments, as exciting as they may be, are at best only a first step towards a quantization of general relativity. It will not be possible to say that this program is firmly on the road to success until a number of additional things have been achieved. At the very minimum, these are:

1) A demonstration of the existence of a consistent operator ordering for the constraints in the context of a regularization procedure. Until this is achieved we cannot be sure that there exist any simultaneous solutions to all of the constraints.

2) Assuming that the ordering that we have used above is shown to be consistent in the context of a suitable regularization procedure, it must then be shown that superpositions of the loop states can be constructed which are annihilated by both the Hamiltonian and the diffeomorphism constraints. These would then constitute examples of physical states of the gravitational field.

3) It will then be important to determine whether the physical states constructed from the loop states contain a basis for all of the physical states of the theory or whether, on the other hand, these

constitute only one sector of the space of physical states. If the latter is the case than we need a way of constructing the other physical states.

4) Once we have a set of physical states which we believe contain a basis for the physical states we must invent an inner product such that the physical states have finite norm and such that the remaining term in the Hamiltonian, the boundary term, is a Hermitian operator.

5) To make contact with physics we will then want to evaluate the expectation value of some physical operators in some physical states. It is quite likely that new issues of regularization will arise at this stage, as it is not clear whether or not the regularization procedures which are sufficient to define the constraint equations are sufficient also to define physical inner products.

6) If all of this can be achieved it will then become important to learn how to construct states which are semiclassical in the sense that smooth metrics, perhaps defined through some procedure of coarse grinding, can be constructed which satisfy a semiclassical extension of the Einstein equations.

The first of these steps, the examination of the operator ordering problem in the context of a regularization, is currently being investigated in the context of the lattice theory [21]. It is probable that we will know shortly whether or not there is a consistent regularized ordering on the lattice. If there is, then at least in principle there ought to exist a space of states which are simultaneous solutions to all of the constraints. However finding them, and then proceeding to the other steps, will not be straightforward tasks, not in the least because of the issues tying together the problems of regularization, the inner product and the spacial diffeomorphisms that we mentioned in Section 2.

Some work on finding superpositions of the loop states which are also annihilated by the diffeomorphism constraints is also underway. The crucial issue in this work seems to be the construction of diffeomorphism invariant measures on loop space. If one treats this problem formally, then one finds results which indicate that knot theory may play an important role in the classification of physical states. This result is not surprising; if one puts the loop states discussed in Section 5 in equivalence classes with respect to the identity connected component of the spacial diffeomorphisms, than these classes are labeled by the link classes of the three manifold Σ. What is difficult about this program is insuring that a diffeomorphism invariant measure can be constructed on loop space of Σ in the presence of a regulator, but in the absence of a background metric on Σ. These ideas are discussed in a preliminary manner in Ref. [23].

At present the situation is, at least to those of us involved with this work, very exciting, as the new variables seem to have given new life to the old program of canonical quantization of general relativity. Whatever the outcome, this must be considered for the good, as there is no other approach in which the conceptual issues which lurk behind any attempt to "quantize gravity" are so intertwined with the day to day technical issues. Thus, whether or not this approach leads to a consistent quantization of general relativity, it seems certain to bring us closer to an understanding of whether the principles of general relativity and quantum mechanics may be consistently combined into a single theory or, if this turns out to be impossible, in what ways we should seek to modify the fundamental principles of physics in order to arrive at a single fundamental theory.

Note added in proof: Very recently, progress has been made on the problem of constructing solutions to all the constraints. The key idea seems to be to construct a new representation for quantum gravity, based directly on functionals of loops [37].

ACKNOWLEDGEMENTS

I would first like to thank Abhay Ashtekar, Ted Jacobson and Paul Renteln for the many, many hours of discussion during which the ideas described here emerged. Most of this work took place at the Institute for Theoretical Physics at the program "Approaches to quantum gravity" and I would like to thank the organizers of the workshop and the administration of the ITP for making the program, which was one of the most exciting experiences of my scientific career, possible. I would also like to thank David Boulware, Louis Crane, Karel Kuchar and Ted Newman for conversations and for their very helpful suggestions during the course of this work.

REFERENCES

1. See, for example, the articles by Boulware, Stelle, Strominger and Tomboulis in *Quantum theory of gravity: Essays in honor of the 60th birthday of Bryce S. DeWitt* ed. S. M. Christensen (Adam Hilger, Bristol, 1984).

2. M. B. Green and J. H. Schwarz, *Phys. Lett.* B149, (1984), 117; B151 (1985), 21.

3. R. Penrose, "Angular momentum: an approach to combinatorial space-time" in *Quantum theory and beyond* ed. T. Bastin, pp. 151-180. (Cambridge University Press, Cambridge, 1971); "Combinatorial quantum theory and quantized directions" in *Advances in twistor theory*, ed. L. P. Hughston and R. S. Ward (Pitman Publishing, London, 1979).

4. R. Penrose, "Twistor theory, its aims and achievements" in *Quantum gravity: an Oxford symposium*, ed. C. J. Isham, R. Penrose and D. W. Sciama (Clarendon Press, Oxford, 1975). See also *Advances in twistor theory* op. cit.

5. D. Finkelstein, "A process conception of nature" in *The Physicist's Conception of Nature* ed. J. Mehra (Reidel, Dordrecht, Holland, 1973); D. Finkelstein and E. Rodriguez, "Quantum time-space and gravity" in *Quantum concepts in space and time* ed. R. Penrose and C. J. Isham (Clarendon Press, Oxford, 1986).

6. B. S. DeWitt, *Phys. Rev. Lett.* 13 (1964), 114.

7. C. J. Isham, A. Salam and J. Strathdee, *Phys. Rev.* D3 (1971), 1805; D5 (1972), 2548.

8. R. Arnowitt, S. Deser and C. W. Misner, *Phys. Rev.* 120 (1960), 313, 321.

9. J. A. Wheeler, in *Relativity, groups and topology*, ed. B. S. DeWitt and C. DeWitt (Gordon and Breach, London, 1964), pp. 317-520 and in *Battelle Recontres*, ed. C. DeWitt and J. A. Wheeler (Benjamin, New York, 1986).

10. S. W. Hawking, *Phys. Rev.* D13 (1976), 2460.

11. L. Crane and L. Smolin, *Nuclear Physics* B267 (1986), 714.

12. C. J. Isham, *Proc. Roy. Soc.* A351 (1976), 209.

13. E. P. T. Liang, *Phys. Rev.* D5 (1972), 2458; C. Teitelboim, in *An Einstein centenary volume* ed. A. Held (New York, Plenum, 1980).

14. O. Klein, in Niels Bohr and the Development of Physics, ed. W. Pauli, L. Rosenfeld and V. Weisskopf (Oxford, Pergamon, 1955); L. Landau, in Niels Bohr and the Development of Physics, op. cit.; W. Pauli, *Helv. Phys. Acta Suppl.* 4 (1956), 69; S. Deser, *Rev. Mod. Phys.* 29 (1957), 417.

15. P. K. Feyerabend, *Against Method* (Verso, London, 1978).

16. A. Ashtekar, *Phys. Rev. Lett.* 57 (1986), 2244.

17. A. Ashtekar, Syracuse preprint (1986), "Einstein constraints in the Yang-Mills form" in *Proceedings of the Florence conference on constrained systems* (World Scientific, Singapore, 1986) (to appear).

18. A. Ashtekar, this volume.

19. A. Ashtekar, T. Jacobson and L. Smolin, "A new characterization of half-flat solutions to Einstein's equations" in *Commun. in Math. Phys.*, to appear.

20. J. Kogut and L. Susskind, *Phys. Rev.* D11 (1975), 395; L. Susskind, "Coarse grained quantum chromodynamics" in *Les Houches XXIX (1976): Weak and Electromagnetic Interactions at High Energy* ed. R. Balian and C. H. Llewellyn Smith (North Holland, Amsterdam, 1977), pp. 207-308.

21. P. Renteln and L. Smolin, ITP Santa Barbara preprint NSF-ITP-86-137, November 1986; P. Renteln, Ph.D. Thesis, Harvard (1987), in preparation.

22. T. Jacobson and L. Smolin, "Nonperturbative quantum geometries, *Nucl. Phys. B*, to appear.

23. L. Smolin, in "New Perspectives in Canonical Gravity", ed. A. Ashtekar (Bibliopolous, Naples, 1988), to appear.

24. The centrality of the role of time as a fundamental issue in quantum gravity has been very well argued by Karel Kuchar in his contribution to *Quantum Gravity II: Second Oxford Symposium* ed. C. J. Isham, R. Penrose and D. W. Sciama, (Clarendon Press, Oxford, 1981).

25. C. J. Isham and K. Kuchar, *Ann. Phys.* (N.Y.), 164 (1985), 288 and 316.

26. P. A. M. Dirac, *Lectures on quantum mechanics* (Yeshiva University Press, New York, 1964). A good review is A. Hanson, T. Regge and C. Teitelboim, *Constrained Hamiltonian systems* (Academia Nazionale Dei Lincei, Rome, 1976).

27. See, for the case of first class constraints, L. D. Faddeev and A. A. Slavnov, *Gauge fields: Introduction to the quantum theory* (Benjamin, New York, 1980).

28. For a review of the standard Hamiltonian formulation of general relativity the reader may consult the discussion in C. W. Misner, K. S. Thorne and J. A. Wheeler, *Gravitation* (W. H. Freeman and Company, San Francisco, 1973) or the recent review, A. Ashtekar, *Notes on quantum gravity* (Bibliopoulis, Naples, in press).

29. For a good discussion of these issues see recent paper by J. Hartle, NSF-ITP-86- . For another view, see the very interesting paper by J. Barbour in *Quantum concepts in space and time*, op. cit.

30. T. Jacobson, this volume.

31. V. Bargmann, *Comm. Pure and Appl. Math.* 14 (1961), 187; *Proc. Nat. Acad. Sci. U.S.A.* 48 (1962), 199.

32. A rather complete discussion of this calculation is in Paul Renteln, "Notes on Ashtekar's new variables", Santa Barbara ITP preprint (1986).

33. A. Sen, *J. Math. Phys.* 22 (1981), 1718; *Phys. Lett.* 119B (1982), 89.

34. E. Witten, *Commun. Math. Phys.* 80 (1981), 381.

35. A. Ashtekar, unpublished.

36. For the Lagrangian approach to Ashtekar's variables, please see J. Samuel, *Pramane-J. Phys.* (28) (1987) L429; and also, T. Jacobson and L. Smolin, Classical and Quantum Gravity, to appear, and *Phys. Lett. B* **196** (1987), 39.

37. Please see, C. Rovelli, "The Loop Representation" in *New Perspectives in Canonical Gravity* op. cit., and C. Rovelli and L. Smolin, "Loop space representation for canonical quantum gravity, in preparation.

LEE SMOLIN
INSTITUTE FOR THEORETICAL PHYSICS
UNIVERSITY OF CALIFORNIA
SANTA BARBARA, CA 93106

(Present address)
DEPARTMENT OF PHYSICS
YALE UNIVERSITY
NEW HAVEN, CT 06511

Contemporary Mathematics
Volume **71**, 1988

SUPERSPACE IN THE SELF-DUAL REPRESENTATION OF QUANTUM GRAVITY

Ted Jacobson*

ABSTRACT. In the self-dual representation of canonical quantum gravity, states are holomorphic functionals $\Psi: \mathcal{A} \to C$ on the space $\mathcal{A}$ of SL(2,C) connections on a 3-manifold Σ. The constraints demand that Ψ be invariant under combined diffeomorphisms and gauge transformations, so Ψ passes to a functional $\Psi: \mathcal{A}/\mathrm{AUT} \to C$ on connections modulo automorphisms of a principal SL(2,C) bundle. $\mathcal{A}/\mathrm{AUT}$ plays the role that the "superspace" of 3-geometries plays in the usual metric representation. The problem of finding such functionals is discussed, and some formal solutions are given.

By a change of variables in the phase space of general relativity, Ashtekar [1,2] has proposed a new representation for the canonical quantization of gravity, in which states are holomorphic functionals of an SL(2,C) connection one-form A_a. In classical solutions A_a is a potential for the self-dual part of the Weyl curvature, hence the A-representation has been called the "self-dual representation".

A principal motivation for this change of variables is the resulting dramatic simplification of the energy constraint, whose quantum form takes the place of the Wheeler-deWitt equation in the usual metric representation. Some initial steps have been taken towards solving this self dual form of the energy constraint in quantum gravity--and some exact solutions are known [3,4]. The subject of this talk however was the remaining constraints in the self dual representation, of which there are six. Three of these correspond to the momentum or diffeomorphism constraint in the metric representation. In addition there are three new "gauge" constraints arising from the local SU(2) invariance introduced in the self dual representation as a result of working with soldering forms (equivalently triads) rather than metric components.

METRIC REPRESENTATION. In the metric representation the states are functionals of the 3-metric q_{ab}, and the meaning of the diffeomorphism constraint is that the functional $\psi[q_{ab}]$ actually depends only on the 3-geometry, i.e., it is in-

1980 Mathematics Subject Classification (1985 Revision). 83C45.
*Supported by NSF grants PHY85-06686 and PHY82-01094.

variant under the action of diffeomorphisms of the metric. Calling the space of 3-geometries "superspace," the states are evidently functionals on superspace.

A handful of superspace functionals can be written as

$$\Psi[q_{ab}] = \psi\,(\textstyle\int dV_q,\ \int dV_q\ R,\ \int dV_q\ R_{ab}R^{ab},\ \ldots) \tag{1}$$

where dV_q is the volume element determined by q_{ab} and R is the curvature scalar of q_{ab}, etc. More generally, one can employ a harmonic coordinate system $X^{\hat{a}}[q]$ determined by the covariant Laplace equation

$$\Delta_q\ X^{\hat{a}} = 0 \quad , \quad \hat{a} = 1,\ 2,\ 3 \tag{2}$$

subject to some boundary conditions at infinity (in the asymptotically flat context). Any functional $\Psi[q_{ab}] = \Psi[q_{\hat{a}\hat{b}}]$ of the components $q_{\hat{a}\hat{b}}$ of q_{ab} in these harmonic coordinates is invariant under diffeomorphisms of q_{ab} which are asymptotically the identity. Conversely any superspace functional can in principle be written in this form. Since it is impossible to solve for the harmonic coordinates explicitly in general ($X^{\hat{a}}[q]$ is a highly non-local functional of the metric) this is probably of limited practical value however.

It is interesting that in 2+1 dimensional quantum gravity the diffeomorphism constraint has actually been solved explicitly [5], by exploiting the fact that every 2-d metric can be brought to a multiple of the Kronecker delta by way of a unique coordinate transformation. The explicit form of the energy constraint as a non-local operator on superspace was obtained in ref. [5].

SELF-DUAL REPRESENTATION. In the self-dual representation the diffeomorphism and new gauge constraints can be written as [1],[2]

$$\frac{d}{d\varepsilon}\ \Psi\ [A_a + \varepsilon\ N^bF_{ab}]\ |_{\varepsilon=0} = 0 \tag{3}$$

$$\frac{d}{d\varepsilon}\ \Psi\ [A_a + \varepsilon\ \mathcal{D}_a\Lambda]\ |_{\varepsilon=0} = 0 \tag{4}$$

Here F_{ab} is the curvature two-form of the connection A_a, N^b is an arbitrary vector field, $\mathcal{D}_a$ is the (gauge) covariant derivative formed with the connection A_a, and Λ is an arbitrary gauge parameter. (I will not be specific about the boundary conditions on N^b and Λ.) Since the original gauge invariance arose from the local rotation group SU(2), Λ is in principle anti-hermitian. This restriction on Λ is immaterial however, because Ψ is required to be holomorphic (otherwise it would depend on too many degrees of freedom [1],[2]).

Equation (4) says that Ψ is a gauge invariant functional of the connection A_a. This is familiar from Yang-Mills theories and corresponds there to the Gauss'-law constraint. To understand the meaning of (3) note that

$$N^aF_{ab} = N^a\ \partial_a\ A_b + (\partial_b\ N^a)A_a - \partial_b\ (N^aA_a) + N^a[A_a,A_b] \tag{5}$$

In a gauge where $N^aA_a = 0$, the last two terms of (5) vanish and N^aF_{ab} becomes just the Lie derivative with respect to N^a of the gauge components of A_a, considered as three ordinary one-forms. Together with the gauge invariance constraint (4), (3) therefore implies that Ψ is invariant under diffeomorphisms connected to the identity.

The expression N^aF_{ab} can be thought of as a gauge-covariant generalization of the Lie derivative for connections, the ordinary Lie derivative being ill-defined because it is not gauge covariant. Thinking of the connection as a Lie algebra valued one form ω on a principal fiber bundle $P \to \Sigma$, the situation can be viewed as follows. The action of a finite diffeomorphism $\phi:\Sigma\to\Sigma$ on the connection is not defined, because ω lives on P, not Σ, and the lift $\tilde{\phi}:P\to P$ of ϕ is defined only up to an arbitrary gauge transformation (automorphism of P that fixes Σ). An infinitesimal diffeomorphism however is given by a vector field N on Σ, and this has a unique <u>horizontal</u> lift $\tilde{N}$. The ordinary Lie derivative with respect to $\tilde{N}$ is the gauge covariant Lie derivative. To see that this is the same as (5), note that

$$L_{\tilde{N}}\omega = \tilde{N} \lrcorner d\omega + d(\tilde{N} \lrcorner \omega) = \tilde{N} \lrcorner \Omega \tag{6}$$

since $\tilde{N} \lrcorner \omega = 0$ ($\tilde{N}$ is horizontal) and $\Omega = d\omega + [\omega,\omega]$. Pulling back by a local section $s:\Sigma\to P$ then yields

$$s^*L_{\tilde{N}}\omega = N \lrcorner F \quad , \tag{7}$$

where $F = s^* \omega$.

We can now state the content of the two constraints (3), (4) in geometric terms. Let P be a principal SL(2,C) bundle over the 3-space Σ, and let $\mathcal{A}$ denote the space of connections on P. If AUT denotes the group of automorphisms of P, then the states in the self dual representation are holomorphic functionals on the complex manifold $\mathcal{A}$ that pass to $\mathcal{A}$/AUT,

$$\Psi: \mathcal{A}/\mathrm{AUT} \to C \tag{8}$$

The space $\mathcal{A}$/AUT therefore plays the role that superspace plays in the metric representation, its elements taking the place of 3-geometries. The complex structure of $\mathcal{A}$ is trivially inherited from the complex structure on the Lie algebra of SL(2,C) in which the connections take their values. The differential structure of $\mathcal{A}$/AUT is complicated due to the presence of singularities. Away from these singularities, it would seem that $\mathcal{A}$/AUT is locally a complex manifold.

Now the question is how do we write down functionals on $\mathcal{A}$/AUT? In the metric representation there are the nice explicit superspace functionals (1), and there is also the general (implicit) solution employing harmonic coordinates (2). Are there analogues of these for functionals on $\mathcal{A}$/AUT?

Our first remark is that without a metric, the only gauge and diffeo invariant holomorphic functionals analogous to (1) appear to be functions of the integral of the Chern-Simons 3-form,

$$\Psi[A] = \psi\left(\int_\Sigma \mathrm{Tr}(A\wedge dA + \frac{2}{3} A\wedge A\wedge A)\right) . \tag{9}$$

This functional is gauge invariant if Σ is compact, or if the gauge transformation is asymptotically the identity.

A large class of formal solutions to the constraints is given by functional integrals of the form

$$\Psi[A] = \int [Dq]\ \Phi[q]\ \chi[A,q] . \tag{10}$$

In (9), $[Dq]$ is a diffeo invariant measure, $\Phi[q]$ is a diffeo invariant functional of the 3-metric q_{ab}, and χ is <u>any</u> holomorphic, gauge invariant functional which is diffeo invariant when q is transformed along with A. For example, one could take

$$\chi[A,q] = \alpha(\int dV_q\ \mathrm{Tr}\ F_{ab}F_{cd}\ q^{ac}q^{bd}) \tag{11}$$

with α any holomorphic function. Of course for (10) to be more than "formal" the functional integral must be given a precise meaning in a manner that preserves the symmetries and holomorphicity.

More generally, by considering the gauge constraint (4) alone, we can reach a powerful conclusion about the form of Ψ. Namely, Ψ must depend on the connection A only through the trace of its holonomy map. (The functionals in (9) and (11) can be written as limits of such functionals.) The argument goes as follows.

The constraint (4) means that Ψ passes to a functional on $\mathcal{A}$/Aut, where Aut is the group of gauge transformations, i.e., automorphisms of P that project to the identity on Σ. The elements of $\mathcal{A}$/Aut are connections up to gauge equivalence, and a characterization of these is provided by holonomy maps. Given a connection A, its holonomy map H^x_A assigns to each loop γ (based at x) the group element $H^x_A(\gamma) = \mathrm{Pexp} \oint A\cdot d\gamma$. ($H^x_A$ is not quite invariant, but transforms as $H^x_A \to g(x)\ H^x_A\ g(x)^{-1}$.) Conversely, H^x_A determines the connection A up to gauge equivalence [6,7]. Thus a gauge invariant functional of A must in fact depend on A only through H^x_A. If Σ is connected, the H^x_A are equivalent for all x, so the x can be dropped.

To eliminate the residual dependence of H^x_A on the gauge at x one can take the trace. The question then arises whether some gauge invariant information is <u>lost</u> upon taking the trace. This actually depends on the group. Let us consider two examples, SU(2) and SL(2,C). Every SU(2) element can be brought by an SU(2) similarity transformation to the form

$$\begin{pmatrix} e^{i\theta} & 0 \\ 0 & e^{-i\theta} \end{pmatrix} .$$

$|\theta|$ thus labels the gauge equivalence class, and this is fully determined from the trace, $2\cos\theta$. Every SL(2,C) element can be brought by an SL(2,C) similarity transformation to one of the forms

$$\begin{pmatrix}\lambda & 0\\ 0 & \lambda^{-1}\end{pmatrix}, \quad \begin{pmatrix}1 & 1\\ 0 & 1\end{pmatrix}, \quad \begin{pmatrix}-1 & 1\\ 0 & -1\end{pmatrix}$$

(these are the Jordan forms of SL(2,C) matrices). The last two correspond to "null rotations". The trace does not distinguish the null rotations from $\pm I$, so here gauge-invariant information is lost by the trace. Thus an SL(2,C) connection cannot in general be reconstructed just from a knowledge of $\mathrm{Tr}H_A$. (For example, a connection with holonomy entirely in a subgroup of null rotations cannot be distinguished from a flat connection.)

In spite of the fact that $\mathrm{Tr}H_A$ fails to determine H_A up to gauge equivalence, any gauge invariant holomorphic functional $\Psi[A]$ can nevertheless be regarded as a functional of $\mathrm{Tr}H_A$. This is because the space $\mathcal{A}$ of SL(2,C) connections is the complexification of the space $\hat{\mathcal{A}}$ of SU(2) connections. Holomorphicity of Ψ therefore implies that Ψ is the analytic continuation of some gauge invariant functional $\hat{\Psi}$ on $\hat{\mathcal{A}}$. By the above argument,* $\hat{\Psi}[\hat{A}] = \hat{\Psi}[\mathrm{Tr}H_{\hat{A}}]$ where $\hat{A} \in \hat{\mathcal{A}}\mathcal{C}\mathcal{A}$. Now extending the domain of Ψ to all of $\mathcal{A}$ we obtain the (unique) analytic continuation of $\hat{\Psi}$, which must be just the original functional Ψ.

In particular, the states in the self-dual representation of quantum gravity can not distinguish between flat connections and those with holonomy entirely in a subgroup of null rotations. Perhaps this is related to the fact that although the little group of a null momentum vector contains a subgroup of null rotations, all known massless particles transform trivially under this subgroup.

An interesting class of functionals of the form (10) can be constructed using the map $\mathrm{Tr}H_A$. The idea is to take

$$\chi[A,q] = \chi\big(\mathrm{Tr}H_A(\gamma_q)\big) , \tag{12}$$

where γ_q is some loop (or set of loops) in Σ determined by the metric q. For example γ_q might be a circuit around a particular cell in the harmonic coordinate system of q - or it might be determined by the geodesic equation of q, together with some boundary conditions. Now it is a result of refs. [3],[4] that the functional $\mathrm{Tr}H_A(\gamma_q)$ satisfies the energy constraint in a regulated form. The upshot is the startling fact that, provided the functional integral (10) can be defined in a manner preserving the constraints (3) and (4), the functional (10),(12) will be a solution to all of the constraints of quantum general relativity in the self-dual representation.

It is a pleasure to thank Abhay Ashtekar, Paul Renteln and Lee Smolin for countless discussions about the self dual representation. Thanks go also to Arthur Fischer and Richard Woodard for useful discussions, and to the Institute for Theoretical Physics and the organizers of the program there on quantum gravity.

BIBLIOGRAPHY

1. A. Ashtekar, Phys. Rev. Lett. 57 (1986) 2244.

2. A. Ashtekar, "A New Hamiltonian Formulation of General Relativity," Syracuse University preprint, submitted to J. Math. Phys.

3. L. Smolin, this volume.

4. T. Jacobson and L. Smolin, in preparation.

5. M. Henneaux, Phys. Lett. 134B (1984) 184.

6. J. Anandan in Conference on Differential Geometric Methods in Theoretical Physics, Trieste, 1981, ed. G. Denardo and H. D. Doebner (World Scientific, Singapore. 1983).

7. J. W. Barrett, Imperial College preprint TP/84-85/9, July 1985.

DEPARTMENT OF PHYSICS
UNIVERSITY OF CALIFORNIA
SANTA BARBARA, CA 93106

DEPARTMENT OF PHYSICS
BRANDEIS UNIVERSITY
WALTHAM, MA 02254 (present address)

*The argument here is incomplete. To complete it, note that H is a unitary representation of the loop group, and invoke the fact that a unitary representation is uniquely determined by its character alone.

Contemporary Mathematics
Volume 71, 1988

SELF-DUAL AND ANTI-SELF-DUAL HERMITIAN METRICS ON COMPACT COMPLEX SURFACES

Charles P. Boyer

The recent symbiosis between mathematics and physics is nowhere more apparent than in quantum gravity. This is especially true in view of the fact that it was Roger Penrose's nonlinear graviton construction [18] that inspired the development of the moduli space of Yang-Mills theory [1] culminating in Donaldson's celebrated use of the moduli space of k=1 SU(2)-instantons to describe the topology of smooth simply-connected four-manifolds with positive definite intersection matrix. The seminal paper of Atiyah, Hitchin and Singer [1] led to many other natural problems. One such problems is: Determine all smooth compact orientable four manifolds that admit a metric with self-dual or anti-self-dual Weyl tensor and describe this metric up to conformal equivalence. In its entirety this is a very difficult problem that involves global existence theorems for nonlinear partial differential equations. One such case is Yau's [2, 25, 26] proof of the Calabi conjecture (in the case of vanishing first Chern class) which implies the existence of an anti-self-dual Ricci flat metric on a K3 surface. Other than Yau's tour de force two lines of attack of the problem have emerged; the first involves giving necessary conditions for existence [4-8, 12, 14, 17] while the second uses twistor methods to prove existence and hopefully explicitly construct the metrics [10, 13, 21, 24]. Actually there are too few known examples that are not conformally flat. Other than the conformally flat structures, the only known self-dual or anti-self-dual metrics on compact 4-manifolds are $\mathbb{P}^2$ and compact quotients of its dual symmetric space HH(2), K3 and its quotients [12, 25], and Poon's [21] recent existence proof by twistor methods of a one-parameter family of self-dual metrics on the connected sum $\mathbb{P}$ # $\mathbb{P}^2$. The purpose of the present paper is to describe some of the author's recent work [6] giving necessary conditions for the existence of anti-self-dual Hermitian metrics on compact complex surfaces and to present some new results concerning self-dual metrics on complex Hermitian surfaces.

1980 Mathematics Subject Classification (1986 Revision). 52C55 (32J15).
[1]Partially supported by NSF grant/contract NO. DMS-8508950.

There is a nice description of our problem in terms of a quadratic variational problem which closely parallels Yang-Mills theory. This is described by the conformally invariant "action"

(1) $$A = \int |W_+|^2 + |W_-|^2$$

where $W_+(W_-)$ denote the self-dual (anti-self-dual) parts of the Weyl conformal tensor. There is a topological invariant playing the role of instanton number, namely the Hirzebruch signature τ, given through Chern-Weil theory by

(2) $$\tau = \frac{1}{12\pi^2} \int |W_+|^2 - |W_-|^2$$

Thus we have $A \geq 12\pi^2 |\tau|$ so

(3a) $$W_+ = 0$$

or

(3b) $$W_- = 0$$

give absolute minimum for A. It is intriguing to conjecture about the existence of solutions to (3a) or (3b) with arbitrarily high instanton number providing infinite hierarchies of conformal gravitational instantons for conformal gravity [22].

Throughout the sequel M will denote a smooth compact orientable manifold usually of dimension four and (M, g) will denote either a Riemannian structure or in the complex case an Hermitian structure. Now on an orientable Riemannian manifold the Hodge * operator $*: \Lambda^p M \to \Lambda^{n-p} M$ on the exterior bundle $\Lambda M = \bigoplus_p \Lambda^p M$ is an involution on $\Lambda^2 M$; thus $\Lambda^2 M$ splits as a direct sum

(4) $$\Lambda^2 M \simeq \Lambda^2_+ M \oplus \Lambda^2_- M$$

where $\Lambda^2_\pm M$ denote the subbundles of $\Lambda^2 M$ whose fibres are the $\pm$ eigenspaces, respectively, of *. Moreover, on $\Lambda^2 M$, * is a conformally invariant operator. It should also be mentioned here that reversing the orientation of M interchanges the eigenspaces Λ^2_+ and Λ^2_- .

Now in addition, suppose that M is almost complex and that g is compatible with the almost complex structure J; i.e. (M, g) is almost Hermitian. The almost complex structure gives another well-known splitting, namely the splitting of the complexified tangent bundle into its holomorphic and anti-holomorphic parts, viz.

(5) $$TM \otimes \mathbb{C} \simeq T^{1,0}M \oplus T^{0,1}M$$

This induces a splitting on $\Lambda^2 M$, namely

(6) $$\Lambda^2 M \otimes C \simeq \Lambda^{2,0}M \oplus \Lambda^{1,1}M \oplus \Lambda^{0,2}M$$

Now the compatibility condition on g relates the two splittings (4) and (6):

$$\Lambda^2_+M \otimes \mathbb{C} \simeq \Lambda^{2,0}M \oplus \Lambda^{0,2}M \oplus \varepsilon$$

(7) $\Lambda^{1,1}M \simeq \Lambda^2_-M \oplus \varepsilon$

where ε denotes the complex trivial line bundle on M generated by the fundamental Hermitian form Ω defined by

(8) $\Omega(X,Y) = g(X,JY)$

where X and Y are any two vector fields on M. In four real dimensions the operator ([11] pg. 111) $L: \Lambda^{p,q}M \to \Lambda^{p+1,q+1}M$ defined by $L\alpha = \alpha \wedge \Omega$ has a special important property: restricted to 1-forms L defines an isomorphism between Λ^1M and Λ^3M. Thus, there is a natural 1-form β and M such that

(9) $d\Omega + \beta \wedge \Omega = 0$

This fact makes Hermitian geometry in two complex dimensions tractable even in the non-Kähler case. In particular, we shall be interested in the case when β is closed (i.e. $d\beta = 0$), and the almost complex structure J is integrable. In this case (M, g) is called locally conformally Kähler [23] since locally $\beta|U_\alpha = d\psi_\alpha$ for some smooth function $\psi_\alpha: U_\alpha \to \mathbb{R}$ where $\{U_\alpha\}$ is an appropriate open cover of M. Then the locally defined metrics $e^{\psi_\alpha} g|U_\alpha$ are Kähler on U_α.

On a complex surface (i.e. a two complex dimensional manifold) the bundle $\Lambda^{2,0}M$ is a complex line bundle, called the canonical line bundle and denoted by K. Thus as real bundles

(10) $\Lambda^2_+M \simeq K \oplus \varepsilon$

where here ε is the real trivial line bundle generated by Ω. An important role is played by the number of linearly independent holomorphic sections $P_\ell(M)$ of the tensor product bundles $K^\ell = K \otimes \overset{\ell \text{ times}}{\cdots} \otimes K$, where $\ell = 0, 1, \ldots,$. The numbers $P_\ell(M)$ are called the plurigenera of M and are bimeromorphic invariants.

In four dimensional Riemannian geometry the curvature tensor splits into four irreducible parts under the action of the orthogonal group SO(4). More explicitly, if $\Gamma = \Gamma_+ + \Gamma_-$ denotes the splitting of the Levi-Civita connection according to the Lie algebra isomorphism $so(4) \simeq su(2) \oplus su(2)$ then the curvature R of Γ splits

(11) $R = R_+ \oplus R_-$

where $R_\pm$ can be viewed as a homomorphism from $\Lambda^2_\pm M \to \Lambda^2 M$. Then using $\mathrm{Hom}(\Lambda^2_\pm M, \Lambda^2 M) \simeq \mathrm{Hom}(\Lambda^2_\pm M, \Lambda^2_\mp M) \oplus \mathrm{End}\,\Lambda^2_\pm M$, the traceless Ricci tensor B is the piece of $R_\pm$ lying in $\mathrm{Hom}(\Lambda^2_\pm M, \Lambda^2_\mp M)$, the scalar curvature R is the part in the image of the trace of the endomorphism of $\Lambda^2_\pm M$, and the self-dual and anti-self dual parts, W_+ and W_-, of the Weyl conformal tensor correspond to the traceless parts of $\mathrm{End}\,\Lambda^2_+M$ and $\mathrm{End}\,\Lambda^2_-M$, respectively.

Now if (M, g) is a Hermitian manifold the orientation is fixed by the complex structure. Thus, for example Λ^2_+M and Λ^2_-M play distinct roles, as evidenced by the relations (7). In fact, the space of self-dual Weyl tensors splits as the direct sum of line bundles $K^{-2} \oplus K^{-1} \oplus \varepsilon_R$, whereas, the space of anti-self-dual Weyl tensors does not. Thus we expect that solutions to $W_+ = 0$ or $W_- = 0$ on a Hermitian manifold have a completely different character.

An important numerical invariant obtained from the curvature of any Hermitian metric on a compact complex surface M is the square of the first Chern class evaluated on the fundamental homology cycle of M. Chern-Weil theory gives

$$(12) \qquad c_1^2 = 2\chi + 3\tau = \frac{1}{4\pi^2} \int [2|W_+|^2 + \frac{R^2}{24} - 2|B|^2]$$

In [6] the author proved the following concerning the $W_+ = 0$ case:

THEOREM 1: Let (M, g) be a compact Hermitian surface with anti-self-dual Weyl tensor, then,

1) If the first Betti number $b_1(M)$ is even, there is a Kähler metric on M conformally equivalent to g with zero scalar curvature, and the plurigenera satisfy $P_\ell(M) \leq 1$ for all ℓ.

2) If $b_1(M)$ is odd, then g is locally conformally Kähler and there is a locally conformally Kähler metric on M conformally equivalent to g with non-negative scalar curvature that is strictly positive almost everywhere. Moreover, all the plurigenera $P_\ell(M)$ vanish.

3) $c_1^2 \leq 0$

In order to find complex surfaces that are candidates for admitting an anti-self-dual Hermitian metric, we briefly recall (cf. [3, 11, 15]) the Enriques-Kodaira classification of compact complex surfaces. First let us recall the simpler case of compact Riemann surfaces. They can be divided into three types of geometries according to the genus g or equivalently the Euler number $\chi = 2(1-g)$:

1) elliptic geometry with g = 0 - the Riemann sphere characterized by a metric of constant positive Gaussian curvature; 2) parabolic geometry with g=1 - the torus characterized by a flat metric: 3) hyperbolic geometry with $g \geq 2$ compact quotients of the unit disc in $\mathbb{C}$ characterized by metrics of constant negative Gaussian curvature.

The situation in two complex dimensions is both cruder and far more complicated. First, in one complex dimension, conformal geometry and complex geometry coincide and this no longer true in higher dimension. Second, in one variable the curvature is determined by one function - the Gaussian curvature, whereas in higher dimension it is given by a fourth rank tensor. Nevertheless, a very rough analogue to the case of Riemann surfaces does exist in two complex variables. The invariant which best generalizes the genus is called the Kodaira dimension and denoted by Kod(M). It measures the growth of the plurigenera and can be defined by

$$\mathrm{Kod}(M) = \begin{cases} -\infty & \text{if } P_\ell(M) = 0 \text{ for all } \ell. \\ 0 & \text{if } P_\ell(M) = 0 \text{ or } 1 \text{ but not all zero.} \\ 1 & \text{if } P_\ell(M) \sim \ell \text{ for } \ell \text{ large.} \\ 2 & \text{if } P_\ell(M) \sim \ell^2 \text{ for } \ell \text{ large.} \end{cases}$$

(Note: In n complex dimensions Kod(M) $\leq$ n). The Kodaira dimension is an example of a bimeromorphic invariant (i.e. an invariant under a bimeromorphic map). A bimeromorphic map is a map between surfaces which is biholomorphic away from certain proper analytic subsets (cf. [3]). For surfaces a bimeromorphic map can be characterized in terms of a process known as blowing-up and its "inverse" blowing-down. Blowing-up can be defined locally in terms of coordinates as follows [3, 11]: Let $x \in M$ be any point. The blow-up at x is given by the quadric

$$\Delta = \{(z,\zeta) \in U \times \mathbb{P}^1 : z_1 \zeta_2 - z_2 \zeta_1 = 0 \}$$

where U is a coordinate neighborhood of x with coordinates (z_1, z_2) centered at x, and (ζ_1, ζ_2) are homogeneous coordinates for $\mathbb{P}^1$. The complex surfaces $\tilde{M}$ defined by replacing U by Δ is called the blow-up of $\tilde{M}$ at x. Notice that away from $(z_1, z_2) = (0, 0)$, U is biholomorphic to Δ, but the point x represented by $(z_1, z_2) = 0$ is replaced by a projective line $\mathbb{P}^1$. Thus $\tilde{M}$ has a copy of $\mathbb{P}^1$ embedded as a curve whose normal bundle is isomorphic to the universal bundle on $\mathbb{P}^1$ often devoted by $\mathcal{O}(-1)$. Such embedded $\mathbb{P}^1$'s are called (-1) curves or exceptional curves [3, 11]. A surface is called minimal [3] if it contains no (-1) curves. The important point is [3]: every compact complex surface can be obtained by blowing-up minimal surfaces. The Enriques-Kodaira classification is a classification of the minimal compact complex surfaces. We give this as a table ([3], pg. 188):

Kod(M)	$-\infty$	0	1	2
	1) rational surfaces including $\mathbb{P}^2$ 2) ruled surfaces of genus ≥ 1 3) surfaces of type VII including Hopf surfaces	1) Tori 2) hyperelliptic surfaces 3) K3 surfaces 4) Enriques surfaces 5) Kodaira surfaces	properly elliptic surfaces	surfaces of general type

REMARKS: Rational surfaces are surfaces that are birationally equivalent to $\mathbb{P}^2$. They include $\mathbb{P}^1 \times \mathbb{P}^1$. Ruled surfaces of genus $g \geq 1$ are holomorphic fibrations of $\mathbb{P}^1$ over a Riemann surface S_g of genus $g \geq 1$. Surfaces of type VII refers to Kodaira's original list as used by Barth et al [3]. They all have $b_1(M) = 1$. Hopf surfaces are surfaces whose universal cover is $\mathbb{C}^2/\{(0, 0)\}$. Hyperelliptic surfaces are certain compact quotients of tori. K3 surfaces are simply connected compact surfaces with trivial canonical bundles. Enriques surfaces are quotients of K3 surfaces by an holomorphic involution. Kodaira surfaces are surfaces with odd first Betti numbers that are fibrations over either an elliptic curve (i.e. a $g = 1$ Riemann surface) or a rational curve with an elliptic curve as its generic fibre. Properly elliptic surfaces are fibrations over a higher genus curve $S_g (g \geq 2)$ with an elliptic curve as generic fibre. Surfaces of general type are all algebraic surfaces. They include complete intersections of hypersurfaces in $\mathbb{P}^n$ of sufficiently high degree, quotients of symmetric domains, and products $S_{g_1} \times S_{g_2}$ of Riemann surfaces with $g_i \geq 2$.

Using theorem 1 one can prove [6].

THEOREM 2: *Let (M, g) be a minimal Hermitian surface with anti-self-dual Weyl tensor; then (M, g) is conformally equivalent to one of the following:*

1) *a torus with its flat metric*
2) *a hyperelliptic surface with a flat metric*
3) *a K3 surface with a Yau metric*
4) *an Enriques surface with a Yau metric*
5) $\mathbb{P}^1 \times \Delta/\Gamma$ *with the conformally flat metric obtained as the product metric of a metric of constant positive Gaussian curvature H on* $\mathbb{P}^2$ *and constant negative Gaussian curvature - H on the unit disc* Δ *where* Γ *is the fundamental group of* $S_g = \Delta/\Gamma$.
6) *a Hopf surface with its conformally flat locally conformally Kähler metric of constant positive scalar curvature.*
7) *a class* VII_0 *surface with positive second Betti number*

REMARK: The corresponding metrics are known to exist in cases 1-6. In case 7 existence is not known.

For nonminimal surfaces we have [6]

THEOREM 3: *Let (M, g) be a nonminimal Hermitian surface with anti-self-dual Weyl tensor; then M can be obtained by blowing-up one of the following minimal surfaces k-times:*

1) *a class* VII_0 *surface,* $k \geq 1$
2) *a ruled surface of genus* $g \geq 1$, $k \geq 1$
3) *a rational surface,* $k \geq 9$
4) $\mathbb{P}^2$, $k \geq 10$

REMARK: Existence is an open problem for all the above cases.

A result which is not given explicitly in [6], but can easily be obtained from the results there and in [16] is:

THEOREM 4: *Let (M, g) be a conformally flat Hermitian surface; then it is conformally equivalent to either case* 1), 2), 5), or 6) *of theorem* 2.

To treat the self-dual case we consider another numerical invariant of the complex structure (in fact a bimeromorphic invariant) coming from the Riemann-Roch theorem, namely the <u>arithmetic genus</u> A defined by

$$A = \frac{c_1^2 + c_2}{12} = \frac{\tau + \chi}{4}$$

Suppose M is an Hermitian surface with self-dual Weyl tensor that is not conformally flat; then an obvious consequence of 1) is $\tau > 0$. But also $c_2 = \chi$ cannot be negative since the only minimal surfaces with $c_2 < 0$ are ruled surfaces of genus $g \geq 2$ which have $\tau = 0$. Furthermore, we can not have a blow-up of a ruled surface or any minimal surface since blowing-up decreases τ and increases c_2 by one [11]. Thus $c_2 \geq 0$, so $c_1^2 = 3\tau + 2c_2 > 0$ and it follows from a result of Kodaira [15] that M is either a rational surface or a surface of general type, and in these cases $\chi = c_2 > 0$. Now since τ and χ are both positive and their sum is divisible by four we must have $\tau \geq 1$ $\chi \geq 3$ or $\tau \geq 3$ $\chi \geq 1$. The latter can be ruled out by Miyaoka's inequality [3] $c_1^2 \leq 3c_2$ which holds on any surface of general type and any rational surface. Thus we must have $\chi = c_2 \geq 3$ and $\tau \geq 1$ in which case $c_1^2 \geq 9$. Now a minimal, rational surface (cf. [11] pg. 513) is either $\mathbb{P}^2$ or a rational ruled surface S_n, $n \neq 1$. But for S_n, $c_2 = 4$ and $c_1^2 = 8$ and so S_n cannot occur. Moreover, since blowing-up decreases c_1^2 no blowing-up of a rational surface can occur. Summarizing our results we have:

<u>THEOREM 5</u>: <u>Let</u> (M, g) <u>be a Hermitian surface with self-dual Weyl tensor that is not conformally flat</u>. <u>Then either</u> M <u>is a surface of general type or</u> M <u>is isomorphic to</u> $\mathbb{P}^2$. <u>Moreover, the following inequalities hold</u>:

i) $\tau \geq 1$

ii) $c_2 \geq 3$

iii) $A \geq 1$

iv) $c_1^2 \geq 9$

v) $c_1^2 \leq 3c_2$, $c_2 \geq 3\tau$

vi) $3A \leq c_2 < 4A$, $8A < c_1^2 \leq 9A$

vii) $A \leq \tau \leq 4A - 3$

REMARK: It is not known whether there are any more self-dual conformal structures on $\mathbb{P}^2$ other than that determined by the standard Fubini-Study metric. However, Singer has shown ([9] pg. 369) that there are no self-dual deformations of the standard structure and Poon [21] has shown that there are no other self-dual conformal structures containing a metric with positive scalar curvature.

Actually for self-dual Hermitian Einstein surfaces more can be said. Recall that $W_{\pm}$ can be represented by 3x3 traceless Hermitian matrices, and so can be diagonalized. If two eigenvalues coincide at $x \in M$ $W_{\pm}$ is called degenerate at x and just degenerate if it is degenerate at all points of M. We give an outline of the proof of the next theorem (details will appear elsewhere).

THEOREM 6: Let (M, g) be an Hermitian Einstein surface with self-dual Weyl tensor. Then if (M, g) is not flat, it is either:

1) $\mathbb{P}^2$ with its standard Fubini-Study metric

or

2) a compact quotient of Hermitian hyperbolic space with its standard Fubini-Study metric

Outline of proof: By the Riemannian version of a well-known theorem in general relativity - the Goldberg-Sachs theorem (cf. [19, 20]), W_+ must be degenerate. So by lemma 1 of [6], (M, g) is locally conformally Kähler. But since g is Einstein but not flat $c_1^2 > 0$ by (9). Thus by a theorem of Kodaira [15] M is algebraic and by the results of [6] g is conformal to a Kähler metric. But then the result follows from the work of Chen [7] or Derdzinski [8].

In conclusion, I would like to mention several open problems the most important of which, as previously mentioned, are the existence proofs of conformal structures with vanishing W_+ or W_-. Better still would be the explicit construction of such structures. Another interesting problem is to determine the moduli space of self-dual or anti-self-dual conformal structures. Little is known here with the exception of K3 surfaces where it is known [9] that the general Ricci flat metric depends on 58 parameters. The noncompact case about which nothing was said here, is also of interest. Important work has already been done by Calabi, Hitchin and the Cambridge relativity group (cf. [9] and references therein).

Acknowledgments: I have had many interesting discussions with many people concerning the work discussed here. I would like to take this opportunity to thank: Eugenio Calabi, Bill Goldman, Nigel Hitchin, Jacques Hurtubise, Claude LeBrun, Ben Mann and Al Vitter.

BIBLIOGRAPHY

1. Atiyah, M. F., Hitchin, N. J., Singer, I. M.: Self-duality in four-dimensional Riemannian Geometry, Proc. Roy. Soc. Lond. A362, 425-461 (1978).

2. Aubin, T.: Nonlinear Analysis on Manifolds, Monge-Ampere Equations, (Springer, New York, 1982).

3. Barth, W. Peters, C., Van der Ven, A.: Compact Complex Surfaces, (Springer, New York, 1984).

4. Besse, A.: Geometrie Riemannienne en dimension 4 (Cedic/Fernand Nathan, Paris, 1981).

5. Bourguignon, J. P.: Les Varietes de dimension 4 a signature non nulle dont la courbure est harmonique sont de'Einstein, Inv. Math. 63, 263-286 (1981).

6. Boyer, C. P.: Conformal duality and compact complex surfaces, Math. Ann. 274, 517-526 (1986).

7. Chen, B. Y.: Some topological obstructions to Bochner-Kähler metrics and their applications, J. Diff. Geom. 13, 547-558 (1978).

8. Derdzinski, A.: Self-dual Kähler manifolds and Einstein manifolds of dimension four. Compos. Math. 49, 405-433 (1983).

9. Eguchi, T., Gilkey, P. B., Hanson, A. J.: Gravitation, gauge theories and differential geometry. Phys. Rep. 66 213-393 (1980).

10. Freidrich, T. and Kurke, H.: Compact four-dimensional self-dual Einstein manifolds with positive scalar curvature. Math. Nachr. 106, 271-299 (1982).

11. Griffiths, P., Harris, J.: Principles of Algebraic Geometry, (John Wiley, New York, 1978).

12. Hitchin, N. J.: Compact four-dimensional Einstein manifolds. J. Diff. Geom. 9, 435-441 (1974).

13. Hitchin, N. J.: Kählerian twistor spaces, Proc. London Math. Soc. 43, 133-150 (1981).

14. Itoh, M.: Self-duality of Kähler surfaces. Compos. Math. 51, 265-273 (1984).

15. Kodaira, K.: On the structure of compact complex analytic surfaces. I. Am. J. Math. 86, 751-798 (1964).

16. Lafontaine, J.: Remarques sur les varietes conformement plates. Math. Ann. 259, 313-319 (1982).

17. LeBrun, C.: On the topology of self-dual 4 manifolds. Stony Brook, preprint.

18. Penrose, R.: Nonlinear gravitons and curved twistor theory, Gen. Rel. Grav. 1, 31-52 (1976).

19. Penrose, R., Rindler, W.: Spinors and Space-time, vol. 2 (Cambridge Univ. Press, Cambridge, 1986).

20. Plebanski, J. F.: Spinors, tetrads, and forms, Cinvestav del I.P.N. monograph, Mexico City.

21. Poon, Y. S.: Compact self-dual manifolds with positive scalar curvature, Oxford thesis (1985).

22. Strominger, A., Horowitz, G. T., Perry, M. J.: Instantons in conformal gravity. Nucl. Phys. B238, 635-664 (1984).

23. Vaisman, I.: Generalized Hopf manifolds. Geom. Dedicata 13, 231-255 (1982).

24. Vitter, A.: Self-dual Einstein metrics. in Nonlinear Problems in Geometry Contemp. Math. 51, 113-120 (1986).

25. Yau, S.-T.: Calabi's conjecture and some new results in algebraic geometry. Proc. Natl. Acad. Sci. U.S.A., 74, 1798-1799 (1977).

26. Yau. S.-T.: On the Ricci curvature of a compact Kähler manifold and the complex Monge-Ampere Equation, I. Commun. Pure. Appl. Math. 31, 339-411 (1978).

DEPARTMENT OF MATHEMATICS AND COMPUTER SCIENCE
CLARKSON UNIVERSITY
POTSDAM, NY 13676

Professor Charles P. Boyer
Department of Mathematics and Computer Science
Clarkson University
Potsdam, NY 13676

Contemporary Mathematics
Volume **71**, 1988

INSTANTONS ON THE QUATERNIONIC SIEGEL SPACE

F. J. Flaherty

ABSTRACT. In this short note I want to sketch a general method of constructing exact solutions to the Yang-Mills equations over a special class of complex surfaces (real four-dimensional manifolds), derived from the quaternionic Siegel-space in the same way that the Riemann surfaces of genus larger than one are gotten from the Poincaré upper-half plane. In addition, the gauge group used will be, for simplicity, Sp(1) = SU(2). The methods used will entail quaternionic analysis, meromorphic complex functions, and a quaternionic version of 't Hooft's ansatz.

INTRODUCTION. Over the last several years there has been increasing interest in the relationship between the Yang-Mills equations and the Einstein equations and as Abhay Ashtekar indicated in his talk, there may be a way of producing solutions to the Einstein equations from solutions to the Yang-Mills equations. This is not as far-fetched as it might appear. Maxwell's equations are very close to the Yang-Mills equations while Kobayashi's notion of Einstein-Hermitian bundles is not too far from the Einstein-Kaehler condition.

1. QUATERNIONS. The quaternions H will be viewed as R = sp(1), in which sp(1) is the algebra of Sp(1) (isomorphic to SU(2)). The product of two quaternions $A_{\bullet}B$ is then defined by using the Killing form, C, and the bracket product [,] on sp(1):

$$A.B = (a,v)(b,w) := (ab + \tfrac{1}{2}C(v,w),\quad aw + bv + [v,w]).$$

Moreover, the inner product on H is defined by $\overline{A}.B + \overline{B}.A = 2\langle A,B\rangle$, in which the conjugate $\overline{A} = (a,-v)$.

The interesting point about the quaternions is that although they form a complex vector space, there are many ways in which they do so. Any pure quaternion (zero real part) has a non-positive square, since

---

1980 Mathematics Subject Classification (1985 Revision). 53C, 81E

$v.v = \frac{1}{2}C(v,v) = -\langle v,v\rangle$. Hence right multiplication by any pure quaternion (real part zero) of unit length yields a complex structure.

The Siegel space, S, in H is then defined as $H-R$, in which R is the axis of reals (pure part zero). Thus an element A of S, which always has its pure part v different from zero, can always be represented in the form:

$$A = (a,v) = U(a + |v|e_3)U^{-1}$$

in which (e_α) form an ortho basis of $sp(1)$ and U is a unitary transformation depending on A. As a result, A has the form: $A = U\,\mathrm{diag}(z,\bar{z})U^{-1}$ in which $z = a + i|v|$. The mappings in $SL(2,R)$ act on the complex z via fractional linear transformation, with real coefficients, and so can be lifted to S. Restricting to the deck transformations of a Riemann surface of genus $g > 1$, the resulting quotient space will be denoted by $S(g)$. In terms of the metric ds^2 on H:

$$ds^2 = (\mathrm{im}\ z)^2((|dz|^2/(\mathrm{im}\ z)^2) + dS^2)$$

in which dS^2 is the standard round metric on the Riemann sphere. It is clear that $S(g)$ is conformally related to the product of a Riemann surface of genus g, $\Sigma(g)$, and the complex projective line $P_2(C) = S^2$, whose metric is then given by:

$$ds^2 = ((|dz|^2/(\mathrm{im}\ z)^2) + dS^2).$$

For complements and details to these ideas the reader is referred to [1] and [2].

2. MEROMORPHIC FUNCTIONS. To construct the proper gauge fields and bundles on $S(g)$ a class of quaternionic meromorphic functions must be found. With the construction from the last paragraph, it is clear that complex meromorphic functions can be lifted to $S(g)$ to yield functions with meromorphic Laplacian in a similar way as for the multiplicative torus, cf. [3]. This entails the construction of complex meromorphic functions on $\Sigma(g)$. Briefly, take the Jacobian variety, J, of $\Sigma(g)$ and construct theta functions on J. From the theta function, prime functions are constructed and thence quotients of products of prime functions yield the desired meromorphic function. Explicit knowledge of the construction is necessary since bundles with prescribed Chern class are ultimately needed. This whole topic is classical but the recent exposition by Mumford [4] is very readable.

3. BUNDLES. The vector bundles on $S(g)$ are of course induced from its universal cover $H-R$. The situation requires some care as there are non-trivial smooth as well as holomorphic vector bundles on $H-R$. Since $S(g)$

has a natural quaternionic structure, being conformally flat, quaternionic line bundles make sense over $S(g)$. In addition, the covering projection π is locally biholomorphic, in the quaternionic sense; so all local questions, for example those involving curvature, can be resolved on the universal cover H-R.

4. SELF-DUAL SOLUTIONS. To construct self-dual solutions for an SU(2)-bundle consider the quantity $e_{\alpha\beta} := \mathrm{vec}(\bar{e}_\alpha . e_\beta)$. It is easy to verify that $(e_{\alpha\beta})$ is self-dual. The connection form or gauge potential from the 't Hooft ansatz is given by $A_\mu := \frac{1}{2}\mathrm{vec}(e_\mu \bar{a})$, in which $a := \partial_{\bar{q}} \ell n\ \sigma$. If in addition σ is given by $\partial_q f$, then the gauge field or curvature form F is self-dual if and only if the function f has holomorphic Laplacian. Returning to the bundle picture, consider a quaternionic line bundle Q over $S(g)$ which when pulled back to H-R has local sections that play the same role as the (e_α) do in H. Using the ansatz, a self-dual connection is constructed using the lift of a meromorphic function on $\Sigma(g)$. There is still freedom in the choice of the meromorphic function f so it can be chosen in such a way that the integral of the second Chern class $c_2(Q)$ over $\Sigma(g)$ yields the desired topological charge.

5. NON-LINEAR SIGMA MODELS. Given the construction of the self-dual solutions to the Yang-Mills equations it is now easy to construct harmonic maps of the Riemann surfaces $\Sigma(g)$ into the complex projective spaces as in [5]. Details will be supplied in the full write-up of this note.

6. ADENDA. There are several unanswered questions arising from the construction. Are the instanton bundles stable? The Mumford-Takemoto definition of stability applies in this case because $S(g)$ is algebraic. It appears that the answer is yes using the work of Donaldson [6]. The whole question of reducibility has been avoided and needs to be carefully examined. In this regard, every holomorphic bundle over $S(g)$ can be pulled back to a holomorphic bundle over H-R, which in turn splits over the $P_1(C)$ in H-R as a direct sum of complex line bundles.

REFERENCES

1. Gursey, F., Tze, H.-C., Complex and quaternionic analyticity in chiral and gauge theories, I, Ann. of Phys., 128(1980), 29-130.

2. ______, Quaternion analyticity and conformally Kaehlerian structure in Euclidean gravity, Lett. in Math. Phys., 8(1984), 387-395.

3. Flaherty, F., Instantons on the multiplicative 4-torus, preprint.

4. Mumford, D., Tata lectures on theta I, Birkhauser Verlag, 1983.

5. Flaherty, F., Doubly periodic non-linear σ models, preprint.

6. Donaldson, S., Anti self-dual Yang-Mills connections over complex algebraic surfaces and stable vector bundles, Proc. London Math. Soc., 50(1985), 1-26.

DEPARTMENT OF MATHEMATICS
OREGON STATE UNIVERSITY
CORVALLIS, OR 97331

Contemporary Mathematics
Volume **71**, 1988

GAUGE THEORIES FOR FIELDS OF SPIN-ONE AND SPIN-TWO

Robert M. Wald[1]

ABSTRACT. We describe an approach which recently has been employed to systematically obtain the types of gauge groups that can occur in theories involving spin-one and spin-two fields.

In this note I shall describe some recent work I have done (partly in collaboration with Curt Cutler) on the types of gauge theories that can exist for fields of spin-one and spin-two. The details of this work have been reported elsewhere [1]-[3], so I will confine myself here to sketching some of the main ideas.

I begin by explaining what I mean by the terms "spin-one and spin-two fields" and "gauge theories." Consider a <u>linear</u> equation for a tensor field on Minkowski spacetime $(\mathbb{R}^4, \eta_{ab})$ which is relativistically invariant, i.e. such that the natural action of the Poincaré group on the tensor field takes solutions into solutions. The vector space of solutions with Poincaré group action thereby yields a representation of the Poincaré group. Of particular interest are equations for which this representation is irreducible. These irreducible representations can be labeled by the squared mass and spin parameters corresponding to the Casimir operators for the Poincaré group. The terms "massless," "spin-one" and "spin-two" refer to this classification. The theory of a massless, spin-one field is given by Maxwell's equations for a dual vector field A_a,

$$(1) \qquad \partial^a \partial_{[a}A_{b]} = 0$$

where square brackets denote antisymmetrization. (We follow the notational conventions of [4].) Here, it is understood that A_a and $A_a' = A_a + \partial_a \chi$ (where χ is an arbitrary function) represent the same physical solution, so that the solution space is actually the equivalence class of solutions of (1) under

$$(2) \qquad A_a \to A_a' + \partial_a \chi$$

Similarly, the theory of a massless spin-two field is given by the linearized Einstein equation, on a symmetric tensor field γ_{ab},

[1]Supported by NSF grant PHY 84-16691 to the University of Chicago.

(3) $$\partial^a \partial_{[a}\gamma_{b]c} - \partial_c \partial_{[a}\gamma_{b]}{}^a = 0$$

where one takes equivalence classes via,

(4) $$\gamma_{ab} \to \gamma_{ab} + \partial_{(a}\chi_{b)}$$

where χ_b is an arbitrary dual vector field and round brackets denote symmetrization.

Although the above theories are, strictly, what one normally means by "massless spin-one and spin-two fields," my use of this terminology shall be somewhat more general. By "massless spin-one field," I shall mean, in addition to Maxwell theory, any nonlinear, relativistically invariant theory of a dual vector field A_a on Minkowski spacetime which reduces in linear order to Maxwell theory. Similarly, by a "massless, spin-two field," I shall mean any relativistically invariant theory of a symmetric tensor field γ_{ab} which reduces to the linearized Einstein theory in linear order.

The transformations (2) and (4) relating physically equivalent solutions are referred to as gauge transformations. I will call a theory a gauge theory if there is a group of transformations of a form similar to (2) or (4) -- in particular, involving an arbitrary function or tensor field on spacetime in a local manner -- taking solutions into physically equivalent solutions. (Below, I will specify precisely the exact form of the transformations to be considered here.) An important example of a nonlinear gauge theory involving spin-one fields is Yang-Mills theory, where the field variable can be interpreted as a connection on a principal fiber bundle over spacetime. (Indeed, in most references, the terminology "gauge theory" is reserved only for theories of this type.) By choosing a cross-section, we can represent the connection as a Lie-algebra valued one-form field $A_a{}^\mu$ on spacetime; the gauge transformations then correspond to changes in the choice of cross-section. An important example of a nonlinear gauge theory (in my sense of the term) for a spin-two field is general relativity, with the field variable γ_{ab} taken as the deviation of the spacetime metric g_{ab} from a flat metric η_{ab}, $\gamma_{ab} = g_{ab} - \eta_{ab}$.

As already indicated by the examples of Yang-Mills theory (which includes Maxwell theory as a special case) and general relativity, gauge theories of spin-one and spin-two fields play a very important role in physics. It is of interest to ask if these known types of gauge theories are the only possible ones. Here, by "types of gauge theories" I do not refer to the specific form of the field equations of the theories. Indeed it is easy to write down numerous mathematically consistent (though not necessarily physically viable) modifications of the standard Maxwell, Yang-Mills, or Einstein field equations. Rather, I am concerned with the types of gauge transformations which can occur. Can one invent a gauge theory of a spin-one field (or a collection of spin-one fields) where the gauge group is not the Yang-Mills gauge group? Can one invent a gauge theory of a spin-two field (or a collection of spin-two fields) where the gauge group is not the diffeomorphism group?

In order to make any progress toward answering this question, it is necessary to specify more precisely the general class of gauge transformations to be considered. Consider, first, the case of a single spin-one field A_a. I shall seek a gauge group such that the infinitesimal action of the group on A_a is of the form

(5) $$\delta A_a = \beta_a{}^b\,(\partial_b\chi + \alpha_b\chi)$$

Here, as in eq. (2), χ is an arbitrary function on spacetime; $\beta_a{}^b$ is required to be locally constructed out of A_a and the flat background metric η_{ab}, with no derivatives of A_a appearing; α_b also is locally constructed out of A_a and η_{ab} but it is also allowed to depend linearly on the first derivative, $\partial_b A_a$, of A_a. Furthermore, in order that eq. (5) agree with what one obtains in Maxwell theory, eq. (2), in the linearized limit, we require $\beta_a{}^b = \delta_a{}^b$ and $\alpha_b = 0$ when $A_a = 0$. I have given plausibility arguments elsewhere [1] that any consistent, nonlinear generalization of Maxwell theory whose field equations are derivable from an action must be a gauge theory with infinitesimal transformations of the form (5) (with the above restrictions on $\beta_a{}^b$ and α_b), so there is some good motivation for restricting consideration to such types of gauge theories. In any case, I have not analyzed more general types of gauge transformations (e.g., ones which involve higher derivatives of χ or allow more derivatives of A_a in $\beta_a{}^b$ and α_b) and do not believe that this would be an easy task.

The key point is that not all transformations of the form (5) can arise as infinitesimal generators of a gauge group. The reason is simply that the commutator of two infinitesimal gauge transformations must itself be an infinitesimal gauge transformation. Thus, the transformation (5) can arise as the infinitesimal generator of a gauge transformation only if $\beta_a{}^b$ and α_b are such that for arbitrary functions ϕ, ψ there exists a function χ such that

(6) $$\delta_\phi(\delta_\psi A_a) - \delta_\psi(\delta_\phi A_a) = \delta_\chi A_a$$

where δ_χ denotes the variation of the quantity under the transformation (5). Equation (6) can be re-written more explicitly as,

(7) $$(\delta_\phi\,\beta_a{}^b)\,(\partial_b\psi + \alpha_b\psi) - (\delta_\psi\beta_a{}^b)\,(\partial_b\phi + \alpha_b\phi) + \beta_a{}^b\,\{(\delta_\phi\,\alpha_b)\psi - (\delta_\psi\alpha_b)\phi\} = \beta_a{}^b(\partial_b\chi + \alpha_b\chi)$$

Remarkably, eq. (7) can be solved if one assumes that $\beta_a{}^b$, α_b, and χ can be expanded as a power series in A_a. The "zeroth order in A_a" part of eq. (7) can easily be solved for the linear order parts of $\beta_a{}^b$, α_b, and χ. Then, by an inductive argument, it can be shown that each linear order solution gives rise to at most one exact solution, i.e., solutions of the higher order equations (if they exist) are uniquely determined by the linear order solution. (Here we eliminate certain trivial solutions arising from freedom to re-write eq. (5) in equivalent forms as explained in [1].) The details of this argument are given in [1], so I will merely state the final result here: The only solution of eq. (7) is the trivial solution $\beta_a{}^b = \delta_a{}^b$ and $\alpha_b = 0$. Thus, the only type of gauge theory (within the class considered here) which can exist for a single spin-one field A_a is one which has the same gauge group (2) as the linear Maxwell theory.

The situation becomes more interesting if one considers a collection of k spin-one fields $A_a{}^\mu$, which is most conveniently viewed as a spin-one field valued in a k-dimensional vector space, $\mathcal{V}$. (Here and below, Latin indices are spacetime indices, while Greek indices refer to the internal vector space.) We now seek gauge theories with gauge group generated by infinitesimal transformations of the general form,

$$\delta A_a{}^\mu = \beta_a{}^b{}_\nu{}^\mu(\partial_b \chi^\nu + \alpha_{b\lambda}{}^\nu \chi^\lambda) \tag{8}$$

As shown in [1], there is now a nontrivial solution of the zeroth order part of the analogous integrability condition for the linear order part, $\alpha^{(1)}{}_{a\mu}{}^\nu$, of $\alpha_{a\mu}{}^\nu$; namely,

$$\alpha^{(1)}{}_{a\mu}{}^\nu = c^\nu{}_{\mu\lambda} A_a{}^\lambda \tag{9}$$

where $c^\nu{}_{\mu\lambda}$ is antisymmetric in its lower indices,

$$c^\nu{}_{\mu\lambda} = - c^\nu{}_{\lambda\mu} \tag{10}$$

The first order part of the integrability condition implies the following additional condition on $c^\nu{}_{\lambda\mu}$,

$$c^\lambda{}_{[\mu\nu} c^\sigma{}_{\rho]\lambda} = 0 \tag{11}$$

Thus, by eqs. (10) and (11), $c^\mu{}_{\nu\lambda}$ endows $\mathcal{V}$ with the structure of a Lie algebra! (Equation (11) is the Jacobi identity.) Again, the full solutions of the integrability conditions are determined by the linear order parts. Such solutions do exist: They are precisely the Yang-Mills gauge transformations for the given Lie algebra, $\mathcal{V}$. Thus, the only types of gauge theories (of the general class considered here) involving a collection of spin-one fields are those having the Yang-Mills gauge group.

Consider, now, the case of a spin-two field γ_{ab}. We consider infinitesimal transformations of the general form

$$\delta\gamma_{cd} = B^{ab}{}_{cd}(\partial_a \psi_b + C^e{}_{ab} \psi_e) \tag{12}$$

where $B^{ab}{}_{cd}$ and $C^e{}_{ab}$ satisfy restrictions analogous, respectively, to those for

$\beta_a{}^b$ and α_b in the spin-one case. In this case, two types of gauge theories are known: those which have the same gauge group (4) as the linear spin-two field (i.e., $B^{ab}{}_{cd} = \delta^{(a}{}_c \delta^{b)}{}_d$, $C^e{}_{ab} = 0$), and those which have the diffeomorphism group as its gauge group (i.e., $B^{ab}{}_{cd} = \delta^{(a}{}_c \delta^{b)}{}_d$, $C^e{}_{ab} = -\Gamma^e{}_{ab}$, where $\Gamma^e{}_{ab}$ is the Christoffel symbol associated with the metric $g_{ab} = \eta_{ab} + \gamma_{ab}$). Do any other solutions of the integrability conditions for $B^{ab}{}_{cd}$ and $C^e{}_{ab}$ exist, i.e., are there other kinds of gauge theories for a spin-two field? I have shown in detail elsewhere [1] that the answer is "no"; the above solutions are the only solutions to the integrability condition. Thus, if one accepts the plausibility arguments mentioned above that any consistent nonlinear theory of a spin-two field with field equations derivable from an action must be a gauge theory with infinitesimal transformations of the general form (12), this means that all such theories must either have the gauge group (4) or they must be "generally covariant" theories, involving a curved spacetime metric g_{ab}.

In the above cases, the approach of solving the integrability condition for infinitesimal gauge transformations produced only previously known types of theories. The only new result in those cases was a proof that no other types of gauge groups (within the class considered) can occur. However, a new type of gauge theory was discovered [2] when this approach was applied to a collection of spin-two fields, which, again, we shall view as a spin-two field $\gamma_{ab}{}^{\mu}$ valued in a vector space $\mathcal{V}$. The candidate infinitesimal gauge transformations now take the form,

$$\text{(13)} \qquad \delta\gamma_{cd} = B^{ab}{}_{cd\nu}{}^{\mu}(\partial_a \chi_b{}^{\nu} + C^e{}_{ab\lambda}{}^{\nu} \chi_e{}^{\lambda}).$$

In close analogy with the spin-one case, the linear order solution, $C^{(1)e}{}_{ab\mu}{}^{\nu}$, for $C^e{}_{ab\mu}{}^{\nu}$ now takes the form [2],

$$\text{(14)} \qquad C^{(1)e}{}_{(ab)\mu}{}^{\nu} = -\frac{1}{2} a^{\nu}{}_{\mu\lambda}(2\partial_{(a}\gamma_{b)}{}^{e\lambda} - \partial^e \gamma_{ab}{}^{\lambda})$$

where $a^{\nu}{}_{\mu\lambda}$ is symmetric in its lower indices

$$\text{(15)} \qquad a^{\nu}{}_{\mu\lambda} = a^{\nu}{}_{\lambda\mu}$$

The first order part of the integrability condition implies the following additional relation on $a^{\nu}{}_{\mu\lambda}$,

$$\text{(16)} \qquad a^{\nu}{}_{\lambda[\tau} a^{\lambda}{}_{\mu]\kappa} = 0$$

Thus, by eqs. (15) and (16), $a^{\nu}{}_{\mu\lambda}$ endows $\mathcal{V}$ with the structure of an associative, commutative algebra! (Equation (15) expresses the commutativity of the multiplication of vectors defined by $a^{\nu}{}_{\mu\lambda}$ and eq. (16) is equivalent [given eq. (15)] to the associative law.) Existence and uniqueness of an exact solution to the integrability condition corresponding to the linear order solution

(14) has been shown [2]. Thus, a new type of gauge theory for a collection of spin-two fields has been obtained. It is similar in many respects to Yang-Mills gauge theory, but it is based upon an associative, commutative algebra rather than a Lie algebra.

I have given elsewhere [3] a detailed discussion of the geometrical interpretation of this new class of gauge theories for a collection of spin-two fields. It turns out that the gauge group can be interpreted as a diffeomorphism group, but for a differential geometry obtained by replacing $\mathbb{R}$ by an arbitrary, irreducible, associative, commutative algebra, $\mathcal{A}$, with identity element. Two examples of such a differential geometry have previously been considered: the well known case $\mathcal{A} = \mathbb{C}$, and the case where $\mathcal{A}$ is the even part of a Grassmann algebra ("bosonic supermanifold theory"). Thus, the new class of gauge theories gives rise to a generalization of these constructions.

I do not know of any reason to believe that this new class of gauge theories for a collection of spin-two fields will play any role in physics. However, the general approach to obtaining new gauge theories (and establishing uniqueness of such theories) described above can be applied to many other types of fields. In particular, work is currently in progress to examine supergravity theories from this point of view. Given the important role played by gauge theories in physics, it is not implausible that any new class of gauge theories that might be discovered in this manner could be of physical relevance.

BIBLIOGRAPHY

1. R. M. Wald, "Spin-two fields and general covariance," Phys. Rev. D 33 (1986) 3613-3625.
2. C. Cutler and R. M. Wald, "A new type of gauge invariance for a collection of massless, spin-two fields. I. Existence and uniqueness," to be published.
3. R. M. Wald, "A new type of gauge invariance for a collection of massless, spin-two fields. II. Geometrical interpretation," to be published.
4. R. M. Wald, General Relativity, University of Chicago Press, Chicago, 1984.

ENRICO FERMI INSTITUTE AND DEPARTMENT OF PHYSICS
UNIVERSITY OF CHICAGO
CHICAGO, IL 60637

Contemporary Mathematics
Volume 71, 1988

EINSTEIN GEOMETRY AND HYPERBOLIC EQUATIONS

S. Klainerman

One of the main goals in the theory of differential equations of mathematical physics is to describe the behavior of general solutions to the Cauchy problem. In a general set-up (in the absence of boundary conditions) this amounts to finding solutions $u = (u^1, \ldots, u^N)$ depending on $x = (x^0, x^1, \ldots, x^n) \in R^{n+1}$ which satisfy a system of N independent equations

$$G(x, J^m u(x)) = 0 \tag{G}$$

and whose partial derivatives of order $\leq m - 1$ are prescribed on a hypersurface $\underline{H}^n \subset \mathbb{R}^{n+1}$. Here $J^m u = (u, \nabla u, \ldots, \nabla^m u)$ denotes all partial derivatives of u of order less than or equal to m and $G = (G^1, \ldots, G^N)$ are smooth, given functions of x, $J^m u$. The basic physical systems are invariant under translations, i.e. G does not depend explicitly on x, and linear in the highest derivatives $\nabla^m u$. Moreover $m = 1$ or 2.

The first $m - 1$ derivatives of u on $\underline{H}^n$ are uniquely determined by the first $m - 1$ normal derivatives i.e.

$$T^j u \big|_{\underline{H}^n} = \psi^j, \quad j = 0,1, \ldots, m-1, \tag{C.D.}$$

0271-4132/88 $1.00 + $.25 per page

with T the derivative in the direction of the unit normal to $\underline{H}^n$. They are called the Cauchy data for u, i.e. $\mathrm{Cauchy}_{\underline{H}^n}(u) = (\psi^0, \psi^1, \ldots, \psi^{m-1})$. The derivatives of order $\geqslant m$ can be formally determined, in a unique fashion, from the Cauchy data provided that $\underline{H}^n$ is *noncharacteristic*. This amounts to a simple algebraic condition on the derivatives of F relative to $\nabla^m u$ on the given Cauchy data. Moreover this formal procedure can be made rigorous provided that we restrict ourselves to real analytic solutions and real analytic F, $\underline{H}^n$ and Cauchy data. The result is the famous theorem of Cauchy-Kowalewski which allows one to find unique real analytic solutions in the neighborhood of a point in $\underline{H}^n$, or a whole strip around it (Fig. 1).

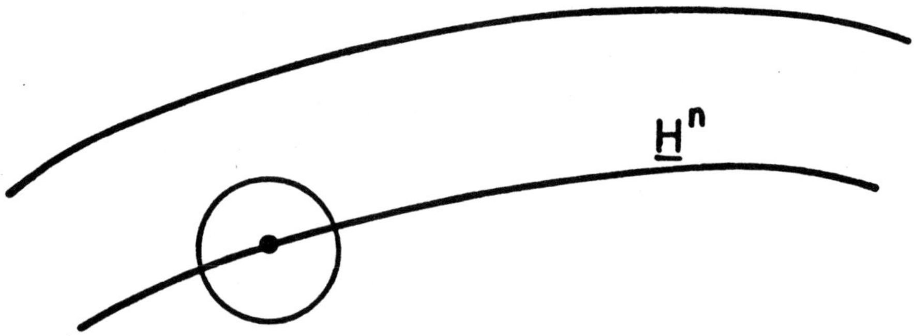

Figure 1

Now the class of real analytic solutions is too limited to describe the evolution of physical systems. Indeed a real analytic function is uniquely determined by the values in any small neighborhood of a point. It is therefore impossible to describe the propagation of local perturbations of the Cauchy data by using real analytic solutions. Moreover Einstein's Principle of Relativity postulates that physical signals cannot propagate with velocity larger than the speed of light. Consequently, acceptable equations, modeling physical systems, have to satisfy the property of *finite speed of propagation.* Roughly, we say that this property is satisfied if, given a solution u and a relatively compact set D in $\underline{H}^n$

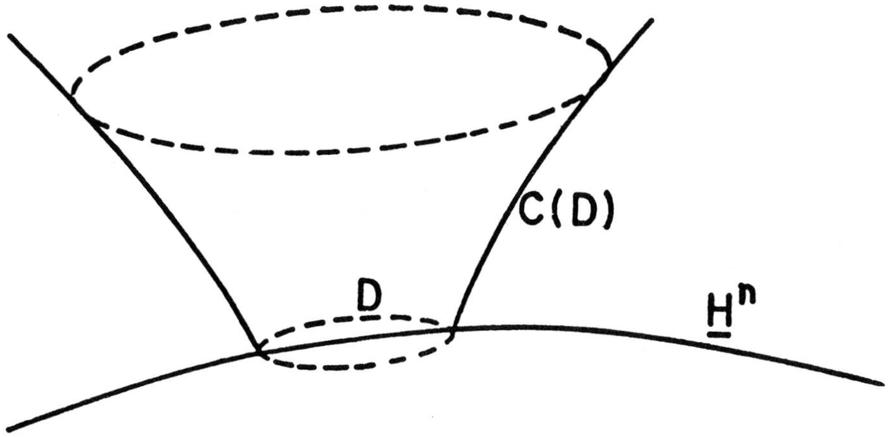

Figure 2

one can find a cone $C(D)$ (as in Fig. 2) such that if v is any other solution with Cauchy(v) = Cauchy(u) on $\underline{H}^n \backslash D$ then $u = v$ in $\mathbb{R}^{n+1} \backslash C(D)$. $\underline{H}^n$ is said to be spacelike relative to u. In particular, the property of finite "speed of propagation" requires that the Cauchy Problem be solved in function spaces with good local properties like C^∞, C^k and so on. If this can be achieved we say that the Cauchy Problem is "well posed". This calls for much more stringent algebraic assumptions on the system. In different formulations they are called hyperbolicity assumptions. For nonlinear equations a good notion of hyperbolicity leads to a local existence and uniqueness theorem for C^∞ or C^k spaces of Cauchy data. One accomplishes this in a neighborhood of a given solution $\bar{u}$. The main step in the proof of such a theorem is

STEP 1. "A-priori L^2 (energy) estimates".

These estimates depend on a splitting of $\mathbb{R}^{n+1}$ (in a neighborhood of $\underline{H}^n$) between a time direction t transversal to $\underline{H}^n$ and spatial directions "parallel" to $\underline{H}^n$. More precisely given $\bar{u}$ and $\underline{H}^n$ noncharacteristic and space-like relative to $\bar{u}$ we can introduce a locally defined time function $t = t(x^0, x^1, \ldots, x^n)$ such that $\underline{H}^n$ is given by $t = 0$ and each level surface $\underline{H}^n_c = \{x \mid t(x) = c\}$ is

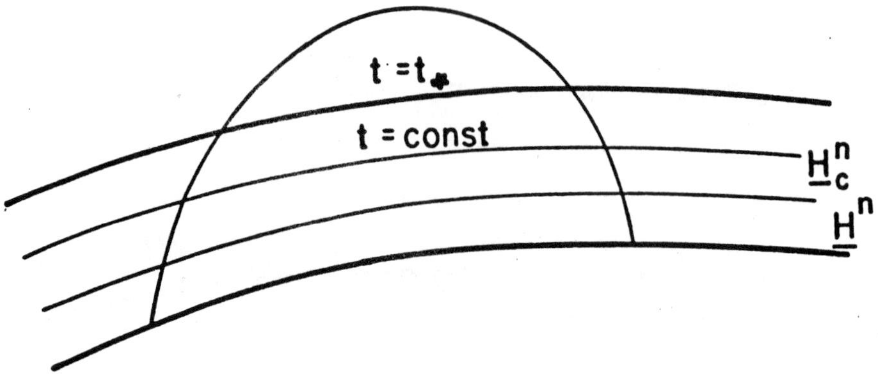

Figure 3

noncharacteristic and space-like relative to all solutions u close to $\overline{u}$,

$$|J^{m-1}u - J^{m-1}\overline{u}| < \delta$$

with $\delta > 0$ sufficiently small, and for $0 \leqslant t \leqslant t_*$ (see Fig. 3).

Given v, a function on $\mathbb{R}^{n+1}$, we define the L^2 norms

$$\|v(t)\|_s = \sum_{0 \leqslant k \leqslant s} \int_{\underline{H}^n_t} |\nabla^k v(x)|^2 d\sigma_t \tag{1}$$

with $d\sigma_t$ the volume element of $\underline{H}^n_t$. Also we define the L^∞ norms,

$$|v(t)|_s = \sum_{0 \leqslant k \leqslant s} \sup_{\underline{H}^n_t} |\nabla^k v(x)|. \tag{2}$$

By so-called energy estimates one establishes an a-priori inequality which has, essentially, the following form

$$\|v(t)\|_s \leqslant \|v(0)\|_s \exp\left[C_s \int_0^t |v(\tau)|_m\right] \tag{3}$$

for all $0 \leqslant t \leqslant t_*$, with $v = u - \overline{u}$ and the constant C depending on

$\sup_{0 \leqslant t \leqslant t_*} |v(t)|_{m-1}$ and certain bounds on $\bar{u}$.*

STEP 2. "Sobolev Inequalities".

The main idea in this step is to try to control the uniform norm $|v(t)|_m$ in terms of the L^2-norms $\|v(t)\|_s$. One has, by the classical Local Sobolev Inequalities

$$|v(t)|_m \leqslant C_m \|v(t)\|_{m+n/2}$$

with $n/2$ the smallest integer > $n/2$. Therefore, inserting this in (3),

$$\|v(t)\|_s \leqslant \|v(0)\|_s \exp\left[c_s \int_0^t \|v(\tau)\|_{m+n/2} d\tau\right].$$

Finally, set $s_0 = m + n/2$ and use a contraction argument relative to the norm $\sup_{0 \leqslant t \leqslant t_*} \|v(t)\|_{s_0}$. This will prove the existence of a local solution u, for any Cauchy data sufficiently close to $\mathrm{Cauchy}_{\underline{H}^n}(\bar{u})$ relative to the Sobolev norm $\| \quad \|_{s_0}$, provided that $t_* \|v(0)\|_{s_0}$ is sufficiently small (see [2], [3], [4], [5]). The critical number s_0 can be made smaller than $m + n/2$ provided that the system is nonlinear in a milder sense.

The concepts of hyperbolicity, finite propagation speed, space-like hypersurfaces and the local existence proof sketched above took many years and many mathematicians to complete. To go beyond them one needs to look much closer to the specific properties of physical systems. In particular most such systems have certain conserved quantities. The one which is very often used is a positive integral on space-like hypersurfaces, depending on (some of) the first $m - 1$ derivatives of u, and which is called

*Estimates of this type were discussed in [1] (for second order equations, or symmetric hyperbolic systems) and are based on the sharp product and chain rule inequalities in Sobolev norms (see [2], [3], [4], [5]).

the (total) energy of the system. The positivity of the energy is usually obvious; however in the particular case of General Relativity it required a nontrivial beautiful theorem proved by Schoen-Yau [1] and Witten [2]. (The concept of energy itself is somewhat more elusive in General Relativity. To make sense of it one needs to introduce the equally elusive notion of an isolated system which will be discussed below.) In any case, the energy allows one to derive a global bound on L^2-norms of some of the first $m - 1$ derivatives of u. When the system is only "mildly nonlinear" this can be used to prove a global existence theorem, in the whole space-time $\mathbb{R}^{n+1}$ and for any smooth Cauchy data. Such results were recently proved for the Yang-Mills equation by Eardley-Moncrief [8], or earlier for semilinear Klein-Gordon equations by Jörgens [9]. Yet for systems with strong nonlinearities the bound provided by the energy is not sufficient to draw any conclusions on the global behavior of solutions. This is the case of the systems describing such fundamental physical theories as Elasticity, Fluid Dynamics or General Relativity. Moreover these systems do not have global smooth solutions for arbitrary data. In fact most solutions, which start by being perfectly smooth on the initial hypersurface, will become singular after only a short lapse of time. The singularities can be so strong as to prohibit their continuation, as classical solutions, as is the case of *shock waves* in Fluid Dynamics, or any continuation at all, as for curvatuve singularities inside *black holes* in General Relativity. So the next fundamental question in the study of the Cauchy Problem is to understand when and how these singularities form, to describe the behavior of solutions near them and find a way, if possible, to continue the solutions beyond the singularities in a physically meaningful fashion. A simpler, related, question is to find under what circumstances singularities may be avoided altogether. Indeed all physical systems have trivial global solutions, corresponding to no action at all like the zero solution $\bar{u} = 0$. If the system G is autonomous and has $\bar{u} = 0$ as a solution we can rewrite it in the form

(F) $$Lu = F(J^m u)$$

with L a linear mth order operator with constant coefficients and F a smooth function of $J^m u$, linear in $\nabla^m u$ and vanishing together with the first derivatives for $J^m u = 0$. For small Cauchy data,

(C.D.) $$\partial_t^j u = \epsilon \psi^j, \qquad j = 0,1, \ldots, m-1$$

the local existence theorem assures the existence of a smooth solution in a time interval $[0,t_*]$ of size $0(1/\epsilon)$. To go beyond this one may want to compare the solutions to (F) with those of the linear problem

(L) $$Lu = 0.$$

Now this approach has been quite successful in the study of boundary value problems for nonlinear elliptic equations where the linear part is manifestly dominant. However, for hyperbolic equations the behavior of (L) and (F) differ even locally (as one can see by comparing their corresponding characteristics). Yet certain global properties of (L) can be used for nonlinear equations. In particular, looking at the energy estimate (3), with $\bar{u} = 0$, we remark that if one could establish that $|u(\tau)|_m$ decays sufficiently rapidly as $\tau \to \infty$ then $\|u(t)\|_s$ is bounded uniformly in $t \geqslant 0$. This would allow us to extend the solutions for all time. Now it is true that most linear equations (L) of importance in mathematical physics have some uniform decay properties. This is a consequence of the spreading of physical signals in the whole space and can be quite easily verified by Fourier methods or, directly, from the form of the explicit solutions. In particular, when L is the usual wave operator, i.e. $L = -\Box = \partial_t^2 - \partial_1^2 - \cdots - \partial_n^2$, in $\mathbb{R}^{n+1}$ and $\Box u = 0$ one has the estimate

(4) $$|u(t,x)| \leqslant C(1 + t)^{-(n+1)/2}$$

uniformly in $x \in \mathbb{R}^n$, $t \geqslant 0$. The constant C depends on the L^1 norm of the first $n/2$ derivatives of the initial data (see [2], [10]). Such estimates have been used to derive global results for scalar nonlinear wave equations. I would like to mention in particular the early work of I. Segal [11] and W. Strauss [12] for semilinear cubic scalar equations (i.e. $F(u) = 0(|u|^3)$ for small u) and by F. John ([13], [14]), J. Shatah [15] and S. Klainerman ([1], [2]) for general classes of scalar wave equations. However, the dependence of the constant C on the L^1-norm of the initial data has seriously restricted the use of (4) to treat quadratic nonlinearities. More recently, in [16], we have presented a method, based on the Lorentz and scale invariance of $\Box$, to derive the same rates of uniform decay in t directly from some generalized version of energy inequalities. In fact the idea was to modify the classical Sobolev inequality used in step 2 of the existence proof outlined below, by a global version of it with $\| \ \|_s$ replaced by a larger weighted Sobolev norm (see (15)) and multiplied by a decay factor of order $0((1+t)^{-(n-1)/2})$ as $t \to \infty$. These new Sobolev norms are bounded when applied to solutions of $\Box u = 0$, which allows us to derive the same rates of decay for u as in (4). Moreover they can be applied to nonlinear equations, $\Box u = F(u,\nabla u,\nabla^2 u)$, to derive estimates similar to those in (3). As a consequence one can follow the same steps as those of the local existence theorem to derive global existence, if $(n-1)/2 > 1$, or "almost global existence", if $(n-1)/2 = 1$, provided that the initial conditions are sufficiently small (see [16], [17], also [18] and the more recent results of [19], [20]).

The aim of this lecture is to discuss the relation between the geometry of the Minkowski space, generalized energy estimates and uniform rates of decay for two other examples of linear field equations in Minkowski space

(M) Maxwell equations

(Sp) Spin-2 field equations.

For completeness we will also consider

($\Box$) Scalar wave equation.

Our main interest is the study of (Sp) and its connection to an outstanding problem in general relativity which we state below. All the results presented in this lecture were derived in collaboration with D. Christodoulou [21].

We first recall a few basic concepts in Einstein Geometry.

A differentiable manifold M^{n+1} of dimension $n+1$ is called an Einstein manifold if it has a nondegenerate metric $\langle\ ,\ \rangle$ of signature $(-1,1,\ldots,1)$. In local coordinates $(x^0,x^1,\ldots,x^n)$,

$$ds^2 = g_{\mu\nu}dx^\mu dx^\nu, \text{ with } g_{\mu\nu} = \langle\frac{\partial}{\partial x^\mu}, \frac{\partial}{\partial x^\nu}\rangle.$$

Denote by $\mathfrak{X}(M^{n+1})$ the set of all vector fields X on M^{n+1},

$$X = X^\mu(x)\frac{\partial}{\partial x^\mu},$$

with

$$\langle X,Y\rangle = g_{\mu\nu}X^\mu Y^\nu = X^\mu Y_\mu.$$

We say that X is time-like, null or space-like depending on whether $\langle X,X\rangle$ is negative, zero or positive. A hypersurface $\underline{H}^n$ is called space-like if its normal is time-like at every point of $\underline{H}^n$. The metric induced by $\langle\ ,\ \rangle$ on a space-like hypersurface is Riemannian.

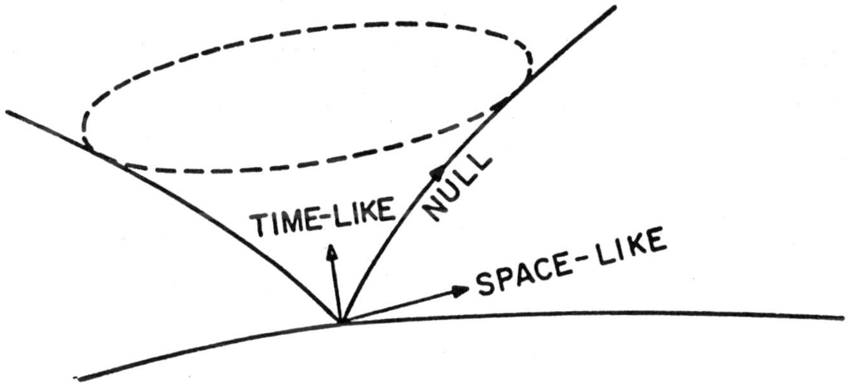

Figure 4

A frame $e_0, e_1, \ldots, e_n$ is called orthonormal if $\langle e_\mu, e_\nu \rangle = \eta_{\mu\nu}$ for $\mu,\nu = 0,1,\ldots,n$. We will also consider null frames i.e. frames $e_1, \ldots, e_{n-1}, e_n, e_{n+1}$ with e_n, e_{n+1} null, $\langle e_n, e_{n+1} \rangle = -1$ and $e_1, \ldots, e_{n-1}$ orthonormal vectors on the space complementary to e_n and e_{n+1}. Relative to an arbitrary frame (e_μ), $\mu = 0, \ldots, n$, the components of a tensor W are $W_{\mu 1 \ldots \mu p}$, and we raise and lower indices with the help of $g_{\mu\nu} = \langle e_\mu, e_\nu \rangle$ and $g^{\mu\nu}$, the matrix inverse of $g_{\mu\nu}$. The Lie derivative of a tensor W relative to a vector field $X \in \mathcal{H}(M)$ will be denoted by $\mathfrak{L}_X W$. As in Riemannian geometry, there exists a unique affine connection compatible with the metric i.e. for $X,Y,Z \in \mathcal{H}(M)$

$$\nabla_X Y - \nabla_Y X = [X,Y]$$

and

$$\nabla_X \langle Y,Z \rangle = \langle \nabla_X Y, Z \rangle + \langle Y, \nabla_X Z \rangle.$$

Relative to a frame $(e_\mu)_{\mu=0,\ldots,n}$ the components of the covariant derivative ∇W of a tensor W are $W_{\mu_1 \ldots \mu_p ; \mu_{p+1}}$. Again as in Riemannian Geometry, the Riemann tensor $R_{\alpha\beta\gamma\delta}$ is obtained by,

$$(\nabla_X \nabla_Y - \nabla_Y \nabla_Y) Z = R(X,Y)Z + \nabla_{[X,Y]} Z$$

for any $X,Y,Z \in \mathcal{H}(M)$.

The basic geometric operators for tensors, like the D'Alembertian $\Box_g$ and Divergence Div_g, are obtained from covariant differentiation and contractions e.g. for vectors $V \in \mathcal{H}(M)$, or symmetric and antisymmetric 2-tensors W,

$$\mathrm{Div}\, V = g^{\alpha\beta} V_{\alpha;\beta}$$

$$(\mathrm{Div}\, W)_\alpha = g^{\beta\gamma} W_{\alpha\beta;\gamma} \qquad *$$

For a scalar u we define the D'Alembertian

$$\Box_g u = g^{\alpha\beta} u_{;\alpha\beta}\,.$$

*We will use the notation div for the induced Riemannian metric on a space-like hypersurface.

The other basic operation is the generalized curl, or the exterior differentiation d for p-forms.

The field equations are systems of differential equations for tensors, obtained by using covariant differentiation, exterior differentiation and contractions. As far as they lead to determined systems of equations, the corresponding Cauchy Problems are, typically, well posed on any "space-like" hypersurface. So in Einstein spaces the differential equations which are tied to the geometry are automatically hyperbolic.

The simplest example is that of the scalar wave equation

$$(\square) \qquad \square_g\, u = 0.$$

The Cauchy problem,

$$u = \psi_{(0)}, \qquad Tu = \psi_{(1)} \quad \text{on} \quad \underline{H}^n$$

is well posed for any space-like hypersurface $\underline{H}^n$.

The Maxwell equations are expressed, in a four-dimensional space-time M^4, with the help of an antisymmetric tensor $F_{\mu\nu}$ called the electromagnetic tensor. Any such F can be decomposed into two vector fields orthogonal to a given direction $X \in \mathfrak{X}(M)$ by forming the contractions

$$(i_X F)_\mu = F_{\mu\nu} X^\nu, \qquad (i_X^* F)_\mu = {}^*F_{\mu\nu} X^\nu,$$

where ${}^*F$ is the Hodge dual to F, i.e. ${}^*F_{\alpha\beta} = \frac{1}{2}\epsilon_{\alpha\beta\mu\nu}F^{\mu\nu}$ and $\epsilon_{\alpha\beta\gamma\delta}$ the volume form on M^4. They determine the full tensor F whenever $\langle X,X\rangle \neq 0$, and are called the *electric* and *magnetic* components of F.

The four dimensional form of the Maxwell equations is

$$(\mathrm{M_i}) \qquad F_{\alpha\beta;\gamma} + F_{\beta\gamma;\alpha} + F_{\gamma\alpha;\beta} = 0$$

$$(\mathrm{M_{ii}}) \qquad F_{\alpha\beta}{}^{;\beta} = 0.$$

The Cauchy problem is given by

$$i_T F = E_{(0)}, \quad i^*_T F = H_{(0)}$$

on any space-like hypersurface $\underline{H}^3$ with unit normal T, and with $E_{(0)}$, $H_{(0)}$ any given vectors in $\underline{H}^3$ satisfying the constraints

$$\text{div}\, E_{(0)} = \text{div}\, H_{(0)} = 0 \quad \text{on} \quad \underline{H}^3.$$

Before defining the Spin-2 equations we will give below a short motivation of their interest for us. They are connected to the Bianchi identities satisfied by the Riemann curvature tensor of Einstein-Vacuum space-times which we describe below.

EINSTEIN EQUATIONS IN VACUUM

The Einstein field equations were proposed by A. Einstein in 1916 as a unified theory of space-time and gravitation. The four manifold M^4 is itself an unknown: one has to find an Einstein metric $g_{\mu\nu}$ such that

$$R_{\mu\nu} - \tfrac{1}{2}\, g_{\mu\nu} R = Q_{\mu\nu}$$

where $R_{\mu\nu}$ is the Ricci tensor of the metric, $R_{\mu\nu} = R^{\alpha}_{\mu\alpha\nu}$, $R = g^{\mu\nu} R_{\mu\lambda}$ the scalar curvature, and $Q_{\mu\nu}$ is the energy momentum tensor of a matter field (e.g. the Maxwell field). Contracting twice the Bianchi identities

$$\nabla_{[\epsilon} R_{\alpha\beta]\gamma\delta} := \nabla_{\epsilon} R_{\alpha\beta\gamma\delta} + \nabla_{\alpha} R_{\beta\epsilon\gamma\delta} + \nabla_{\beta} R_{\epsilon\alpha\gamma\delta} = 0$$

we derive $\nabla^{\nu}(R_{\mu\nu} - \frac{1}{2}\, g_{\mu\nu} R) = 0$ which are equivalent with the divergence equations of the matter field:

$$\text{Div}\, Q = 0.$$

In the simplest situation (i.e. $Q_{\mu\nu} = 0$), of physical vacuum, the Einstein equations take the form

(E.V.) $$R_{\mu\lambda} = 0$$

called the Einstein Vacuum Equations. Written explicitly, in an arbitrary system of coordinates, they lead to a degenerate system of equations. However, since the equations are invariant under any diffeomorphisms of M^4 it suffices to find coordinate conditions called also gauge conditions, which lead to a well posed Cauchy problem. This was achieved by Y. Choquet-Bruhat in the harmonic gauge [22], yet, as she pointed out later [23], the harmonic gauge is unstable in the large. This problem of finding a globally stable, "well-posed", gauge condition (whether it exists at all) is the first major difficulty one has to overcome when trying to construct global space-times.

One particular solution of (E.V.) is the Minkowski space $\mathbb{R}^{3+1}$ i.e. $\mathbb{R}^4$ with a given canonical coordinate system (x^0x^1,x^2,x^3) and metric $\langle \ , \ \rangle$ such that

$$\langle \partial_\mu, \partial_\nu \rangle = \eta_{\mu\nu} = \mathrm{Diag}(-1,1, \ldots, 1).$$

So we would like to ask whether Cauchy developments, (i.e. solutions to Cauchy data on a given 3-manifold) which are small in an appropriate sense, lead to global, complete, smooth, solutions to the Einstein vacuum equations. We expect these solutions to be close, in a certain sense, to the Minkowski space. More precisely they should be "asymptotically flat space-times" as they are known in the Physics literature. The study of such space-times, as stressed by Geroch [24], is the study of isolated systems. They are of great interest in General Relativity since "it is only through a suitable notion of an isolated system that one acquires an ability at all to deal individually with various subsystems in the universe; in particular to assign to subsystems such physical attributes as mass, angular momentum, character of emitted radiation, etc." Yet it is not at all obvious how to even formulate the notion that a space time is close to the Minkowski space. (This task is comparatively easier in Riemannian Geometry due to the directional uniformity of the Euclidean space.) However a global notion of an asymptotically flat manifold has been developed in the last 20 years (see [25] for a survey) beginning with the work of Bondi [26], [27] (see also Sachs [28]) who introduced the idea of analyzing solutions of the field equations along null, or characteristic, surfaces of a space-time. The present state of understanding was set by Penrose ([29], [30]) who formalized the idea of asymptotic flatness by adding a boundary, attached by the process of a conformal compactification. In fact, Penrose defined a space-time to be asymptotically flat if this boundary could be attached in a regular fashion.

With all the appealing features of the idea of regular, conformal compactification it is not at all clear that there are any, nontrivial solutions (involving gravitational radiation) satisfying the Penrose requirements. We believe that a more complete understanding of asymptotically flat space-times can only be accomplished by constructing them from given initial data. Our aim is to achieve that for sufficiently small data.

To make the problem more precise one has to introduce the notion of "space-like asymptotically flat" manifolds. We say that a Riemannian 3-manifold $\underline{H}^3$, diffeomorphic to $\mathbb{R}^3$, is asymptotically flat if there exists a coordinate system (x^1,x^2,x^3) globally defined, with the possible exception of a relatively compact set, such that the metric has the asymptotic properties

$$g_{ij} = \left(1 + \frac{M}{2r}\right)\delta_{ij} + o(r^{-\alpha}) \tag{5}$$

as $r = [\Sigma_{i=1}^3 (x^i)^2]^{1/2} \to \infty$, with M a constant, called the mass of $\underline{H}^3$, and with $\alpha > 1$.

So we will try to construct global (E-V) 4-manifolds which are asymptotically flat on any space-like hypersurface. We will call them space-like asymptotically flat or simply asymptotically flat, though one should be careful not to confuse the definition here with the much more delicate global one discussed above. (For a survey on this topic see [31].)

We are going to look at asymptotically flat 4-manifolds which can be foliated by maximal hypersurfaces (i.e. hypersurfaces whose

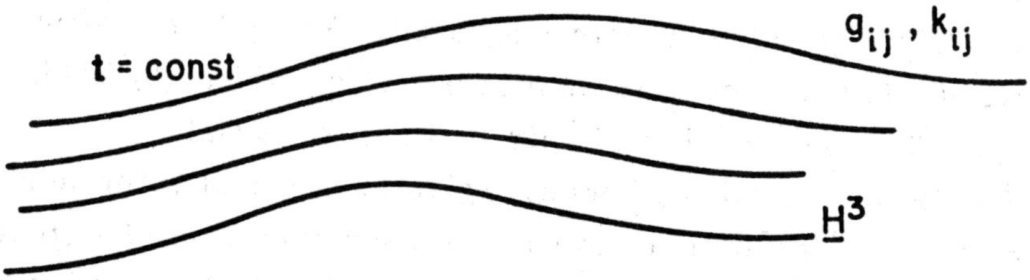

Figure 5

second fundamental form has zero trace). More precisely we want ${}^4M$ to be a product manifold $\mathbb{R} \times \underline{H}^3$, with $\underline{H}^3$ diffeomorphic to $\mathbb{R}^3$ and, with the metric given by

(6) $$ds^2 = -\alpha^2(t,x)dt^2 + g_{ij}(t,x)dx^i dx^j$$

where (x^1,x^2,x^3) are coordinates in $\underline{H}^3$. The hypersurfaces t = const. are space-like, asymptotically flat and maximal i.e.

Maximal Foliation: $tr_g k := g^{ij}k_{ij} = 0.$

The possibility that the gauge condition (6), with tr $k = 0$, is the right one to use in the study of global space-times was suggested to us by the work of Bartnik [32].

Applying the Gauss and Gauss-Codazzi equations on each slice, we derive from (E.V.) the following

Constraint Equations: tr $k = 0$, div $k = 0$, $R = |k|^2$

Lapse Equation: $\Delta\alpha - |k|^2\alpha = 0$

Evolution Equations: $\dfrac{\partial g_{ij}}{\partial t} = -2\alpha k_{ij}$

$$\frac{\partial}{\partial t}k_{ij} = -\nabla_i\nabla_j\alpha + \alpha(R_{ij} - 2k_i^{\ell}k_{\ell j})$$

where Δ, div, tr and the operation of raising and lowering indices are taken relative to the induced metric g_{ij}. R is the scalar curvature of g_{ij}.

In particular we see from the constraint equations that $R \geq 0$. Now the Positive Mass Theorem of Schoen-Yau [6] and Witten [7] assures us that, in any coordinate system for which (5) holds, we have $M \geq 0$, and $M = 0$ if and only if the space is flat.

The above discussion allows us to give a more precise formulation of the problem we want to solve. We say that a triplet $(\underline{H}^3, g, k)$ formed by a 3-manifold $\underline{H}^3$ diffeomorphic to $\mathbb{R}^3$, a Riemannian metric g and a symmetric 2-tensor k is *"asymptotically flat" admissible Cauchy data if*

$$\mathrm{tr}_g k = 0, \quad \mathrm{div}_g k = 0, \quad R = |k|_g^2$$

and there exists a coordinate system (x^1, x^2, x^3), in a neighborhood of infinity, such that

$$g_{ij} = \left(1 + \frac{M}{2r}\right)\delta_{ij} + O(r^{-3/2-\epsilon}), \quad M > 0 \tag{7}$$

$$k_{ij} = o(r^{-5/2-\epsilon}), \quad \text{as} \quad r := (\Sigma\,(x^i)^2)^{1/2} \to \infty$$

and

$$\nabla^\ell g_{ij} = o(r^{-1-\ell}) \tag{7'}$$

$$\nabla^\ell k_{ij} = O(r^{-5/2-\epsilon-\ell}) \quad \text{as} \quad r \to \infty$$

for any partial derivatives of order $0 \leq \ell \leq 4$, and some small $\epsilon > 0$.

CONJECTURE. Any system of asymptotically flat, small, admissible Cauchy data, leads to a globally smooth, complete solution (M^4, g) of the Einstein vacuum equations admitting a foliation (6) (with $\inf \alpha > \delta > 0$) and a smooth embedding of $\underline{H}^3$ in M^4 such that g_{ij}, k_{ij} are the first and second fundamental forms of $\underline{H}^3$ to M^4.

In collaboration with D. Christodoulou we have developed a

method which, if successful, will solve the conjecture by also deriving the correct asymptotic behavior of the Riemann curvatue tensor along null and time-like directions. One essential part of our strategy is to use the Bianchi identities, which for an (E-V) space M^4 read

$$\begin{aligned} &\nabla_{[\epsilon} R_{\alpha\beta]\gamma\delta} = 0 \\ &R_{\alpha\beta\gamma\delta}{}^{;\alpha} = 0 \, , \end{aligned} \tag{8}$$

as evolution equations and then derive g_{ij}, k_{ij}, α by solving purely elliptic problems. Indeed, let the dual of R be ${}^*R_{\alpha\beta\gamma\delta} = \frac{1}{2}\epsilon_{\alpha\beta}{}^{\mu\nu} R_{\mu\nu\gamma\delta}$. Given $X \in \mathfrak{X}(M)$, define the 2-tensors

$$(ii_X R)_{\alpha\gamma} = R_{\alpha\beta\gamma\delta} X^\beta X^\delta, \quad (ii^*_X R)_{\alpha\gamma} = {}^*R_{\alpha\beta\gamma\delta} X^\beta X^\delta.$$

Both $ii_X R$, and $ii^*_X R$ are symmetric and trace-less and together determine the tensor R, provided that $\langle X,X \rangle \neq 0$. In analogy to the electromagnetic field, they can be called the *electric* and *magnetic* decomposition of R. In particular let $E = ii_T R$, $H = ii^*_T R$. Then

$$\begin{aligned} &\mathrm{Ric} - k^2 = E \\ &\mathrm{curl}\ k = H \end{aligned} \tag{9}$$

where Ric denotes the Ricci curvature of the slice, and $(\mathrm{curl}\ k)_{ij} = \epsilon_i^{st} k_{js;t}$. Therefore our strategy is to use the Bianchi identifies (8) to estimate $R_{\alpha\beta\gamma\delta}$, with its electric and magnetic components E, H, and then use (9) together with the *constraint* and *lapse* equations to control k, g and α.

Notice that, on the initial slice,

$$E,H = 0(r^{-3}) \quad \text{as} \quad r \to \infty$$

However, note that the asymptotic behavior of the Lie derivatives $\mathfrak{L}_0 E$, $\mathfrak{L}_0 H$ relative to the vector fields $O_1 = x_2\partial_3 - x_3\partial_2$, $O_2 = x_3\partial_1 - x_1\partial_3$, $O_3 = x_1\partial_2 - x_2\partial_1$ are better by a factor of r^{-1}. Thus we can verify that

$$(10) \qquad \sum_{i=1}^{3} \sum_{\ell=0}^{s} \int (1 + |x|)^{4+2\ell} \left(|\nabla^\ell \mathcal{L}_{O_i} E(x)|^2 + |\nabla^\ell \mathcal{L}_{O_i} H(x)|^2 \right) dx < \infty$$

on the initial slice, where s is some positive integer and ∇^ℓ the covariant derivative of order ℓ relative to the Riemann metric on the slice.

Given an Einstein space M^4 and a tensor $W_{\alpha\beta\gamma\delta}$ which satisfy all the symmetries of the Riemann tensor, i.e.

$$W_{\alpha\beta\gamma\delta} = - W_{\beta\alpha\gamma\delta} = W_{\gamma\delta\alpha\beta}$$

$$W_{\alpha\beta\gamma\delta} + W_{\alpha\gamma\delta\beta} + W_{\alpha\delta\beta\gamma} = 0$$

and also

$$W^{\alpha}{}_{\beta\alpha\delta} = 0$$

we say that W satisfies the *spin*-2 *equations* if

$$(\mathrm{Sp}_i) \qquad W_{\alpha\beta\gamma\delta;\epsilon} + W_{\alpha\beta\delta\epsilon;\gamma} + W_{\alpha\beta\epsilon\gamma;\delta} = 0$$

$$(\mathrm{Sp}_{ii}) \qquad W_{\alpha\beta\gamma\delta}{}^{;\delta} = 0.$$

Given a space-like hypersurface $\underline{H}^3$ with unit normal T we prescribe the initial conditions on $\underline{H}^3$

$$ii_T W = E_{(0)}, \qquad ii^*_T W = H_{(0)}$$

with $ii_T W$, $ii^*_T W$ defined exactly as above for $R_{\alpha\beta\gamma\delta}$. The tensors $E_{(0)}$, $H_{(0)}$ are symmetric and traceless in $\underline{H}^3$, satisfying the constraint equations

$$\mathrm{div}\, E_{(0)} = 0, \quad \mathrm{div}\, H_{(0)} = 0.$$

With the aim of understanding the asymptotic behavior of the curvature tensor R of solutions to the E-V equations it is natural to first ask the same question for the spin-2 equations (they are also called the linearized Einstein vacuum field equations) in Minkowski space, subject to initial conditions $E_{(0)}$, $H_{(0)}$ which verify the assumption (10). To achieve this we use a method which is similar to that mentioned above for solutions to ($\Box$) and which also works for (M), based on generalized Sobolev norms and energy estimates. The latter are connected to the concepts of Killing or conformal Killing vector fields, and of the energy-momentum tensors which we discuss below. We say that

$$(11)\qquad \begin{gathered} X \in \mathfrak{X}(M^{n+1}) \quad \text{is conformal Killing if} \\ X_{\mu;\nu} + X_{\nu;\mu} := \mathfrak{L}_X g_{\mu\nu} = \chi g_{\mu\nu} \end{gathered}$$

for some scalar χ. If $\chi = 0$, X is called simply, Killing.

Now define the following tensors corresponding to solutions u, F, W of ($\square$), (M), (Sp),

$$(Q_\square)\qquad Q_{\alpha\beta} = u_{,\alpha}u_{,\beta} - \tfrac{1}{2}\, g_{\alpha\beta}(g^{\mu\nu}u_{,\mu}\, u_{,\nu})$$

$$(Q_M)\qquad Q_{\alpha\beta} = F_{\alpha\lambda}F_{\beta}{}^{\lambda} + {}^*F_{\alpha\circ}{}^*F_{\beta}{}^{\lambda}$$

$$(Q_{Sp})\qquad Q_{\alpha\beta\gamma\delta} = W_{\alpha\mu\beta\nu}W_{\gamma}{}^{\mu}{}_{\delta}{}^{\nu} + {}^*W_{\alpha\mu\beta\nu}{}^*W_{\gamma}{}^{\mu}{}_{\delta}{}^{\nu}.$$

They are the energy momentum tensors of the field equations; $Q_{(Sp)}$ is the so-called Bell-Robinson tensor [33]. One can show that they are symmetric, traceless (with the exception of $(Q_\square)$) in all pairs of indices and have zero divergence. Moreover they satisfy the *positive energy condition* which states that $Q(X_1,X_2)$ (or $Q(X_1,X_2,X_3,X_4)$ for $Q_{(Sp)}$) is positive for all time-like vector fields X_1, X_2 (or X_1,X_2,X_3,X_4) with the same time-like orientation.

Now let X be an arbitrary vector field and Q the energy-momentum tensor of either ($\square$) or (M). Take

$$P^{\mu} = Q^{\mu\nu}X_{\nu}.$$

Then,

$$P^{\mu}{}_{;\mu} = \tfrac{1}{2}\, Q^{\mu\nu}(X_{\mu;\nu} + X_{\nu;\mu}).$$

Consequently, if X is conformal Killing

$$(12)\qquad P^{\mu}{}_{;\mu} = \tfrac{1}{2}\,\chi\, \mathrm{tr}(Q).$$

In particular, if $\mathrm{tr}(Q) = 0$ (or X is Killing),

$$P^{\mu}{}_{;\mu} = 0. \tag{12'}$$

The same holds true for $Q_{(\mathrm{Sp})}$ by considering $P^{\mu} = Q^{\mu\beta\gamma\delta}X_{\beta}Y_{\gamma}Z_{\delta}$ for any X, Y, Z conformal Killing. Assume that the divergence equation (12') holds true and that M^{3+1} is foliated by space-like hypersurfaces t = const with future oriented unit normal T. Then, for (□) and (M)

$$\int_{H_t^3} Q(T,X) = \text{const in } t \tag{13}$$

and for (Sp)

$$\int_{H_t^3} Q(T,X,Y,Z) = \text{const in } t. \tag{13'}$$

By the positive energy condition the integrands are positive provided that X (and resp. X,Y,Z) are unit time-like, future oriented. Thus, to derive energy identities, with positive integrands, we need conformal Killing time-like vector-fields. Such vector fields exist in the Minkowski space to which we now restrict our attention.

First recall that the Minkowski space, introduced by H. Minkowski as the natural geometric space of Special Relativity, is the simplest example of an Einstein manifold i.e. the space diffeomorphic to $\mathbb{R}^{n+1}$ and with a canonical coordinate system $(x^0, x^1, \ldots, x^n)$ for which

$$\langle \partial_{\mu}, \partial_{\nu} \rangle = \eta_{\mu\nu} = \mathrm{Diag}(-1, 1, \ldots, 1).$$

By abusing our previous notation we will denote x^0 by t and $x = (x^1, x^2, \ldots, x^n)$. The Minkowski space possesses a large family of Killing and conformal Killing vector fields. The first ones exhibited below are the generators of the groups of translations and Lorentz transformations (i.e. the Poincare group) and they are all Killing,

$$(14_{\rm i}) \qquad T_\mu = \partial_\mu; \qquad \mu = 0,1, \ldots, n$$

$$(14_{\rm ii}) \qquad L_{\mu\nu} = x_\mu \partial_\nu - x_\nu \partial_\mu; \qquad 0 \leqslant \mu < \nu \leqslant n$$

with $x_\mu = g_{\mu\nu} x^\nu$. In particular for $i,j = 1, \ldots, n$, $L_{ij} = x_i \partial_j - x_j \partial_i$ are the angular momentum operators corresponding to purely spatial rotations on the space-like slices t = const. The vector fields $L_i = L_{i0} = x_i \partial_t + t\partial_i$ correspond to space-time rotations (i.e. "boosts").

The next two groups of vector fields are conformal Killing and correspond to scaling transformations

$$(14_{\rm iii}) \qquad S = x^\mu \partial_\mu$$

and conformal transformations

$$(14_{\rm iv}) \qquad K_\mu = -2x_\mu S + \langle S,S \rangle \partial_\mu.$$

Denote by K the Lie algebra generated by T_μ, S, $L_{\mu\nu}$ for $0 \leqslant \mu < \nu \leqslant n$. Given any tensor W, we define

$$(15) \qquad \begin{aligned} \|W(t)\|^2_{\mathbf{K},k} &= \sum_{0 \leqslant \ell \leqslant k} \int_{\mathbb{R}^n} |\mathcal{L}_{\Gamma_1} \cdots \mathcal{L}_{\Gamma_\ell} W(t,x)|^2 dx \\ \|W\|_{\mathbf{K},k} &= \sup_{t \geqslant 0} \|W(t)\|_{\mathbf{K},k} \end{aligned}$$

where $\Gamma_1, \ldots, \Gamma_\ell$ are any of the generators of K and $|\mathcal{L}_{\Gamma_1},\ldots \mathcal{L}_{\Gamma_\ell} W|$ is the Euclidean length, in $\mathbb{R}^{n+1}$, of the tensor $\mathcal{L}_{\Gamma_1} \ldots \mathcal{L}_{\Gamma_\ell} W$.

The linear vector fields S and $L_{\mu\nu}$ are dimensionless relative to scaling transformations $x \to \lambda x$ and span all directions in $\mathbb{R}^{n+1}$ except along the null cone

$$-t^2 + (x^1)^2 + \cdots + (x^n)^2 = 0$$

where S is null. In fact at every point in $\mathbb{R}^{3+1}$ we have the formulas

$$(16) \qquad \partial_\mu = \langle S,S \rangle^{-1} (x_\mu S + x^\nu L_{\mu\nu})$$

for all $\mu = 0,1, \ldots, n$, where $\langle S,S\rangle = -t^2 + (x^1)^2 + \cdots + (x^n)^2$.

Similar formulas, which degenerate for $\langle S,S\rangle = 0$, express an electromagnetic tensor F in terms of $i_S F$, $i_S^* F$, and a spin-2 field W in terms of $ii_S W$, $ii_S^* W$. They imply in particular,

$$(16') \qquad |F(t,x)| \leq c \frac{1}{|t - |x||^2}(|i_S F(t,x)| + |i_S^* F(t,x)|)$$

$$(16'') \qquad |W(t,x)| \leq c \frac{1}{|t - |x||^2}(|ii_S W(t,x)| + |ii_S^* W(t,x)|)$$

at all points $(t,x) \in \mathbb{R}^{3+1}$ with $t \neq |x|$, for some constant c.

The degeneracy of the formulas above along the null cone $\langle S,S\rangle = 0$ is responsible for the different asymptotic behavior of solutions to field equations in Minkowski space, away from and near null directions. To describe what happens along the characteristic directions, it is useful to introduce the following null frame.

$$(\mathrm{N}) \qquad E_+ = E_{n+1} = \frac{1}{\sqrt{2}}\left(\partial_t + \Sigma_i \frac{x^i}{|x|}\, \partial_i\right)$$

$$E_- = E_n = \frac{1}{\sqrt{2}}\left(\partial_t - \Sigma_i \frac{x^i}{|x|}\, \partial_i\right).$$

Then pick $E_1, \ldots, E_{n-1}$ unit space-like vector fields orthogonal to each other and to E_+, E_-. Assuming them to be parallel transported in $\mathbb{R}^{n+1}$, we have

$$\nabla_{E_+} E_A = \nabla_{E_-} E_A = 0$$

for all $A = 1,2, \ldots, n-1$.

We are now ready to return to the field equations ($\Box$), (M), (Sp). Assume that the initial data $\psi_{(0)}$, $\psi_{(1)}$, for ($\Box$); $E_{(0)}$, $H_{(0)}$, for (M); and $E_{(0)}$, $H_{(0)}$, for (Sp), given at $t = 0$, satisfy the conditions

$$I_k^{(\Box)}(\psi_{(0)},\psi_{(1)}) < \infty$$

$$I_k^{(M)}(E_{(0)},H_{(0)}) < \infty$$

$$I_k^{(Sp)}(E_{(0)},H_{(0)}) < \infty$$

where, $I^{(\Box)}$, $I^{(M)}$, $I^{(Sp)}$ are the weighted norms

$$(17) \qquad I_k^{(\Box)} = \Big[\sum_{\ell=0}^{k+1} \int_{\mathbb{R}^n} (1 + |x|^2)^\ell \, |\nabla^\ell \psi_{(0)}(x)|^2 dx + \sum_{\ell=0}^{k} \int_{\mathbb{R}^n} (1 + |x|^2)^\ell \, |\nabla^\ell \psi_{(1)}(x)|^2 dx \Big]^{1/2}$$

$$(17') \qquad I_k^{(M)} = \Big[\sum_{\ell=0}^{k} \int_{\mathbb{R}^n} (1 + |x|^2)^{\ell+1} (|\nabla^\ell E_{(0)}(x)|^2 + |\nabla^\ell H_{(0)}(x)|^2) \, dx \Big]^{1/2}$$

$$(17'') \qquad I_k^{(Sp)} = \Big[\sum_{\ell=0}^{k} \int_{\mathbb{R}^3} (1 + |x|^2)^{\ell+2} (|\nabla^\ell E_{(0)}(x)|^2 + |\nabla^\ell H_{(0)}(x)|^2) dx \Big]^{1/2}$$

with ∇^ℓ the ℓ-th spatial covariant derivatives on $t = 0$.

Now notice that among the vector fields displayed in (14) the only ones which are globally non space-like are T_0 and K_0. Since K_0 becomes null along $t^2 - |x|^2 = 0$ we replace it by the strictly time-like conformal Killing vector field

$$(14_v) \qquad K = \partial_t + K_0 = (1 + t^2 + |x|^2)\partial_t + 2t\, x^i \partial_i.$$

Now let $X = T = T_0$ in the integrand in (13). Note that T_0 is the forward unit normal to the canonical foliation $t =$ const. We find,

$$Q_{(\Box)}(T,T) = \tfrac{1}{2}[(\partial_t u)^2 + \sum_i (\partial_i u)^2]$$

$$Q_{(M)}(T,T) = |i_T F|^2 + |i_T^* F|^2$$

which lead by integration to the usual energy identities for $(\Box)$ and (M). In the same way, if we take $X = Y = Z = T$ in the integrand of (13') we infer that

$$Q_{(Sp)}(T,T,T,T) = |i i_T F|^2 + |i i_T^* F|^2.$$

Now take $X = K$ in (13); we claim that

$$Q_{(M)}(T,K) = |i_S F|^2 + |i_S^* F|^2 + |i_T F|^2 + |i_T^* F|^2$$

with S the scaling vector field in (14_{iii}). This leads to

$$(18_M) \quad \|i_S F(t)\| + \|i_S^* F(t)\| + \|i_T F(t)\| + \|i_T^* F(t)\| \leqslant I_0^{(M)}(E_{(0)}, H_{(0)})$$

for all $t \geqslant 0$. Similarly, taking $X = T$, $Y = Z = K$ in (13') we find, for all $t \geqslant 0$

$$(18_{Sp}) \quad \|ii_S W(t)\| + \|ii_S^* W(t)\| + \|ii_T W(t)\| + \|ii_T^* W(t)\| \leqslant I_0^{(Sp)}(E_{(0)}, H_{(0)})$$

for all $t \geqslant 0$. By modifying somewhat the argument leading to (13'), to take into account the fact that $Q_{(\Box)}$ is not traceless, we derive, for solutions u of ($\Box$) (see [34], [35])

$$(18_\Box) \qquad \|u(t)\|_{\mathbf{K},1} \leqslant I_0^{(\Box)}(\psi_{(0)}, \psi_{(1)})$$

for any $t \geqslant 0$.

Notice that, together with (16), (16'), and (16''), the estimates (18) provide us already with some averaged decay properties of solutions to ($\Box$), (M) and (Sp). Indeed one can show, in the particular case of $\Box u = 0$, that the local energy norm

$$\frac{1}{2} \int_\Omega \left[(\partial_t u)^2 + \sum_{i=1}^n (\partial_i u)^2 \right] dx^1 dx^2 dx^3$$

decays like t^{-2} for any compact set $\Omega \in \mathbb{R}^n$. These types of estimates were derived first by C. Morawetz [36], [37] for solutions of the wave equation in the exterior of a star shaped domain in $\mathbb{R}^n$. Similar local results were obtained by Costa-Strauss [38] for the Maxwell equations.

Next we note that if u, F, W are solutions to the corresponding equations in Minkowski space and X is a generator of the Lie algebra $\mathbf{K}$ then the Lie derivatives $\mathfrak{L}_X u$, $\mathfrak{L}_X F$, $\mathfrak{L}_X W$ are again solutions. This implies,

$$(19_{\Box}) \qquad \|u\|_{\mathbf{K},k+1} \leqslant C I_k^{(\Box)}(\psi_{(0)},\psi_{(1)})$$

$$(19_{M}) \qquad \|i_S F\|_{\mathbf{K},k} + \|i_S^* F\|_{\mathbf{K},k} + \|i_T F\|_{\mathbf{K},k} + \|i_T^* F\|_{\mathbf{K},k} \leqslant I_k^{(M)}(E_{(0)},H_{(0)})$$

$$(19_{Sp}) \qquad \|ii_S W\|_{\mathbf{K},k} + \|ii_S^* W\|_{\mathbf{K},k} + \|ii_T W\|_{\mathbf{K},k} + \|i_T^* W_{\mathbf{K},k} \leqslant I_k^{(Sp)}(E_{(0)},H_{(0)}).$$

These generalized energy estimates are the key to our derivation of the asymptotic properties of solutions to the linear equations (□), (M) and (Sp). To go from the weighted Sobolev norms displayed above to uniform norms we rely on the following (see [16], [34] and also [18], [39]).

GLOBAL SOBOLEV LEMMA. Let $u = u(t,x)$ be a smooth scalar function in $\mathbb{R}^{n+1}$ with $x = (x^1, \ldots, x^n) \in \mathbb{R}^n$, $|x| = [\Sigma_i(x^i)^2]^{1/2}$. Assume $\|u\|_{\mathbf{K},k} < \infty$ for some $k > n/2$, with k a given positive integer. Then, given any generators $\Gamma_1, \ldots, \Gamma_\ell$ of $\mathbf{K}$, $0 \leqslant \ell < k-n/2$, we have

$$|\Gamma_1 \cdots \Gamma_\ell\, u(t,x)| \leqslant C_k(1 + |t - |x||)^{-1/2}(1 + t + |x|)^{-(n-1)/2}\|u\|_{\mathbf{K},k}$$

uniformly in $t \geqslant 0$, $x \in \mathbb{R}^n$.

As a consequence of the G.S.L. and the formula (16) we derive

COROLLARY. Under the same assumptions as above

$$|\nabla^\ell u(t,x)| \leqslant C_k(1 + |t - |x||)^{-1/2-\ell}(1 + t + |x|)^{-(n-1)/2}\|u\|_{\mathbf{K},k}$$

uniformly for $t \geqslant 0$, $x \in \mathbb{R}^n$. Here $|\nabla^\ell u|$ denotes the euclidean norm in $\mathbb{R}^{n+1}$ of the ℓ-th covariant derivatives of u.

Finally, we state below the results which we can derive by using the estimates (18) and the G-S-L.

THEOREM 1 ([34], [35]). Let u be a solution of (□) subject to the initial conditions

$$u = \psi_{(0)}, \qquad u_t = \psi_{(1)}, \quad \text{at} \quad t = 0$$

and assume that $I_k^{(\Box)}(\psi_{(0)},\psi_{(1)}) < \infty$ for some $k > n/2$. Then, for all $0 \leqslant \ell < k - n/2$,

$$|\nabla^\ell u(t,x)| \leqslant C_k(1 + |t - |x||)^{-1/2-\ell}(1 + t + |x|)^{-(n-1)/2}, \quad \text{for } t \geqslant 0,\ x \in \mathbb{R}^n.$$

(ii) Moreover let $(E_1,\ldots, E_{n-1},E_n,E_{n+1})$ be the null frame discussed above; then

$$|E_{\mu_1}\cdots E_{\mu_\ell}\, u(t,x)| \leqslant C(1 + |t - |x||)^{-11/2-\ell_*}(1 + t + |x|)^{-(n-1)/2-\ell+\ell_-}$$

for $t > 0$, $x \neq 0$ in $\mathbb{R}^n$, $|t - |x|| < 1/2(t + |x|)$ with ℓ_* the number of times $E_- = E_n$ appears in the sequence $E_{\mu_1}, \ldots, E_{\mu_\ell}$.

THEOREM 2 ([21]). Let F be a solution of the Maxwell equations (M), in $\mathbb{R}^{3+1}$, subject to the initial conditions

$$i_{\partial_t}F = E_{(0)}, \quad i_{\partial_t}^{*}F = H_{(0)} \quad \text{at } t = 0$$

where $E_{(0)}$, $H_{(0)}$ are vectors in $\mathbb{R}^3$ satisfying the constraints div $E_{(0)}$ = div $H_{(0)} = 0$ and such that $I_k^{(M)}(E_{(0)},H_{(0)}) < \infty$ for some integer $k > 3/2$. Then for every $0 \leqslant \ell < k - 3/2$

(i) $$|\nabla^\ell F(t,x)| \leqslant C(1 + |t - |x||)^{-3/2-\ell}(1 + t + |x|)^{-1}$$

uniformly in $t \geqslant 0$, $x \in \mathbb{R}^3$.

Moreover, relative to the null frame E_1, E_2, $E_3 = E_-$, $E_4 = E_+$

(ii) $$|F_{A3}(t,x)| \leqslant C(1 + |t - |x||)^{-3/2}(1 + t + |x|)^{-1}$$

$$|F_{12}(t,x)| \leqslant C(1 + |t - |x||)^{-1/2}(1 + t + |x|)^{-2}$$

$$|F_{34}(t,x)| \leqslant C(1 + |t - |x||)^{-1/2}(1 + t + |x|)^{-2}$$

$$|F_{A4}(t,x)| \leqslant C(1 + t + |x|)^{-5/2}$$

for all $A = 1,2$, $t \geqslant 0$, $x \neq 0$, in $\mathbb{R}^3$ and $|t - |x|| \leqslant 1/2|t + |x||$.

THEOREM 3 ([21]). Let W be a solution of the Spin-2 equations (Sp) subject to the initial conditions

$$ {}^{i}{}_{i}\partial_t W = E_{(0)}, \quad {}^{i}{}_{i}\partial_t W = H_{(0)}, $$

for given symmetric, traceless 2-tensors $E_{(0)}$, $H_{(0)}$ in $\mathbb{R}^3$ satisfying the constraints div $E_{(0)}$ = div $H_{(0)}$ = 0 and assume that $I_k^{(\mathrm{Sp})}(E_{(0)}, H_{(0)}) < \infty$ for some integer $k > 3/2$. Then, for any $0 < \ell < k - 3/2$,

(i) $$|\nabla^{\ell} W(t,x)| \leqslant C(1 + |t - |x||)^{-5/2-\ell}(1 + t + |x|)^{-1}$$

uniformly in $t \geqslant 0$, $x \in \mathbb{R}^3$.

Moreover relative to the null frame E_1, E_2, E_3, E_4,

(ii) $$|W_{A3B3}(t,x)| \leqslant C(1 + |t - |x||)^{-5/2}(1 + t + |x|)^{-1}$$

$$|W_{A343}(t,x)| \leqslant C(1 + |t - |x||)^{-3/2}(1 + t + |x|)^{-2}$$

$$|W_{1212}(t,x)| \leqslant C(1 + |t - |x||)^{-1/2}(1 + t + |x|)^{-3}$$

$$|W_{1234}(t,x)| \leqslant C(1 + |t - |x||)^{-1/2}(1 + t + |x|)^{-3}$$

$$|W_{A434}(t,x)| \leqslant C(1 + |t + |x||)^{-7/2}$$

$$|W_{A4B4}(t,x)| \leqslant C(1 + |t + |x||)^{-7/2}$$

for all $A,B = 1,2$, $t \geqslant 0$, $x \neq 0$ in $\mathbb{R}^3$ and $|t - |x|| \leqslant 1/2(t + |x|)$.

In both Theorems 2 and 3, the Lie derivatives relative to E_1, E_2, E_4 improve by a factor of $(1 + t + |x|)^{-1}$ while the Lie derivative relative to E_3 improves only by $(1 + |t - |x||)^{-1}$. However, the derivatives in the direction E_3 of certain components of F and W improve also by $(1 + t + |x|)^{-1}$.

At the end of this lecture I want to discuss the relation between the initial condition $I_k^{(\mathrm{Sp})}(E_{(0)},H_{(0)}) < \infty$ and condition (10). In fact one can prove that, if $E_{(0)}$, $H_{(0)}$ are initial data for (Sp), satisfying div $E_{(0)}$ = div $H_{(0)}$ = 0 and (10), and some $s = k > 0$, then also $I_k^{(\mathrm{Sp})}(E_{(0)},H_{(0)}) < \infty$.* This is an interesting consequence of the

*This is not quite true for the Einstein Vacuum equations where one component of the curvature tensor will behave differently, due to the mass term in (7). Yet the arguments presented above can be modified to take the mass term into account.

Poincare inequality on the unit sphere S^2 in $\mathbb{R}^3$. Moreover we note here that the initial asymptotic condition $I_k^{(Sp)}(E_{(0)},H_{(0)}) < \infty$ is singular relative to the Penrose conformal compactification [30]. Thus the uniform decay rates for (Sp), which could be obtained in principle, by using that method, do not seem justified. The conformal compactification method was used by D. Christodoulou ([40], [41]) to prove global existence results for certain classes of nonlinear field equations for which the problem of slow decay of infinity of the initial data does not occur. A similar method was proposed by H. Friedrich [42] in an attempt to prove global existence results for the Einstein equations. He has been able to circumvent the difficulty of lack of conformal invariance of the Einstein equations and obtain a symmetric hyperbolic system for the transformed variables. The corresponding initial data are singular however at space-like infinity. The question whether that singularity propagates (in the radiative case) to a bona fide singularity of the compactified picture remains wide open, though certain recent results [43], [44] suggest that this is what happens.

The methods presented here have to be modified considerably for the (E-V) equations. In fact the mass term in the expansion of the induced metric on slices has the effect of distorting the asymptotic behavior of the light cones. To take this effect into account one has to modify the Killing and conformal Killing vector fields of the Minkowski space. This can be done [45] by a systematic study of "optical geometry" of a space-time whose curvature components behave similarly to those given by Theorem 3. Finally, to derive a global existence theorem for the Einstein equations one needs to treat the nonlinear terms which appear each time when we differentiate the Bianch identities. For scalar wave equations it is known ([46], [47]) that quadratic nonlinearities could lead to formation of singularities for arbitrary small data unless the "null condition" is satisfied ([48], [40], [34], [35]). An appropriate version of the null condition, which takes into account the tensorial nature of the equations is satisfied by the Einstein vacuum equations.

REFERENCES

[1] S. Klainerman, G. Ponce, "Global Small Amplitude Solutions to Nonlinear Evolution Equations," Comm. Pure Appl. Math. 36, 1983, 133-141.

[2] S. Klainerman, "Global Existence for Nonlinear Wave Equations," Comm. Pure Appl. Math. 33, 43-101, 1980.

[3] S. Klainerman, A. Majda, "Compressible and Incompressible Fluids," Comm. Pure Appl. Math. 35, 1982, 629-651.

[4] A. Majda, "Compressible Fluid Flow and Systems of Conservation Laws in Several Space Variables," Appl. Math. Sci. 53, Springer-Verlag.

[5] S. Klainerman, "Lecture Notes on Nonlinear Hyperbolic Systems," in preparation.

[6] R. Schoen-S. T. Yau, "On the Proof of the Positive Mass Conjecture in General Relativity," Comm. Math. Phys. 65, 1979, 45-76.

[7] E. Witten, "A Simple Proof of the Positive Energy Theorem," Comm. Math. Phys. 80, 1981, 381-402.

[8] D. M. Eardley - V. Moncrief, "The Global Existence of Yang-Mills-Higgs Fields in 4-Dimensional Minkowski Space," Comm. Math. Phys. 83, 171-191, and 193-212, 1982.

[9] K. Jörgens, Math. Z. 77, 295, 1961.

[10] W. von Wahl, "L^p Decay Rates for Homogeneous Wave Equations," Math. Z. 120, 1971, pp. 93-106.

[11] I. E. Segal, "Dispension of Nonlinear Relativistic Equations," Ann. Sci. Ecole Norm. Sup. 4^e Ser., 459-497, 1968.

[12] W. Strauss, "Decay and Asymptotics for $\Box u = f(u)$," J. Funct. Anal. 2, 1968, pp. 409-457.

[13] F. John, "Delayed Singularity Formation in Solutions of Nonlinear Wave Equations in Higher Dimensions," Comm. Pure Appl. Math. 29, 649-681, 1976.

[14] F. John, "Lower Bounds for the Life-Span of Solutions of Nonlinear Wave Equations, in Three Dimensions," Comm. Pure Appl. Math. 36, 1-36, 1983.

[15] J. Shatah, "Global Existence of Small Solutions to Nonlinear Evolution Equations," J. Diff. Eqns. 78, 73-98, 1982.

[16] S. Klainerman, "Uniform Decay Estimates and the Lorentz Invariance of the Classical Wave Equation," Comm. Pure Appl. Math. 38, 321-332, 1985.

[17] F. John, S. Klainerman, "Almost Global Existence to Nonlinear Wave Equations in Three Space Dimensions,"

[18] L. Hörmander, "On Sobolev Spaces Associated with Some Lie Algebras," Inst. Mittag-Leffler, Report No. 4, 1985.

[19] F. John, "A Lower Bound for the Life-Span of Solutions of Nonlinear Wave Equations in Three Space Dimensions," preprint.

[20] L. Hörmander, "The Life-Span of Classical Solutions of Nonlinear Hyperbolic Equations," Inst. Mittag-Leffler, Rpt. #5, 1985.

[21] D. Christodoulou, S. Klainerman, "Einstein Geometry and Linear Field Equations in Minkowski Space," in preparation.

[22] Y. Choquet-Bruhat, "Theoreme d'Existence Pour Certains Systemes d'Equations aux Derivees Partielles Nonlineaires," Acta Mathematica 88, 141-225.

[23] Y. Choquet-Bruhat, "Un Theoreme d'Instabilite Pour Certaines Equations Hyperbolique Nonlineaires," C. R. Acad. Sci. Paris 276A, 281.

[24] R. Geroch, "Asymptotic Structure of Space-Time," P. Esposito and L. Witten (eds.) Plenum, New York, 1976.

[25] E. T. Newman - K. P. Todd, "Asymptotically Flat Space-Times," General Rel. and Gravitation, Vol. 2, A. Held, Plenum, 1980.

[26] H. Bondi, M. G. J. van der Burg, A. W. K. Metzner, Proc. R. Sol, London Series, A269, 21, 1962.

[27] H. Bondi, Nature 186, 535, 1960.

[28] R. K. Sachs, "Gravitational Waves in General Relativity VIII. Waves in Asymptotically Flat Space-Times," Proc. R. Soc. A270, 103-126 (1962).

[29] R. Penrose, Proc. R. Soc. London Ser. A284, 159, 1965.

[30] R. Penrose, "Conformal Treatment of Infinity," In Relativity, Groups and Topology, B. DeWitt and C. DeWitt (eds.), Gordon and Breach, 1963.

[31] A. Ashtekar, "Asymptotic Structure of the Gravitational Field at Spatial Infinity in General Relativity and Gravitation," Vol. 2, edited by A. Held at Plenum Publ. Corp., 1980.

[32] R. Bartnik, "Existence of Maximal Surfaces in Asymptotically Flat Space-Times," Comm. Math. Phys. 94, 155-175, 1984.

[33] L. Bel, "Introduction d'un Tensur du Quatrieme Ordre," Computes Rendus 247, p. 1094, p. 1297, 1978.

[34] S. Klainerman, "The Null Condition and Global Existence to Nonlinear Wave Equations," Lectures on Appl. Math., Vol. 23, 293-326, 1986.

[35] L. Hörmander, "On Global Existence of Solutions of Nonlinear Hyperbolic Equations in $\mathbb{R}^{1+3}$," Inst. Mittag-Leffler Report No. 9, 1985.

[36] C. Morawetz, "The Limiting Amplitude Principle," Comm. Pure Appl. Math. 15, 1962, pp. 349-362.

[37] P. Lax, R. Phillips, "Scattering Theory for Automorphic Functions," Princeton Univ. Press, 1976.

[38] D. Costa, W. Strauss, "Energy Splitting," Quaterly of Appl. Math., 351-361, 1981.

[39] S. Klainerman, "Remarks on the Global Sobolev Inequalities in Minkowski Space," to appear in Comm. Pure. Appl. Math.

[40] D. Christodoulou, "Global Solutions of Nonlinear Hyperbolic Equations for Small Data," to appear in Comm. Pure Appl. Math., 1986.

[41] D. Christodoulou, "Solutions Globeles des Equations des Equations des Champs de Yang-Mills," C.R. Acad. Sci. Paris, 293, Series A, pp. 39-42 (1981).

[42] H. Friedrichs, "On the Hyperbolicity of Einstein and Other Field Equations," Comm. Math. Phys. 100, 525, 1985.

[43] B. G. Schmidt, J. M. Stewart, "The Scalar Wave Equation in a Schwarzschild Spare-Time," Proc. R. Soc., London Series, A367, 503, 1979.

[44] J. Porrill, J. M. Stewart, "Electromagnetic and Gravitational Fields in a Schwarzschild Space-Time," Proc. R. Soc., London Series, A376, 451, 1981.

[45] D. Christodoulou, S. Klainerman, "Optical Geometry and Quasiconformal Transformations of an Einstein Space-Time,"

[46] F. John, "Blow-up for Quasilinear Wave Equations in Three Space Dimensions," Comm. Pure Appl. Math. 34, 29-51, 1981.

[47] F. John, "Blow-up of Radial Solutions of $\Box u = (\partial/\partial t)F(u_t)$," preprint.

[48] S. Klainerman, "Long-Time Behavior of Solutions to Nonlinear Wave Equations," Proc. of Int. Congress for Mathematicians, Warsaw, 1982.

The work in this paper was supported by the National Science Foundation, Grant No. DMS-8504033

S. KLAINERMAN
COURANT INSTITUTE OF MATHEMATICAL SCIENCES
NEW YORK UNIVERSITY
NEW YORK, NEW YORK

Contemporary Mathematics
Volume **71**, 1988

THE WELL-POSEDNESS OF ROSEN'S FIELD EQUATIONS

R. J. Knill and A. P. Whitman

Cosmologically, general relativity leads to a satisfying theory of space-time which attributes the curvature of space to the gravitational influence of the matter in it. One relativistic model describes space-time as a sphere expanding as the result of the "big bang" and then contracting again into the so-called "big crunch." The theory possesses internal consistency and is readily modified to take into account various possible densities for the universe, considered as a cloud of dust.

Less successful is the application of general relativity to the description of localized phenomena involving more than one relatively large body or gravitational source. All of the explicit solutions to Einstein's field equations are for one gravitational source. That is to say, they describe the curvature of space induced by only one object at a time. Although relativity is supposed to be a refinement of Newtonian mechanics, it does not treat the two gravitational source problem with as much ease or explicitness. This is related to another difficulty. A desirable property of Newtonian mechanics is the ease of combining two gravitational fields to obtain another in the process called "superposition." This property is essential, for example, for one to be able to treat the Newtonian gravitational field of a body (such as the earth), which has spherical symmetry, as if it were from a point source. This is a considerable simplification. In relativity such superposition is not possible at all. This is no mystery, relativists will tell you, because the field equations of general relativity are nonlinear. In fact, in Newtonian mechanics the equations of motion are nonlinear as well so that the potential of an orbiting system does not depend linearly on the static potentials of its component bodies.

One reason that superposition cannot be done in general relativity is that the geometry of space-time is not completely specified by the field

---

1986 Mathematics Subject Classification Numbers. Primary 58E20, Secondary 58G16, 53C80

equations and by so-called initial conditions. While relativists frequently will invoke the Cauchy-Kowalewsky theorem to obtain the existence (and uniqueness) of analytic solutions, as one writer (John in Bers et al., 1964) puts it, "...the theorem of Cauchy and Kowalewsky only fits a universe which can be constructed completely by analytic continuations from its smallest portion." Even the initial data on a Cauchy surface can be deduced from its values on a small open subset, so that one is not free to specify initial data everywhere else on the Cauchy surface.

On the other hand complete specification of solutions by global initial data should be possible and is so desirable that scientists have come to say, after Hadamard, that when equations and initial conditions continuously specify a unique solution to a problem then that problem is "well-posed." In the absence of analyticity general relativity lacks this well-posedness. (It is well-posed from another point of view (Y. Bruhat, 1962).) Although this is no drawback in cosmology nor in the problem of the curvature of space near a single isolated gravitational source, it is a problem when one tries to compare (for the purpose of superposition) the geometries generated by two distinct gravitational sources. This is due in part to the fact that solutions to Einstein's field equations are not uniquely determined by initial data. Relativists can avoid this problem by regarding as the same, two solutions which are diffeomorphically equivalent via a diffeomorphism which leaves the Cauchy data fixed (Hawking and Ellis, 1973). Thus one may give Cauchy data for each of two gravitational sources and be unable to intelligently superpose one set of equivalent solutions onto the other simply because there is no way to select representative solutions to each set of Cauchy data.

For these reasons the authors have turned to a viable alternative to general relativity called "Rosen's bi-metric theory of gravitation," introduced by physicist Nathan Rosen (1973). Although it has recently come under criticism for its explanation of the binary pulsar (Will, 1981), by compelling solar system tests of a gravitational theory, for example in terms of such phenomena as the bending of light rays near the sun and the red shift, it is widely recognized that Rosen's theory gives as good an agreement with experiment as does general relativity. Furthermore, as will be seen here in several cases, the field equations of Rosen are well-posed. While Rosen's theory might be questioned as a cosmology, it is hoped that its study will simplify the problem of dealing with more localized relativistic phenomena, such as the n-body problem.

In four previous papers (Stoeger, 1983; Stoeger, Whitman and Knill, 1985; Whitman, Knill and Stoeger, in press; and Knill, Whitman and Stoeger, in press) the following results have been obtained. Rosen's field equations are shown to be those of a harmonic map into a suitably defined symmetric space, a means of decomposing certain solutions into solutions of simpler equations has been given and Rosen's one solution to his field equations has been expanded to four distinct classes of solutions, each depending on various constants of integration. It remains however, to determine when Rosen's field equations are well-posed and have global solutions. It is the purpose of this paper to give a partial answer to this problem and to derive the gravitational field of an extended body by invoking a form of superposition.

It is with great pleasure that we dedicate this paper on the occasion of his 70th birthday to Professor Ky Fan, whose steadfast mathematical taste and broad contributions have set a high standard.

1. ROSEN'S BIMETRIC THEORY OF GRAVITATION

Let M be the space of 4×4 symmetric matrices with one negative and three positive eigenvalues. The metric defined on M by Stoeger (1983) is an indefinite metric $\langle\ ,\ \rangle$ given by the rule: for a matrix σ in M and for symmetric matrices A and B, regarded as tangent vectors at σ, define

$$\langle A,B\rangle = 1/4\ \mathrm{tr}(\sigma^{-1}A\sigma^{-1}B) - 1/8\ \mathrm{tr}(\sigma^{-1}A)\mathrm{tr}(\sigma^{-1}B)\ .$$

We note here that Misner (1978) has observed that M is a homogeneous space. It is in fact a symmetric space when equipped with the Stoeger metric (Stoeger, Whitman and Knill, 1985).

Let U be an open subset of Minkowski space-time, equipped with a flat metric γ. Its line element is

$$ds^2 = \gamma_{ij}dx^i dx^j\ .$$

By a result of Stoeger (1983), a solution to Rosen's empty space field equations is a harmonic map g from U,γ to M (equipped with the metric $\langle\ ,\ \rangle$ of Stoeger). This harmonic map is used to assign to each point x of U the Lorentzian metric whose line element is

$$ds^2 = g_{ij}(x)dx^i dx^j\ .$$

Here i and j range over 0 to 3, with $x^0 = t$ and x^1, x^2 and x^3 denoting x, y and z. This metric is a gravitational potential tensor satisfying the empty space field equation of Rosen's bimetric theory of gravitation. These field equations have the form

$$(1) \qquad \gamma^{ij}(g_{mn|ij} - g^{pq}g_{mp|i}g_{nq|j}) = 0$$

which imply the harmonic mapping nature of g as a map $g : G,\gamma \to M,\langle\cdot,\cdot\rangle$. Here g^{pq} is the pq entry of the inverse matrix to g and the vertical bar denotes covariant differentiation relative to γ and with respect to the space-time variables whose indices follow the bar. Repeated indices are summed using the Einstein convention. As one sees from the form of this partial differential equation, it is quasilinear.

Rosen's bimetric theory makes the correct predictions with respect to the classical solar system experiments, but our interest here is in the fact that it makes use of harmonic maps, and that geodesics in M are so simple to describe that they are useful in obtaining explicit solutions to Rosen's field equations. This description is given in the following theorem.

THEOREM 1 (W,K,S). Let σ be a matrix in M and let A be a symmetric matrix regarded as a tangent to M at σ. Then the geodesic through σ in M in the direction of A is

$$\tau(t) = \exp(tA\sigma^{-1})\sigma$$

where $\exp$ denotes matrix exponentiation, and t is a real number.

It is possible for $A\sigma^{-1}$ to have imaginary eigenvalues in which case the geodesic is periodic and we write $\tau(\theta)$ instead of $\tau(t)$, where θ is an angle from zero to 2π, i.e., we think of τ as having the unit circle for domain.

The usefulness of geodesics stems from the following factorization theorem.

THEOREM 2 (W,K,S). Suppose that a harmonic map $g : U \to M$ is such that its differential dg has rank one at every point of its domain U in $\mathbb{R}^4_1$ and such that dg is non-null in the sense that at each point of U there is a cotangent vector A at $g(x)$ such that the dual homomorphism dg^* takes A to a non-null cotangent vector of U at x. Then there is a harmonic function $\alpha : U \to \mathbb{R}^1$ or $U \to S^1$ and geodesic τ in M such that $g = \tau\alpha$.

Of course, since the metric on U is Lorentzian, a harmonic function α on U, which by definition is a zero of the Laplace-Beltrami operator, is a solution to a linear hyperbolic partial differential equation, not Laplace's equation.

2. WELL-POSEDNESS OF ROSEN'S FIELD EQUATIONS

The notion of a well-posed problem goes back to Hadamard. Suppose that there is given data D for a system of differential equations. (The data D may include the exact domain of the system of initial conditions.) The problem of solving the system of equations for a function satisfying D is "well-posed" if a solution exists, is unique and depends continuously on the data D. It is "locally" well-posed if such solutions are only proven to exist in a neighborhood of D. Certain problems are known to be locally well-posed. Let us recall one such situation which will be of use to us.

A system of partial differential equations is said to be quasilinear if it is of the form

$$\sum_{|\alpha| = n} A_\alpha D^\alpha f = B .$$

Here α is a multiindex $\alpha = (\alpha_0,\dots,\alpha_k)$, $|\alpha| = \alpha_0 +\dots+ \alpha_k$, f is a function of k variables, A_α is an $N\times N$ matrix function of the independent variable u of f as well as of the partial derivatives of f of order less than n, B is an N dimensional vector function of u and of the partial derivatives of f of orders less than n, and D^α is the partial differential operator formed from the 0-th partial derivative to order α_0 times...times the k-th partial derivative to the order α_k.

Consider the system (1). Indeed since γ is flat, we may assume with no loss of generality that coordinates have been chosen so that $\gamma = \mathrm{diag}(-1,1,1,1)$. With such coordinates physicists are accustomed to write η for γ. Since η is constant, then the covariant differentiation of formula (1) becomes partial differentiation. Then (1) is evidently quasilinear. Let S be the hyperplane $t = x^0 = 0$ in $\mathbb{R}^4$. Then S is a free surface in the sense of (Bers, John, and Schechter, 1964) and (1) is a quasilinear hyperbolic partial differential equation with respect to which S is space-like. We shall assume that U is in the domain of determinacy of $S \cap U$, that is, for any point u in U the set of points w in S such that $\langle w-u,w-u\rangle_\eta > 0$ is in U. If $w-u = (x^0,x^1,x^2,x^3)$, then this is the statement that $\eta_{ij}x^ix^j > 0$.

Cauchy data for the field equations of Rosen consist of initial values $g(w) = f(w)$ and the initial wave velocity $g_t(w) = v(w)$, where g_t denotes the first partial of g with respect to $t = x^0$, w is an arbitrary point in S, and f and v are C^∞ functions taking their values in M, and TM, respectively.

We have no need of any more generality than that given above for hyperbolic partial differential equations, for Cauchy data, nor for domain of determinacy. However, the following theorem of Leray holds in somewhat more general cases than we are considering. The interested reader is referred to (Leray, 1953).

THEOREM (Leray). Suppose that we are given the quasilinear system of hyperbolic differential equations (1) and Cauchy data on a free surface $S \subset U$. Then there exists a neighborhood V of $S \cap U$ in U and a C^∞ solution g of (1) on V satisfying the Cauchy data. Furthermore any solution to (1) on U is uniquely determined throughout U by the initial data.

As Leray pointed out, the proof of the above theorem depends on a priori bounds for the derivatives of g of high order and there are no examples where such a priori bounds are known in the large, so Leray's theorem is strictly local as it applies to existence of solutions. Uniqueness, of course, is global in the domain of determinacy of $S \cap U$.

DEFINITION. A map G of $\mathbb{R}^n$ into M is "totally geodesic" if for any affine function $\alpha : \mathbb{R}^1 \to \mathbb{R}^n$ the composition $G\alpha$ is a geodesic.

Recall (Eells-Lemaire, 1983) that the composition of a harmonic map $H : U \to \mathbb{R}^n$ with a totally geodesic map G is a harmonic map $GH : U \to M$. Again recall that from the signature of η and the fact that both $\mathbb{R}^4, \eta$ and $\mathbb{R}^n$ (equipped with the Euclidean metric $|\cdot|$) are flat, a harmonic map H of $U \subset \mathbb{R}^4$ into $\mathbb{R}^n$ is a solution to a system of wave equations

$$(2) \qquad \eta^{ij} H_{,ij} = 0 \ , \ k = 1,\ldots,n \ .$$

Thus H is harmonic if and only if each coordinate function H^k is a solution to the wave equation. Here a comma before the subscripts i and j denotes second partial differentiation with respect to the i and j coordinates.

DEFINITION. Cauchy data $g(w) = f(w)$ and $g_t(w) = v(w)$ is said to be "totally geodesic" if there is a totally geodesic embedding $G : R^n, |\cdot| \to M, \langle\cdot,\cdot\rangle$, for some $n > 1$, such that for each w in $S \cap U$, $f(w)$ is in the image of G and $v(w)$ is tangent to the image of G.

THEOREM 3. Suppose that the given Cauchy data on $S \cap U$ is totally geodesic. Then there exists a unique solution g to Rosen's field equations satisfying the Cauchy data. Furthermore g factors into a composition $g = GH$ of the totally geodesic map G with a harmonic function H.

PROOF. G is an embedding so one may use it to pull back the Cauchy data to Cauchy data for the system (2). Since the system (2) is linear hyperbolic there exists a solution H to (2) on U satisfying the pulled back data. Thus, (Eells and Lemaire, 1983, Proposition 2.20 et seq.) the composition $g = GH$ is a harmonic map. It satisfies Rosen's field equations and since H satisfies the pulled back data, g satisfies the given Cauchy data. By the uniqueness part of Leray's theorem g is the only solution on U satisfying the given Cauchy data.

There is an instance in which local existence of a solution g in a neighborhood V of $S \cap V$ implies the existence of a (unique) extension of g to all of U. We take this up now.

DEFINITION. Let $g : U \to M$ be a C^∞ function. We say that the "rank" of g is n if the differential dg has rank n at each point x of U as a linear map of the tangent space U over x.

The notion of rank has broader significance than that discussed here. See (Whitman, Knill and Stoeger, in press). For the moment we are just concerned about rank one maps in the context of Leray's theorem. The following is a corollary to theorem 2.

COROLLARY 4. Suppose that a solution g to Rosen's field equations exists in a neighborhood V in U of $S \cap U$, and that solution has rank 1 and that for each point x of V the differential dg^* takes some covariant vector of $TM^*(g(n))$ to a covariant vector of nonzero norm in the manifold V, equipped with the pseudo-Riemannian metric η. Then g extends uniquely to a solution on U to Rosen's field equations.

PROOF. The conditions imply, by theorem 2.1 of (Whitman, Stoeger and Knill, preprint) that g factors as $g = GH$ when the domain of G (= range of H) is $R^1, |\cdot|$, and where G is a geodesic in $M, \langle\cdot,\cdot\rangle$, and G is a solution to the wave equation with initial data given as the pull back under M of the initial data of g. By the theorem, g extends uniquely to all of U.

3. APPLICATION: A SUPERPOSITION ARGUMENT

As an application suppose that there is given a massive point gravitational source. (The solution to his field equations given by Rosen assume such a source.) This solution is the Lorentzian metric with matrix representation

(3) $$g(u) = \mathrm{diag}(-\exp(-m_1/r),\ \exp(m_2/r),\ \exp(m_2/r),\ \exp(m_2/r))$$

where $u = (t,x,y,z)$ is in U and $r = r(u) = (x^2 + y^2 + z^2)^{1/2}$. The solution is said to be static since (in part) it is time independent. Using superposition we give here an extension of this solution to an extended body with spherically symmetric density. This superposition will be well defined by virtue of the uniqueness of solutions discussed in the previous paragraph.

First we comment on an elementary form of superpostion mentioned previously in (Knill, Whitman and Stoeger) but not justified by well-posedness arguments until now. Suppose that we write $r = R(u)$ for r defined as in Formula (3). Let

$$u_1 = (0,x_1,y_1,z_1), \quad \text{and} \quad u_2 = (0,x_2,y_2,z_2)$$

be the locations of two gravitational potential sources of Rosen type metrics g_1 and g_2 so that

$$g_i(u) = \mathrm{diag}(-\exp(-m_1 r(u-u_i)),\exp(m_2 r(u-u_i)),\exp(m_2 r(u-u_i),\exp(m_2 r(u-u_i)).$$

Then it is clear that g_i is the composition

$$g_i = GH_i \quad , \quad i = 1,2$$

where

$$G(s) = \tau(s) = \exp(s\ \mathrm{diag}(m_1,m_2,m_2,m_2)\eta^{-1})\eta$$

is the geodesic τ of theorem 1 defined in terms of $\sigma = \eta$ and $A = \mathrm{diag}(m_1,m_2,m_2,m_2)$, and H_i is the solution to the wave equation satisfying the Cauchy data

(4) $$H_i(0,x,y,z) = 1/r((0,x,y,z)-u_i), \qquad i = 1,2$$

(5) $$D^\alpha H_i = 0 , \qquad i = 1,2 ; \alpha = (1,0,0,0) .$$

Here $r((0,x,y,z)-u_i)$ is the space distance from u_i to $(0,x,y,z)$. H_i is uniquely determined as a solution to the Cauchy problem specified by (4) for the values of H_i on $S - \{(0,0,0,0)\}$ and by (5) for the values of $D^\alpha H_i$ on $S - \{(0,0,0,0)\}$. The domain of uniqueness of H_i is the set U_i of points u such that

$$\eta_{jk}(u-u_i)^j(u-u_i)^k \geq 0 \qquad i = 1,2 \quad (i \text{ is not summed}).$$

Here the subscripts refer to coordinates and repeated indices j and k are to be summed via the Einstein convention. In order to circumvent the mismatch between U_1 and U_2 we specify that there is given a ball $B: r \leq r_0$ which contains both u_1 and u_2 and on $U = \text{exterior}(B)$ of which H_i is specified. Then H_i is the unique solution to the Cauchy problem on the domain of determinacy of U for both $i = 1$ and 2.

Since the wave equation is linear, $H_0 = H_1 + H_2$ is the solution to the Cauchy problem using the sum over $i = 1,2$ of the Cauchy data in (4) and (5). Furthermore, $G_0 = GH_0$ is a solution to Rosen's field equations. Theorem 3 assures us of the uniqueness of this solution on U.

Now suppose one is given a sphere of radius less than or equal to r_0, of spherically symmetric density, centered at the origin. Then by integrating the scalar potentials (4) over points u_1 of the sphere one obtains a scalar potential

$$H = m/r \tag{6}$$

on the exterior $r \geq r_0$ of the sphere $r = r_0$. This specifies a solution

$$g(u) = GH(u) ,$$

of Rosen's field equations which may reasonably be called the superposition of all the gravitational potential tensors sourced at points in the sphere, and which is the unique solution having the Cauchy data (5) and (6) on $S - \{r \leq r_0\}$. Such a superposition argument has not heretofore been solidly based on the well-posedness of the field equations involved, and although the argument was elementary, it seemed worthwhile to sketch it to emphasize the need for well-posed field equations.

Our long-range goal, however, is to solve the n-body problem, and in particular the time dependent two body gravitational source problem. It is unclear how to do this at this point; perhaps by replacing η with the gravitational potential tensor of one body and then deriving a solution to Rosen's field equations in this case of a background metric with nonzero curvature. However, the arguments above depend on the flatness of η. New techniques will need to be developed to handle the case of curved background metrics. These will probably involve harmonic sections of fibre bundles, rather than harmonic maps as discussed in this paper.

BIBLIOGRAPHY

L. Bers, F. John and M. Schechter, Partial Differential Equations, Lectures in Applied Mathematics, vol. 3, John Wiley and Sons, N.Y., 1964.

Y. Bruhat, The Cauchy Problem, Gravitation: An introduction to current research (Ed.: L. Witten) Wiley, 1962, pp. 130-168.

P. Dionne, Sur les problemes de Cauchy bien poses, J. Anal. Math. Jerusalem, 10, pp. 1-90, 1962.

J. Eells and L. Lemaire, A report on harmonic maps, Bull. London Math. Soc., 10, (1978), pp. 1-68.

J. Eells and L. Lemaire, Selected Topics in Harmonic Maps, Tulane (1980), CBMS Regional Conf. Series #50.

S. W. Hawking and G.F.R. Ellis, The large scale structure of space time, Cambridge University Press, Cambridge, 1973.

R. Knill, A. P. Whitman and W. R. Stoeger, The decomposition and superposition of rank-one gravitational potential tensors, Contemporary Mathematics, 51, (1986), pp.81-91.

M. Krzyzanski and J. Schauder, Quasilinear Differentialgleichungen zweiter ordnung von hyperbolischen Typos. Gemischte Randwertaufgaben, Studia Math., 6 (1936), pp. 162-189.

J. Leray, Hyperbolic Differential Equations, Institute for Advanced Study, Notes, 1953.

C. W. Misner, Harmonic maps as models for physical theories, Phys. Rev. D, 18 (1978), 4510-4524.

N. Rosen, A Bi-metric theory of gravitation, Gen. Rel. and Grav., 4 (1973), pp. 435-447.

N. Rosen, A theory of gravitation, Ann. Phys. (N.Y.), 84 (1974), pp. 455-473.

N. Rosen, A Bi-metric theory of gravitation, II, Gen. Rel. and Grav., 6 (1975), pp. 259-268.

J. H. Sampson, Some properties and applications of harmonic mappings, Ann. Scient. Ec. Norm. Sup., 11 (1978), pp. 211-228.

W. R. Stoeger, Rosen's bi-metric theory of gravity as a harmonic map, Proceedings of the Third Marcel Grossman Meeting on Recent Developments in General Relativity, Shanghai, ed. Hu Ning (North Holland Publishing Co., 1983), Part B, pp. 921-925.

W. Stoeger, A. Whitman and R. Knill, The harmonic mapping character of Rosen's bi-metric theory of gravity and the geometry of its harmonic mapping space, J. of Math. Phys., 26 (1985), pp. 2032-2038.

A. Whitman, R. Knill and W. Stoeger, Some harmonic maps on Pseudo-Riemannian manifolds, Int. J. of Theoretical Phys., in press.

C. M. Will, Theory and experiment in gravitational physics, Cambridge University Press, (1981), p. 131.

DEPARTMENT OF MATHEMATICS
TULANE UNIVERSITY
NEW ORLEANS, LA 70118

DEPARTAMENTO DO MATEMATICA
PUC/RJ
R. MARQUES S. VINCENTE 293
22451 RIO DE JANEIRO
BRAZIL

Contemporary Mathematics
Volume 71, 1988

DYNAMICS OF QUADRATIC LAGRANGIANS IN GRAVITY
Fairchild's theory

Victor Szczyrba[1]

ABSTRACT. In the framework of SO(3,1)-gauge theories of gravity we investigate the dynamics of the Fairchild theory, which is based on a lagrangian quadratic in components of the Riemann tensor and linear in the Ricci scalar. The dynamical variables of the theory are defined and the complete system of 13 constraints for their initial values on a spacelike surface is presented. It is shown that the dynamical equations give rise to a consistent evolution picture. In particular, the constraints are preserved in time. The gauge structure of the theory is investigated and the number of independent degrees of freedom is determined.

1. INTRODUCTION. In the recent past quadratic lagrangians in gravity have attracted considerable interest on both clasical and quantum levels. There exist, in principle, two different approaches to such theories. In the first one connections on spacetime M are (pseudo)riemannian, constructed from (pseudo)riemannian metrics $g_{\mu\nu}$ on M by means of the Levi-Civita formula. In such a scheme lagrangians quadratic in components of the Riemann tensor give rise to fourth order systems of field equations for the metric. Thio method is very popular in quantum gravity cf. papers by Fradkin and Tseytlin [1], Barth and Christensen [2], Tomboulis [3], Boulware [4], Antoniadis and Tomboulis [5] and references therein. Classical aspects of these theories have recently been discussed by Boulware [4] and Szczyrba [6].

The other conceivable scheme is usually called the SO(3,1)- (or SL(2,C)-) framework for gravity. It admits non-riemannian connections and is based on

---

1980 Mathematics Subject Classification [1985 Revision] 83-02, 83D05.
1 Supported by NSF grant PHY 8503072.

the Einstein-Palatini variational principle with components of tetrads $e^{(\alpha)}{}_{\lambda}$ and anholonomic components of connections $\Gamma_{\lambda}{}^{(\alpha)(\beta)}$ treated as independent variational potentials. Many authors analyzed the field equations for a general SO(3,1)- (SL(2,C)-) covariant lagrangian in this scheme and tried to extract as much information as possible from general geometric principles cf. papers by Trautman [7], Hehl at al [8], Szczyrba [9], Tseytlin [10], Blagoević, Nicolić and Vasilić [11], Antonowicz and Szczyrba [12]. On the other side several particular lagrangians were investigated Yang [13], Fairchild [14], Sezgin and Nieuwenhuizen [15], Rauch and Nieh [16], Hehl et al [17], from the classical and quantum points of view.

As far as the dynamics of SO(3,1)-gauge theories is concerned, it is clear that for different gravitational lagrangians we may expect completely different time evolutions. Therefore it is natural that we restrict ourselves to gravitational lagrangians at most quadratic in components of the curvature and torsion tensors. This class embodies the Einstein-Cartan-Sciama-Kibble theory, whose SO(3)-covariant dynamical analysis has recently been presented in Ref. [18,19,12]. (For a more traditional treatment of the ECSK theory see Ref.[7,20].)

Let $(e^{(\alpha)}{}_{\lambda})^{3}_{\alpha=0}$ be a Minkowski orthonormal tetrad field on spacetime , that is,

(1.1) $$e^{(\alpha)}{}_{\mu}\, e^{(\beta)}{}_{\nu}\, \eta_{(\alpha)(\beta)} = g_{\mu\nu},$$

where $\eta_{(\alpha)(\beta)}$ is the constant diagonal Minkowski metric with signature (-1,1,1,1). Connection coefficients in an anholonomic basis of tetrads are given by an so(3,1)-valued one form on M $\Gamma_{\lambda}{}^{(\alpha)(\beta)}dx^{\lambda}$. The metricity of the connection is assured by the relations $\Gamma_{\lambda}{}^{(\alpha)(\beta)} = -\Gamma_{\lambda}{}^{(\beta)(\alpha)}$ that follow directly from the properties of the Lie algebra so(3,1).

The torsion and curvature are defined by the following formulas

(1.2) $$Q^{(\alpha)}{}_{\mu\nu} = \partial_{\mu} e^{(\alpha)}{}_{\nu} + \Gamma_{\mu}{}^{(\alpha)}{}_{(\tau)} e^{(\tau)}{}_{\nu} - (\,\mu \leftrightarrow \nu\,)\ ,$$

(1.3) $$R^{(\alpha)(\beta)}{}_{\mu\nu} = \partial_\mu \Gamma_\nu{}^{(\alpha)(\beta)} + \Gamma_\mu{}^{(\alpha)}{}_{(\epsilon)}\Gamma_\nu{}^{(\epsilon)(\beta)} - (\mu \leftrightarrow \nu) .$$

We denote $e = \det[e^{(\alpha)}{}_\mu]$, and $[e^\mu{}_{(\alpha)}]$ is the inverse matrix of the tetrad matrix $[e^{(\alpha)}{}_\mu]$. Then $g^{\mu\nu} = e^\mu{}_{(\alpha)} e^\nu{}_{(\beta)} \eta^{(\alpha)(\beta)}$.

The Ricci tensor, Ricci scalar and Einstein tensor are

(1.4) $$R^\mu{}_{(\alpha)} = R^{(\epsilon)(\beta)}{}_{\tau\nu} e^\nu{}_{(\beta)} e^\tau{}_{(\alpha)} e^\mu{}_{(\epsilon)}, \qquad R = R^\mu{}_{(\alpha)} e^{(\alpha)}{}_\mu$$

and $G^\mu{}_{(\alpha)} = R^\mu{}_{(\alpha)} - 1/2 e^\mu{}_{(\alpha)} R$, respectively.

(An)holonomic indices $\mu, \nu, \ldots$ $((\alpha),(\beta),\ldots)$ are raised and lowered by means of the (an)holonomic metrics $g^{\mu\nu}$ and $g_{\mu\nu}$ ($\eta^{(\alpha)(\beta)}$ and $\eta_{(\alpha)(\beta)}$).

A general at most quadratic gravitational lagrangian can be built from one linear and 8 quadratic invariants of curvature and torsion [15]. In the present paper, however, we investigate only the Fairchild subclass of quadratic lagrangians [14]

(1.5) $$L = AeR + (1/4)BeR^{(\alpha)(\beta)}{}_{\mu\nu} R_{(\alpha)(\beta)}{}^{\mu\nu} ,$$

where A and B are non-vanishing constants.

Two limit cases of the lagrangian (1.5) give rise to the ECSK theory (B = 0) and to the Stephenson-Kilmister-Yang theory (A = 0). The latter case has recently been investigated profoundly in Ref. [21]. This paper clarifies the gauge structure of the SKY theory that is much more richer than that of a generic SO(3,1)-gauge theory of gravity.

Our goal is to find the complete system of constraints for Fairchild's theory, its dynamical variables and their time evolution, to prove the time-maintenance of the constraints, explain the role of gauge tranformations and gauge variables as well as to count independent degrees of freedom. In the result we get 16 independent degrees of freedom (in the phase space) for the Fairchild theory, comparing with 10 degrees of freedom in SKY gravity and 4 in the Einstein theory.

2. FIELD EQUATIONS IN THE FAIRCHILD THEORY. For the lagrangian (1.5) the tetrad and connection momenta are

(2.1a) $U^{\lambda\mu}{}_{(\alpha)} = 0$,

(2.1b) $P^{\lambda\mu}{}_{(\alpha)(\beta)} = Ae(e^{\lambda}{}_{(\alpha)}e^{\mu}{}_{(\beta)} - e^{\lambda}{}_{(\beta)}e^{\mu}{}_{(\alpha)}) + BeR_{(\alpha)(\beta)}{}^{\lambda\mu}$.

We have the following set of field equations

(2.2) $(E1)^{\mu}{}_{(\alpha)} = \delta L/\delta e^{\mu}{}_{(\alpha)} = - 2AeG^{\mu}{}_{(\alpha)} +$
$+ Be[\ 1/4\ e^{\mu}{}_{(\alpha)}R^{(\rho)(\sigma)}{}_{\varepsilon\omega}R_{(\rho)(\sigma)}{}^{\varepsilon\omega} - R^{(\rho)(\sigma)\mu\varepsilon}R_{(\rho)(\sigma)(\alpha)\varepsilon}\ ] = 0$,

(2.3) $(E2)^{\mu}{}_{(\alpha)(\beta)} = - D_{\lambda}P^{\lambda\mu}{}_{(\alpha)(\beta)} = 0$.

We observe that the second term in (2.2) is a symmetric traceless quantity. Therefore for solutions of the Fairchild theory the Ricci and Einstein tensors are symmetric and traceless.

For a given slicing of spacetime into a family of 3 dimensional surfaces of constant x^{o} we introduce the caret variables according to the rules presented in Ref. [9] and [18]. We have from (2.1)

(2.4a) $\hat{P}^{ok}{}_{(o)(b)} = A\hat{e}\hat{e}^{k}{}_{(b)} - Be\hat{\hat{R}}_{(o)(b)oq}\bar{g}^{qk}$,
$\hat{P}^{ok}{}_{(a)(b)} = - Be\hat{\hat{R}}_{(a)(b)oq}\bar{g}^{qk}$;

(2.4b) $\hat{P}^{mn}{}_{(o)(b)} = Be\hat{\hat{R}}_{(o)(b)}{}^{mn}$,
$\hat{P}^{mn}{}_{(a)(b)} = Ae(\hat{e}^{\hat{m}}{}_{(a)}\hat{e}^{n}{}_{(b)} - \hat{e}^{m}{}_{(b)}\hat{e}^{n}{}_{(a)}) + Be\hat{\hat{R}}_{(a)(b)}{}^{mn}$.

According to the general canonical analysis of SO(3,1)-gauge theories of gravity presented in Ref.[9] we have the following set of canonical variables

(2.5) positions $\hat{e}^{(a)}{}_{k}$, $\hat{\Gamma}_{k}{}^{(\alpha)(\beta)}$, $n^{(a)}$;
and momenta $\hat{U}^{ok}{}_{(a)}$, $\hat{P}^{ok}{}_{(\alpha)(\beta)}$, $M_{(a)}$.

The primary constraints (2.1a) eliminate the momenta canonically conjugate to the triads $\hat{e}^{(a)}{}_{k}$ and these quantities drop out from the expression for the symplectic 2-form Ω . We have (cf. [9])

(2.6) $\Omega(X_1,X_2) =$
$= \int_{\sigma}[\ \delta_1\hat{P}^{ok}{}_{(\alpha)(\beta)}\wedge\delta_2\hat{\Gamma}_{k}{}^{(\alpha)(\beta)} + \delta_1M_{(a)}\wedge\delta_2n^{(a)}\ ]dx^1\wedge dx^2\wedge dx^3$.

Taking into account that $n^{(a)}$ are the gauge variables of boost transformations and that the dynamics of their conjugate momenta $M_{(a)}$ is trivial, that is ,

(2.7) $M_{(a)} = 0$ (cf [9])

we have for the F-theory only 36 genuine dynamical symplectic variables

(2.8) $\hat{P}^{ok}{}_{(\alpha)(\beta)}, \quad \hat{\Gamma}_k{}^{(\alpha)(\beta)}$.

From relations (2.4) we can compute the time-components of the Riemann tensor. Moreover, these components represents the time-derivatives of the connection coefficients $\hat{\Gamma}_k{}^{(\alpha)(\beta)}$ (cf. [18]). Therefore we have the following dynamical equations for the canonical positions

(2.9)
$$\hat{D}_o\hat{\Gamma}_k{}^{(o)(b)} = (\hat{D}_k + \partial_k \ln N)\hat{\Gamma}_o{}^{(o)(b)} - (1/B\hat{e})\hat{P}^{on(o)(b)}\bar{g}_{nk} - (A/B)\hat{e}^{(b)}{}_k,$$
$$\hat{D}_o\hat{\Gamma}_k{}^{(a)(b)} =$$
$$= - \hat{\Gamma}_o{}^{(a)}{}_{(o)}\hat{\Gamma}_k{}^{(o)(b)} + \hat{\Gamma}_k{}^{(a)}{}_{(o)}\hat{\Gamma}_o{}^{(o)(b)} - (1/B\hat{e})\hat{P}^{on(a)(b)}\bar{g}_{nk} .$$

The dynamics of the canonical momenta is determined by the equations $(\hat{E2})^k{}_{(\alpha)(\beta)} = 0$. In the explicit form they read

(2.10) $$- \dagger\hat{D}_o\hat{P}^{ok}{}_{(\alpha)(\beta)} - (\dagger\hat{D}_s + \partial_s \ln N)\hat{P}^{sk}{}_{(\alpha)(\beta)} = 0 ,$$

where the dagger-derivatives $\dagger\hat{D}_\lambda$ are defined in the Appendix.

In the present Section we have explained the role of the spatial components of the connection $\hat{\Gamma}_k{}^{(\alpha)(\beta)}$, they are the dynamical symplectic variables. On the other hand, we know from the results of Ref. [9],[12],[18] that the quantities $\hat{\Gamma}_o{}^{(a)(b)}$ are SO(3)-gauge variables. Therefore we have to elucidate the meaning of three remaining connection coefficients $\hat{\Gamma}_o{}^{(a)(o)}$ as well as the role of triads components $\hat{e}^{(a)}{}_k$. These questions are answered in subsequent Sections.

3. TRANSLATIONAL FIELD EQUATIONS IN THE (3+1)-DECOMPOSITION.

The translational field equations (2.2) written in terms of caret variables split in three subsets

(3.1) $(\hat{E1})^o{}_{(\alpha)} = 0$,

(3.2) $(\hat{E1})^k{}_{(a)} = 0$,

(3.3) $(\hat{E1})^k{}_{(o)} = 0$.

Similarly as in the SKY theory [21] it is very convenient to introduce the following quantities

$$(3.4)\qquad X^{k}{}_{(\alpha)(\beta)} = \hat{P}^{ok}{}_{(\alpha)(\beta)}\,, \qquad Y^{k(\alpha)(\beta)} = 1/2\ B\varepsilon^{kuv}\hat{R}^{(\alpha)(\beta)}{}_{uv}$$

resembling the electric and magnetic fields of Yang-Mills theories. Also the Einstein-Cartan triad vector density [18]

$$(3.5)\qquad \zeta^{k}{}_{(a)} = 2\hat{e}\hat{e}^{k}{}_{(a)}$$

is very useful in subsequent considerations. The quantities (3.4) can be expressed by means of the symplectic variables (2.8) and their spatial derivatives but so far the role of the triad density (3.5) remains unclear.

Let us observe that instead of the system consisting of the first equation in (3.1) and 9 equations in (3.2) we may consider the equivalent system

$$(3.6a)\qquad (\hat{E1})^{o}{}_{(o)} = 0\,,$$

$$(3.6b)\qquad \hat{B}e[\ (\hat{E1})^{ks} + \bar{g}^{ks}(\hat{E1})^{o}{}_{(o)} - 1/2\ \bar{g}^{ks}(\hat{E1})^{\lambda}{}_{\lambda}\ ] = 0\,.$$

Taking into account (2.4) and (3.4) we rewrite (3.6b) in the following elegant form

$$(3.6b')\qquad (\ X^{k}{}_{(\alpha)(\beta)}X^{s(\alpha)(\beta)} + Y^{k(\alpha)(\beta)}Y^{s}{}_{(\alpha)(\beta)}\ ) + A(\ X^{k}{}_{(o)}{}^{(c)} + (1/2)Y^{k}{}_{(m)(n)}\varepsilon^{(m)(n)(c)}\ )\zeta^{s}{}_{(c)} = 0\,.$$

Now we assume that on the initial surface σ the 3×3 matrix

$$(3.7)\qquad I^{k(c)} = X^{k}{}_{(o)}{}^{(c)} + (1/2)Y^{k}{}_{(m)(n)}\varepsilon^{(m)(n)(c)}$$

is invertible. Moreover, we assume that the following orientability condition holds

$$(3.8)\qquad -A^{-3}(\det[I^{k(c)}]^{-1}\cdot\det[X^{k}{}_{(\alpha)(\beta)}X^{s(\alpha)(\beta)} + Y^{k}{}_{(\alpha)(\beta)}Y^{s(\alpha)(\beta)}] = \det[\zeta^{k}{}_{(c)}] > 0\,.$$

In such a situation (3.6b') gives us a system of algebraic equations for $\zeta^{k}{}_{(c)}$ and we write

$$(3.9)\qquad \zeta^{k}{}_{(a)} = -(1/A)I^{-1}{}_{(a)r}(\ X^{k}{}_{(\alpha)(\beta)}X^{r(\alpha)(\beta)} + Y^{k}{}_{(\alpha)(\beta)}Y^{r(\alpha)(\beta)}\ ).$$

By virtue of the relation $\hat{e} = (1/2\sqrt{2})(\ \det[\zeta]^{k}{}_{(a)}\ )^{1/2}$ we can compute $\hat{e}^{k}{}_{(a)}$ and $\hat{e}^{(a)}{}_{k}$ from (3.9). Therefore in the F-theory triads of vectors (covectors) on the surfaces of a slicing are algebraic functions of the symplectic variables (2.8) and their spatial derivatives.

The translational hamiltonian constraints read

$$(3.10)\quad \hat{e}B(\hat{E}1)^{o}{}_{(o)} = - [\, 1/2(X^{k(\alpha)(\beta)}X_{k(\alpha)(\beta)} + Y^{k(\alpha)(\beta)}Y_{k(\alpha)(\beta)}) + 3A^{2}\hat{e}^{2}\,] = 0\ ,$$

$$(3.11)\quad B(\hat{E}1)^{o}{}_{q} = - \varepsilon_{qmn}X^{m}{}_{(\alpha)(\beta)}Y^{n(\alpha)(\beta)} = 0\ .$$

Equations (3.3) read

$$(3.12)\quad B(\hat{E}1)^{k}{}_{(o)} = - B(\hat{E}1)^{ok} + A\bar{g}^{kr}\varepsilon_{rpq}J^{p(a)}\zeta^{q}{}_{(a)} = 0\ .$$

where

$$(3.13)\quad J^{p(a)} = Y^{p(o)(a)} + (1/2)X^{p}{}_{(m)(n)}\varepsilon^{(m)(n)(a)}\ .$$

It follows from (3.11) and (3.12) that

$$(3.14)\quad J^{pq} - J^{qp} = 0 \quad \text{or equivalently} \quad \varepsilon_{rpq}J^{p(a)}\zeta^{q}{}_{(a)} = 0\ .$$

In Section 5 we show that symplectic constraints (3.14) are the time-conservation conditions for second class constraints $(\hat{E}2)^{o}{}_{(a)(o)} = 0$.

4. ROTATIONAL FIELD EQUATIONS IN THE (3+1)-DECOMPOSITION. We have two subsystems of rotational equations

$$(4.1a)\quad (\hat{E}2)^{o}{}_{(a)(b)} = 0,$$

$$(4.1b)\quad (\hat{E}2)^{o}{}_{(o)(b)} = 0,$$

and

$$(4.2)\quad (\hat{E}2)^{k}{}_{(\alpha)(\beta)} = 0\ .$$

Equations (4.2) are the dynamical equations for $X^{k}{}_{(\alpha)(\beta)}$. They read

$$(4.2')\quad \dagger\hat{D}_{o}X^{k}{}_{(\alpha)(\beta)} = \varepsilon^{kuv}(\dagger\hat{D}_{u} + \partial_{u}\ln N)(Y^{n}{}_{(\alpha)(\beta)}\bar{\underset{\sim}{g}}_{nv}) + - \hat{e}A(\dagger\hat{D}_{s} + \partial_{s}\ln N)(\hat{e}^{s}{}_{(\alpha)}\hat{e}^{k}{}_{(\beta)} - \hat{e}^{s}{}_{(\beta)}\hat{e}^{k}{}_{(\alpha)}).$$

Here $\bar{\underset{\sim}{g}}_{pq} = (1/\sqrt{g})\bar{g}_{pq}$ is a tensor density of weight -1 .

Analogous dynamical equations hold for the quantities $Y^{k(\alpha)(\beta)}$. Taking into account (2.9) we get

$$(4.3)\quad \dagger\hat{D}_{o}Y^{k(\alpha)(\beta)} = - \varepsilon^{kuv}(\dagger\hat{D}_{u} + \partial_{u}\ln N)(X^{n(\alpha)(\beta)}\bar{\underset{\sim}{g}}_{nv}) + + A\varepsilon^{kuv}(\dagger\hat{D}_{u} + \partial_{u}\ln N)(\hat{e}^{(\alpha)}{}_{v}\hat{e}^{(\beta)}{}_{o} - \hat{e}^{(\beta)}{}_{v}\hat{e}^{(\alpha)}{}_{o}).$$

Equations (4.1a) read

(4.1a') $\hat{\dagger D}_k X^k{}_{(a)(b)} = 0$.

They are SO(3)-constraints (cf.[9],[12]).

Equations (4.1b) give rise to 3 additional symplectic constraints

(4.1b') $\hat{\dagger D}_k X^k{}_{(a)(o)} = 0$.

We would like to emphasize that (4.1a') are not the boost constraints. We know from Ref.[9] and [12] that those are the equations (2.7), which in our discussion are solved. That is to say, the momenta $M_{(a)}$ are eliminated from the dynamical picture.

Constraints (4.1b') are not generic for an arbitrary SO(3,1)-gauge theory of gravity. For instance, equations (4.1b) are not symplectic constraints in the ECSK theory [18]. However, for gravitational lagrangians purely quadratic in curvature (no torsion terms) we always have 3 additional symplectic constraints of type (4.1b').

Now we have four translational constraints, three SO(3)-constraints and six additional constraints (3.14) and (4.1b') for the reduced system of 36 symplectic variables $X^k{}_{(\alpha)(\beta)}$, $\hat{\Gamma}_k{}^{(\alpha)(\beta)}$ (or for $X^k{}_{(\alpha)(\beta)}$, $Y^{k(\alpha)(\beta)}$). In the next Section we will prove that this set gives us all constraints and that they are preserved during the evolution.

Let us observe that the quantities $Y^{k(\alpha)(\beta)}$ satisfy the system

(4.4) $\hat{\dagger D}_k Y^{k(\alpha)(\beta)} = 0$.

These relations, however, follow from the Bianchi identities $D_{[\lambda} R^{(\alpha)(\beta)}{}_{\mu\nu]} = 0$ and so do not bring any new information.

5. TIME CONSERVATION OF CONSTRAINTS AND INDEPENDENT DEGREES OF FREEDOM. Thirteen symplectic constraints in the F-theory can be divided in three subsets.

(i) four translational constraints (3.10) and (3.11),

(ii) three rotational constraints (4.1a'),

(iii) six additional constraints (3.14) and (4.1b').

Constraints (3.10), (3.11) and (4.1a') are related to the invariance of the theory with respect to the actions of Diff M and the local SO(3)-group, respectively. Therefore they are preserved in time. As a matter of fact, if we $\dagger\hat{D}_o$-differentiate the left-hand sides of these constraints and make use of the dynamical equations (4.2) and (4.3) then we obtain linear combinations of the left hand sides of (3.10), (3.11) and (4.1a').

If we time-differentiate the left-hand sides of (4.1b') then we are led to constraints (3.14). This property becomes clear if we recall that for a general SO(3,1)-theory the time-conservation conditions for (4.1b) are (3.3) (cf.[9],[12]).

Now we investigate the conditions

$$\hat{D}_o(J^{pq} - J^{qp}) = 0. \tag{5.1}$$

This time-differentiation leads to linear algebraic equations for unknown quantities $\hat{\Gamma}_o{}^{(c)}{}_{(o)}$. Therefore these 3 quantities can be computed as algebraic functions of the symplectic variables and their spatial derivatives. In an explicit form (5.1) read

$$[\,\delta^{(c)}{}_{(j)}\mathrm{tr}Z - Z^{(c)}{}_{(j)}\,]\hat{\Gamma}_o{}^{(o)}{}_{(c)} = \tag{5.2}$$
$$= \text{expression of } X^k{}_{(\alpha)(\beta)}\,,\ Y^{k(\alpha)(\beta)}\,,\ \text{their } x^k\text{-derivatives, and } \partial_k \ln N\,,$$

where

$$Z^{(i)}{}_{(j)} = 4A\hat{e}(X^{\bullet}{}_{(\alpha)(\beta)}X^{\bullet(\alpha)(\beta)} + Y^{\bullet}{}_{(\alpha)(\beta)}Y^{\bullet(\alpha)(\beta)})^{-1}_{pq} \cdot(\,I^{p(i)}I^q{}_{(j)} + J^{p(i)}J^q{}_{(j)}\,). \tag{5.3}$$

Z is a symmetric matrix. If its eigenvalues satisfy the conditions $\lambda_1 + \lambda_2 \neq 0$, $\lambda_1 + \lambda_3 \neq 0$, and $\lambda_2 + \lambda_3 \neq 0$, then equations (5.3) can be uniquely solved for $\hat{\Gamma}_o{}^{(o)}{}_{(c)}$ and these quantities are algebraic functions of $X^k{}_{(\alpha)(\beta)}$, $Y^{k(\alpha)(\beta)}$ and their spatial derivatives.

Thus, in Fairchild gravity 36 dynamical variables $X^k{}_{(\alpha)(\beta)}$, $\hat{\Gamma}_k{}^{(\alpha)(\beta)}$ are subject to 13 constraints. The dynamics is consistent with constraints. The triad coefficients $\hat{e}^{(a)}{}_k$ as well as the quantities $\hat{\Gamma}_o{}^{(o)}{}_{(c)}$ are algebraic functions of the dynamical variables and their spatial derivatives (up to second order).

The constraints (i) are generators of an action of the diffeomorphism group of spacetime in the set of field variables. We refer the reader to Ref.[12], where a discussion of the standard and an SO(3,1)-covariant action of DiffM has been presented. The constraints (ii) generate the standard action of SO(3)-rotations (cf.[9],[12]). These seven constraints are first class in Dirac's classification. The remaining six constraints (iii) are second class. It is to be expected because their time-conservation conditions lead either to new constraints or to algebraic equations (5.2) for the quantities $\hat{\Gamma}_o{}^{(o)}{}_{(c)}$. Thus, 36 dynamical variables (2.8) are constrained by 13 equations and subject to an action of a seven parameter group. Therefore we expect to have 36 - (13+7) = 16 independent degrees of freedom in the phase space.

More precise analysis of the space of independent degrees of freedom requires methods of symplectic geometry and the theory of elliptic differential operators on compact (non-compact) 3-manifolds. It can be performed adopting results of Ref.[22-25] and [19].

In some sense Fairchild's theory gives us a generic example in the class of theories with quadratic gravitational lagrangian without torsion terms. As it has recently been shown by the present author (unpublished) for such lagrangians we have typically 16 degrees of freedom in the phase space. Of course, this statement does not exclude other more degenerate cases with smaller numbers of degrees of freedom. One example is provided by the ECSK theory with four degrees of freedom, another one by SKY gravity with ten. This theory has been investigated in Ref.[21], where it has been proven that the connection components $\hat{\Gamma}_{o}{}^{(o)}{}_{(c)}$ cannot be determined from the field equations and that they are gauge variables corresponding to a non-standard action of Lorentz boosts. This additional gauge invariance causes that equations (4.1b') are first class constraints. Therefore in the set of 36 dynamical variables (2.8) we have an action of a 10-parameter group (4 parameters of DiffM, 3 of standard SO(3)-rotations and 3 of the non-standard action of boosts). Moreover, in the SKY gravity equations (3.6b') do not determine the triad densities $\zeta^{k}{}_{(a)}$ but they give us six symplectic constraints, which include the energy constraint (3.10). Together we have 15 constraints, 10 of them ((3.11),(4.1a',b') and the trace of (3.6b')) are first class and five contraints given by the traceless part of (3.6b') are second class. The odd number of second class constraints could suggest contradiction but there is another sixtienth (second class) constraint, which ensures consistency of the dynamical picture. We have 36 - (16+10) = 10 independent degrees of freedom in the phase space. More about SKY gravity and a method of how to determine the triads (or 3-metric) by means of the dynamical variables can be found in Ref.[21].

Acknowledgments: The author thanks Professor V.Moncrief for his continuous support and the organizers of the AMS gravitational conference for an opportunity to present these results.

APPENDIX: SO(3)-COVARIANT DAGGER-DERIVATIVES. The (3+1)-decomposition of SO(3,1)-object-valued tensor fields on spacetime has been presented in Ref. [9], [18] and in Ref. [12] for the spinor language. In the present paper we follow the notation of Ref.[18]. In particular SO(3)-covariant derivatives $\hat{D}_k$ and $\hat{D}_o$ are defined by formulas (D.34-D.37) of Ref. [18], where they have been denoted by Script characters. In the present paper we also use the dagger SO(3)-covariant derivatives $^{\dagger}\hat{D}_{\lambda}$ that differ essentially from $\hat{D}_{\lambda}$. In particular, for an SO(3,1)-tensor-valued spacetime tensor density of weight r

$F^{(\alpha)}{}_{(\beta)}{}^{\mu}{}_{\nu}$ we have in terms of its caret components

$$(A.1)\quad {}^{\dagger}\hat{D}_{\lambda}\hat{F}^{(\alpha)}{}_{(\beta)}{}^{\mu}{}_{\nu} = \hat{D}_{\lambda}\hat{F}^{(\alpha)}{}_{(\beta)}{}^{\mu}{}_{\nu} + \delta^{(\alpha)}{}_{(o)}\hat{\Gamma}_{\lambda}{}^{(o)}{}_{(c)}\hat{F}^{(c)}{}_{(\beta)}{}^{\mu}{}_{\nu} + \delta^{(\alpha)}{}_{(a)}\hat{\Gamma}_{\lambda}{}^{(a)}{}_{(o)}\hat{F}^{(o)}{}_{(\beta)}{}^{\mu}{}_{\nu} + - \delta^{(o)}{}_{(\beta)}\hat{\Gamma}_{\lambda}{}^{(d)}{}_{(o)}\hat{F}^{(\alpha)}{}_{(d)}{}^{\mu}{}_{\nu} - \delta^{(b)}{}_{(\beta)}\hat{\Gamma}_{\lambda}{}^{(o)}{}_{(b)}\hat{F}^{(\alpha)}{}_{(o)}{}^{\mu}{}_{\nu}.$$

We observe that for objects with purely holonomic indices the ${}^{\dagger}\hat{D}_{\lambda}$ and $\hat{D}_{\lambda}$ operators coincide. The importance of the dagger derivatives lies in their elegant commutation rules

$$(A.2)\quad [{}^{\dagger}\hat{D}_i, {}^{\dagger}\hat{D}_j]\,\hat{F}^{(\alpha)}{}_{(\beta)}{}^{\mu}{}_{\nu} = \hat{R}^{(\alpha)}{}_{(\tau)ij}\hat{F}^{(\tau)}{}_{(\beta)}{}^{\mu}{}_{\nu} - \hat{R}^{(\sigma)}{}_{(\beta)ij}\hat{F}^{(\alpha)}{}_{(\sigma)}{}^{\mu}{}_{\nu} + + \delta^{\mu}{}_{m}\,{}^{R}\bar{R}^{m}{}_{pij}\hat{F}^{(\alpha)}{}_{(\beta)}{}^{p}{}_{\nu} - \delta^{n}{}_{\nu}\,{}^{R}\bar{R}^{q}{}_{nij}\hat{F}^{(\alpha)}{}_{(\beta)}{}^{\mu}{}_{q}.$$

Here ${}^{R}\bar{R}^{m}{}_{pij}$ is curvature of the Riemannian connection $\bar{\gamma}_k{}^p{}_q$ on σ.

$$(A.3)\quad [{}^{\dagger}\hat{D}_o, {}^{\dagger}\hat{D}_j]\,\hat{F}^{(\alpha)}{}_{(\beta)}{}^{\mu}{}_{\nu} = = \partial_j \ln N\, {}^{\dagger}\hat{D}_o\hat{F}^{(\alpha)}{}_{(\beta)}{}^{\mu}{}_{\nu} + \hat{R}^{(\alpha)}{}_{(\tau)oj}\hat{F}^{(\tau)}{}_{(\beta)}{}^{\mu}{}_{\nu} - \hat{R}^{(\sigma)}{}_{(\beta)oj}\hat{F}^{(\alpha)}{}_{(\sigma)}{}^{\mu}{}_{\nu} + - r\lambda_j{}^p{}_p\,\hat{F}^{(\alpha)}{}_{(\beta)}{}^{\mu}{}_{\nu} + \delta^{\mu}{}_{m}\lambda_j{}^{m}{}_{p}\hat{F}^{(\alpha)}{}_{(\beta)}{}^{p}{}_{\nu} - \delta^{n}{}_{\nu}\lambda_j{}^{q}{}_{n}\hat{F}^{(\alpha)}{}_{(\beta)}{}^{\mu}{}_{q}.$$

Here $\lambda_j{}^q{}_p$ are the following Christofel-like tensors

$$(A.4)\quad \lambda_j{}^q{}_p = = 1/2\bar{g}^{qu}[(\hat{D}_j + \partial_j \ln N)\hat{D}_o\bar{g}_{up} + (\hat{D}_p + \partial_p \ln N)\hat{D}_o\bar{g}_{uj} - (\hat{D}_u + \partial_u \ln N)\hat{D}_o\bar{g}_{jp}].$$

A complete description of dagger-differentiations can be found in Ref.[21].

BIBLIOGRAPHY

1. E.S.Fradkin and A.A.Tseytlin, "Renormalizable asymptotically free quantum theory of gravity," Nucl.Phys. B, 201 (1982),469-481.

2. N.H.Barth and S.M.Christensen, "Quantizing fourth-order gravity theories: The functional integral," Phys.Rev. D, 28 (1983), 1876-1893.

3. E.T.Tomboulis, "Renormalization and asymptotic freedom in quantum gravity," in Quantum Theory of Gravity, edited by S.M Christensen, A.Hilger, Bristol, 1984.

4. D.G.Boulware, "Quantization of higher derivative theories of gravity," in Quantum Theory of Gravity, edited by S.M. Christensen, A.Hilger, Bristol, 1984.

5. I.Antoniadis and E.Tomboulis, "Gauge invariance and unitarity in higher-derivative quantum gravity," Phys.Rev. D, 33 (1986), 2756-2779.

6. V.Szczyrba, "Hamiltonian dynamics of higher-order theories of gravity," J.Math.Phys. (to appear February 1987).

7. A.Trautman, "On the structure of the Einstein-Cartan equations," Symposia Mathematica, 12 (1973), 139-162;
"Yang-Mills theory and gravitation. A comparision," in Geometric Techniques in Gauge Theories, edited by M.Martini and E.De Jager, Lecture Notes in Mathematics 926, Springer, Berlin, 1982.

8. F.W.Hehl, J.Nitsch and P.von der Heyde, "Gravitation and the Poincaré gauge theory with quadratic lagrangian," in General Relativity and Gravitation, edited by A. Held, Plenum, New York, 1980.

9. W.Szczyrba, "Hamiltonian dynamics of gauge theories of gravity," Phys.Rev. D, 25 (1982), 2548-2568.

10. A.Tseytlin, "Poincaré and De Sitter gauge theories with propagating torsion," Phys.Rev. D, 26 (1982), 3327-3341.

11. M.Blagoević and I.Nicolić, "Hamiltonian dynamics of Poincaré gauge theory I," Phys.Rev. D, 28 (1983), 2455-2463;
I.Nicolić, "Dirac hamiltonian structure of $R + R^2 + T^2$," Phys.Rev. D, 30 (1984), 2508-2520;
M.Blagoević and M.Vasilić, "Hamiltonian analysis of extra gauge symmetries in $R + T^2$ theory of gravity," Phys.Rev. D, 34 (1986), 357-366.

12. M.Antonowicz and W.Szczyrba, "Spinor matter fields in SL(2,C)-gauge theories of gravity," Phys.Rev. D, 31 (1985), 3104-3129.

13. C.N.Yang, "Integral formalism for gauge fields," Phys.Rev.Lett., 33 (1974), 445-447.

14. E.E.Fairchild, Jr, "Gauge theory of gravitation," Phys.Rev. D, 14 (1976), 384-391, Errata 2833;
"Yang-Mills formulation of gravitational dynamics", Phys.Rev. D, 16 (1977), 2438-2447.

15. E.Sezgin and P.van Nieuwenhuizen, "New ghost-free gravity lagrangians with propagating torsion," Phys.Rev. D, 21 (1980), 931-933.

16. R.Rauch and H.T.Nieh, "Birkhoff's theorem for general Riemann-Cartan type $R + R^2$ theories of gravity," Phys.Rev. D, 24 (1981), 2029-2048.

17. P.Baekler, F.W.Hehl and E.W.Mielke, "Vacuum solutions with double duality properties of a quadratic Poincaré gauge field theory," in Proceedings of the Second Marcel Grossmann Meeting on General Relativity, edited by R. Ruffini, North-Holland, Amsterdam, 1982;
P.Baekler, F.W.Hehl and H.J.Lenzen, "Vacuum solutions with double duality properties of the Poincaré gauge field theory II," in Proceedings of the Third Marcel Grossmann Meeting on General Relativity, edited by Hu Ning, Science Press, North-Holland, Beijing, Amsterdam, 1983.

18. W.Szczyrba, "The dynamical structure of the Einstein-Cartan-Sciama-Kibble theory of gravity," Ann.Phys.(N.Y), 158 (1984), 320-373.

19. M.Antonowicz and W.Szczyrba, " An SU(2)-covariant dynamical formulation of the Einstein-Cartan-Dirac theory," Class.Quant.Grav., 2 (1985), 515-533.

20. F.W.Hehl, P.von der Heyde, G.D.Kerlick, J.M.Nester, "General

relativity with spin and torsion," Rev.Mod.Phys., 48 (1976), 393-416.

21. V.Szczyrba, "Stephenson-Kilmister-Yang theory of gravity and its dynamics," preprint, Yale University, 1986.

22. V.Moncrief, "Decomposition of gravitational perturbations," J.Math.Phys., 16 (1975),1556-1560.

23. W.Szczyrba, "A symplectic structure of the set of Einstein metrics," Comm.Mathh.Phys. 51 (1976), 163-182.

24. A.Fischer and J.Marsden, "The initial value problem and the dynamical formulation of general relativity," in General Relativity - An Einstein Centenary Survey, edited by S.W. Hawking and W. Israel, Cambridge U.P., Cambridge, 1979.

25. M.Cantor, "Elliptic operators and decomposition of tensor fields," Bull.Am.Math.Soc., 5 (1981), 235-262.

Victor Szczyrba
Department of Physics
Yale University
New Haven, Connecticut 06511

Contemporary Mathematics
Volume **71**, 1988

STABILITY IN DISSIPATIVE RELATIVISTIC FLUID THEORIES

William A. Hiscock[1] and Lee Lindblom[2]

ABSTRACT. This paper examines the problem of finding a theory that describes the effects of dissipation (viscosity and thermal conductivity) in a fully relativistic fluid. Several of the proposed theories, including those of Eckart, Landau-Lifshitz, Havas-Swenson, and Israel-Stewart, are examined. A number of difficulties have been identified with these theories in the literature: non-causal propagation of signals, poorly posed dynamical evolution of initial data, and generic instability of the equilibrium states. This paper describes how the stability of the equilibrium states can be analyzed in this entire class of theories. It is shown that all of the "first-order" theories (Eckart, Landau-Lifshitz, and Havas-Swenson) have very short timescale instabilities in every equilibrium state. These first-order theories are consequently inadequate. The second-order theories (Israel-Stewart) in contrast can have stable equilibria. Furthermore, it is shown that the conditions needed for these theories to have stable equilibrium states are equivalent to the conditions needed to guarantee that perturbations propagate causally via hyperbolic differential equations. Thus the second-order theories appear to be promising candidates for an acceptable theory of dissipative relativistic fluids.

1980 Mathematics Subject Classification (1985 Revision). 83C55, 76E99, 82A35, 76Y05.

[1]Supported by the National Science Foundation Grant #PHY 85-05484.

[2]Supported by the National Science Foundation Grant #PHY 85-18490.

1. INTRODUCTION

As is well-known, in the Fourier theory of heat flow, temperature fluctuations propagate via a parabolic equation. In a non-relativistic (Newtonian) theory this is acceptable, if somewhat of a curiosity (Maxwell[1] was already concerned about the resulting infinite propagation velocity in 1867), since Newtonian physics does not possess a maximum velocity for the transmission of information. It does hint, however, that it may be difficult to create an acceptable theory of dissipative relativistic fluids, in which all fluid variables obey hyperbolic equations, and all transmission of information through the fluid occurs inside the light cone (i.e., at velocities less than the speed of light). The purpose of this paper is to review recent work aimed at finding the simplest acceptable truly relativistic theory of dissipative fluids. The unifying theme of the present discussion is the question of the stability of the equilibrium states in various proposed theories. Many of the theories have *no* stable equilibrium states at all!

The first attempts at creating a theory of relativistic dissipative fluids are now called "first-order" theories; in these theories the definition of the entropy current contains no terms of higher than first order in the deviations from equilibrium (heat flow, viscous stresses, etc.). The simplest such theories are those of Eckart[2] and Landau and Lifshitz[3], which are presented in many textbooks on relativistic physics[3,4,5]. These theories are the simplest covariant generalizations of the Navier-Stokes-Fourier theory of Newtonian dissipative fluids. It is also possible to construct more complicated theories which are still first-order in this sense; the theory created by Havas and Swenson[6] may be the most general such theory possible.

It is also possible to create either Newtonian or relativistic second-order theories, where the entropy current definition is extended to include terms quadratic in the deviations from equilibrium. The kinetic theory version of such a Newtonian second-order theory was first studied in 1949 by Grad[7]; the associated second-order Newtonian phenomenological fluid theory was later developed by Müller[8]. Around 1970, a number of workers began to extend Grad's work to the relativistic regime[9-11]. The first relativistic second-order phenomenological fluid theory was described by Israel[12]; later work of Stewart[13] and Israel[14-16] clarified the relation between the kinetic theory and phenomenological approaches, and extended the theories to more complicated situations. Section 2 of this paper reviews how the first- and second-order theories are constructed.

It has been known for some time that there are definite problems with the first-order theories. Under certain special conditions, a parabolic equation may be obtained for the propagation of thermal fluctuations in the simpler (Eckart and Landau-Lifshitz) first-order theories[16]. This raised serious doubts as to whether the first-order theories should be considered truly "relativistic", but in itself is not conclusive, for the following reasons. First, an analysis of the propagation of fluctuations for the complete theory (with both viscosity and thermal conductivity nonzero) has not been completed. Second, in the more complicated, but still first-order, Havas-Swenson theory, thermal fluctuations obey a hyperbolic equation[17], at least in the Newtonian limit.

The question of the stability of equilibrium states in the Eckart theory has also been raised[18]. It was shown that instabilities driven by thermal conductivity exist in the Eckart theory, but only for perturbations with length scales much smaller than is physically meaningful (i.e., smaller than the interparticle separation in the fluid).

While the results mentioned above are suggestive of serious problems in the first-order theories, they are not conclusive. These potential problems (the stability of equilibrium states, and the causality of linear wave propagation) may be addressed by studying the linearized equations of motion for small perturbations about an equilibrium state of the fluid. Determining the speed at which information can be transmitted through the fluid is a very complicated problem. For all of the first-order theories, the system of equations which govern the evolution of linear perturbations is neither purely hyperbolic, nor parabolic, nor elliptic. Little seems to be known about the properties of such complicated systems[19]. In particular, it is not clear whether a well-posed initial value problem exists for this type of system of equations. As a result, the analysis of signal propagation to determine the causal properties of the first-order fluids is even more complicated than in the dispersive, dissipative electromagnetic case[20]. The stability of equilibrium states, on the other hand, can be studied by a simple plane-wave normal mode analysis of the perturbation equations. Such an analysis has been performed and it has been found that all of these first-order theories predict rapid evolution away from an arbitrary equilibrium state[21,22]. In other words, *every* equilibrium state in *every* first-order theory is unstable in the sense that small spatially bounded departures from equilibrium at one instant in time grow exponentially with time. The time scales for these instabilities can be ridiculously short: for example, water at room temperature and pressure has an instability with a growth time scale of about 10^{-34} seconds in these theories. Since these theories predict nonsensical behavior for phenomena which should be well within their range of applicability, we feel that these first-order theories are unacceptable. Section 3 of this paper reviews the stability analysis of these theories.

The stability of equilibrium states and the causality of signal propagation in the second-order Israel-Stewart[23] theory have also been investigated. The major result of this study was that the conditions necessary for equilibrium states to be stable in the second-order theories are equivalent to the conditions needed to guarantee that perturbations propagate causally and obey a hyperbolic set of equations. Thus, stability implies causality and hyperbolicity for perturbative waves. The converse is also true (although a particular definition of hyperbolicity must be imposed); any Israel-Stewart fluid with perturbations which propagate causally and obey a set of hyperbolic equations will possess stable equilibrium states. Section 4 of this paper reviews the stability and causality analysis of the second-order theories.

In addition, several new results not contained in Ref. (23) are discussed in Section 4 and an accompanying appendix. In Section 4 we show that causality and a naive version of hyperbolicity do not imply stability. In particular, it is not sufficient to simply assume that the characteristic velocities are all real (a more naive notion of hyperbolicity) and less than the speed of light; this is demonstrated by explicitly constructing a counterexample in which all the characteristic velocities are real and bounded between zero and one (the speed of light in our units), yet the equilibrium states are unstable. In an appendix we present a new result which widens the class of equilibrium states of Israel-Stewart fluids for which stability can be rigorously analyzed. Stability is analyzed via an energy functional, quadratic in the perturbation variables, which has a non-positive time derivative. In the appendix, we show that any perturbation which conserves the overall particle number of the fluid, and the total momenta associated with any background Killing vector fields of the spacetime, will grow without bound if the energy functional for that perturbation is negative on some spacelike surface. This result eliminates some assumptions about the properties of the equilibrium states made in Ref. (23).

The combination of the main results, that the first-order theories are necessarily unstable, while the second-order theories can be both stable and causal, strongly suggests to us that the second-order theories should replace the first-order theories as the standard theory of relativistic dissipative fluid mechanics. The first-order theories can probably not be made causal; their equilibrium states are all unstable; and it appears that the first-order theories have no well-posed initial value problem. The Israel-Stewart theory, on the other hand, possesses stable equilibrium states, and has perturbations which propagate causally according to hyperbolic equations.

2. THEORIES OF RELATIVISTIC DISSIPATIVE FLUIDS

(a) *Constructing the Theories.* The fundamental variables of a relativistic theory of fluids are the stress-energy tensor, T^{ab}, and the particle number current N^a. The fundamental equations of motion are the conservation laws for these quantities:

(1) $$\nabla_a T^{ab} = 0,$$

(2) $$\nabla_a N^a = 0.$$

The derivative operators in Eqs. (1-2) are four-dimensional covariant derivatives; no specialization to a fixed background spacetime is to be assumed. In an equilibrium state these fundamental tensors can be decomposed in terms of other familiar fluid variables,

(3) $$T^{ab} = \rho u^a u^b + p q^{ab},$$

(4) $$N^a = n u^a,$$

where u^a is the four-velocity of the fluid, ρ is the energy density (as measured by an observer co-moving with the fluid), p is the pressure, n is the number density, and q^{ab} is the projection tensor orthogonal to u^a:

(5) $$q^{ab} = g^{ab} + u^a u^b.$$

For a fluid which is not in an equilibrium state, it is also customary to introduce a four-velocity vector field u^a and a set of thermodynamic variables ρ, p, n, s (entropy per particle), T (temperature), μ (chemical potential), etc. which are used to supplement the description of the fluid contained in the fundamental tensors T^{ab} and N^a. There are, however, a variety of ways of introducing these auxiliary fields. Different theories adopt different rules for identifying these extra fields.

The four-velocity u^a of a fluid which is not in equilibrium can be defined in terms of the fundamental tensors of the theory in a variety of reasonable ways. In the Eckart[2] theory the four-velocity is identified with the direction in spacetime in which the particles of the fluid move, i.e., the four-velocity is parallel to the particle number current:

(6) $$u^a = (-N^b N_b)^{-1/2} N^a.$$

In the Landau-Lifshitz[3] theory, on the other hand, the four-velocity is identified with the spacetime direction of energy flow. Thus u^a is taken as the timelike eigenvector of T^{ab} in the Landau-Lifshitz theory:

$$q_{ab}T^{bc}u_c = 0. \tag{7}$$

Other choices are possible. In the Havas-Swenson theory[6], and also in a general class of theories which we have studied[21], the initial choice of u^a is only restricted by demanding that it be a unit timelike vector field.

A great deal of freedom is also available in introducing the thermodynamic variables ρ, p, n, etc. into the description of a fluid which is not in equilibrium. It is customary to constrain the thermodynamic variables by the "first law of thermodynamics",

$$d\rho = nTds + \frac{\rho + p}{n}\, dn \tag{8}$$

where $d\rho$, ds, and dn are one-forms. It is also customary to take over to the nonequilibrium theory the equation of state which describes the equilibrium states of the fluid. The equation of state defines the value of one of the thermodynamic variables (often s, which is not generally considered to be directly observable) as a function of two other observable thermodynamic variables, e.g., $s = s(\rho,n)$. The given equation of state and the first law of thermodynamics [Eq. (8)] then reduce the thermodynamic variables to a space containing only two independent functions. These two independent thermodynamic variables may be identified with the fundamental tensors of the nonequilibrium theory in a variety of ways. The thermodynamic energy density, ρ, could be defined as the physical energy density,

$$\rho = T^{ab}u_a u_b, \tag{9}$$

the thermodynamic pressure, p, could be defined as the average physical pressure,

(10) $$p = \frac{1}{3} T^{ab} q_{ab},$$

and the thermodynamic particle number density, n, could be defined as the physical particle number density:

(11) $$n = - N^{a} u_{a}.$$

Since the space of thermodynamic variables is only two-dimensional, one cannot make all of these identifications [Eqs. (9)-(11)] for an arbitrary nonequilibrium state of the fluid. The most common choice, made in the Eckart and Landau-Lifshitz theories, takes ρ and n to be related to the fundamental tensors by Eqs. (9) and (11). The thermodynamic pressure, p, is then determined by Eq. (8) and the equation of state. It will in general not satisfy Eq. (10). A nonequilibrium scalar stress, τ, may then be defined as the difference between the average physical pressure and the thermodynamic pressure:

(12) $$\tau = \frac{1}{3} T^{ab} q_{ab} - p.$$

This is, however, not the only possible choice. One could take ρ and p, or n and p, as being defined by Eqs. (9)-(11) and then introduce a nonequilibrium correction field to modify the third equation [Eq. (11) or (9) respectively]. The Havas-Swenson theory allows an even more general identification of the thermodynamic variables with the fundamental tensors. In this type of theory one assumes that a procedure exists to measure two of the thermodynamic functions (not necessarily n, p, or ρ), for example the temperature and the chemical potential, in an arbitrary nonequilibrium state of the fluid. Given the values of these two functions, the rest of the thermodynamic variables (in particular n, p, and ρ) are determined by the equation of state and Eq. (8). In this case additional nonequilibrium correction terms must be added to each of Eqs. (9)-(11). Finally, note that once a set of rules for identifying the thermodynamic variables with the fundamental tensors is chosen, the particular values of the thermodynamic variables for a given nonequilibrium state still depend strongly on the choice of four-velocity definition.

To simplify the presentation in this paper, we will restrict further consideration to theories of the Eckart type. This implies that the four-velocity of the fluid be identified with the particle number current vector as in Eq. (6). It further implies that ρ and n be chosen as the fundamental observable thermodynamic variables, to be identified with the physical energy density and particle number density as in Eqs. (9) and (11), even for states which are not in equilibrium. Readers desiring more details concerning the Landau-Lifshitz theory, or other more general first-order theories, such as the Havas-Swenson theory, are referred to Refs. (3), (6), (21), and (22). The results concerning the stability of equilibrium states are not qualitatively different for those other theories.

Using the Eckart choice of four-velocity, u^a, and thermodynamic variables ρ and n, the stress-energy tensor and the number current vector can be decomposed in the following manner for an arbitrary nonequilibrium state

$$T^{ab} = \rho u^a u^b + (p + \tau)q^{ab} + q^a u^b + q^b u^a + \tau^{ab}, \tag{13}$$

$$N^a = nu^a, \tag{14}$$

where

$$u^a q_a = u^a \tau_{ab} = \tau^a{}_a = \tau^{ab} - \tau^{ba} = 0. \tag{15}$$

The three fields τ, q^a, and τ^{ab} describe the deviations from local equilibrium in the fluid. The vector field q^a describes the heat flow, and τ and τ^{ab} are nonequilibrium stresses in the fluid.

To complete the construction of the theory, equations to determine τ, q^a, and τ^{ab} must be given. The definitions of these variables will be based on the need to satisfy the second law of thermodynamics. The total entropy associated with a spacelike surface Σ (i.e., at one instant in time) is obtained by integrating the entropy current vector field over the surface:

$$S(\Sigma) = \int_\Sigma s^a \, d\Sigma_a. \tag{16}$$

The second law of thermodynamics requires this total entropy to be a non-decreasing function of time for isolated systems. If another spacelike surface Σ' lies to the future of Σ, then the second law requires that

$$S(\Sigma') - S(\Sigma) = \int \nabla_a s^a \, dV \geq 0, \tag{17}$$

where the two surface integrals implied by Eq. (16) have been converted into a volume integral by Gauss' theorem. If the second law holds in the form of Eq. (17) for all surfaces Σ' to the future of Σ, then it is clear that the following inequality (a local form of the second law) must also hold:

$$\nabla_a s^a \geq 0. \tag{18}$$

The first-order (Eckart) theory of relativistic dissipative fluids is obtained by modeling the entropy current, s^a, by a sum of terms no higher than first order in the deviations from equilibrium, τ, q^a, and τ^{ab}. The entropy current must therefore have the form

$$s^a = sn\, u^a + \beta q^a, \tag{19}$$

where β is an as yet unconstrained (zeroth-order) thermodynamic function. A term linear in τ multiplying u^a has been omitted as it inevitably leads to a defining equation for τ which is nonlinear. The divergence of this entropy current can now be evaluated, using the equations of motion for the fluid [Eqs. (1)-(2)] to simplify the resulting expressions; the following expression results:

$$\begin{aligned} T\nabla_a s^a = -\tau\nabla_a u^a + q^a(T\nabla_a\beta - u^b\nabla_b u_a) - \tau^{ab}\langle\nabla_a u_b\rangle, \\ + (T\beta - 1)\nabla_a q^a, \end{aligned} \tag{20}$$

where the brackets $\langle\ \rangle$ which appear in Eq. (20) are defined by:

$$\langle A_{ab}\rangle = \tfrac{1}{2}\, q_a{}^c q_b{}^d (A_{cd} + A_{dc}) - \tfrac{1}{3}\, q_{ab} q^{cd} A_{cd} \tag{21}$$

for any second rank tensor. The simplest way to guarantee that

Eq. (20) is consistent with the second law, Eq. (18), is to require that the deviations from equilibrium be defined as follows:

(22) $$\beta = 1/T,$$

(23) $$\tau = -\zeta\nabla_a u^a,$$

(24) $$q^a = -\kappa q^{ab}(\nabla_b T + Tu^c\nabla_c u_b),$$

(25) $$\tau_{ab} = -2\eta\langle\nabla_a u_b\rangle.$$

With these definitions the divergence of the entropy current takes on its familiar quadratic form:

(26) $$T\nabla_a s^a = \frac{\tau^2}{\zeta} + \frac{q^a q_a}{\kappa T} + \frac{\tau^{ab}\tau_{ab}}{2\eta}.$$

which is manifestly non-negative if the three thermodynamic coefficients ζ, κ, and η are required to be positive. These coefficients may be identified with the familiar Newtonian dissipation coefficients by examining the theory in the Newtonian limit; one finds that ζ is the bulk viscosity, η is the shear viscosity, and κ is the thermal conductivity. The far more general Havas-Swenson theory results from assuming that all of the non-equilibrium fields (including non-equilibrium contributions to the energy and number densities) are allowed to depend on all first derivatives of the background equilibrium fields which have the correct tensor rank. This results in eleven additional coefficients in the theory; the constraints on these coefficients which result from enforcing the second law of thermodynamics are quite complicated and are described in the paper of Havas and Swenson[6].

Equations (1), (2), and (22)-(25) form a complete set of equations of motion for the dynamical variables n, ρ, u^a, τ, q^a, and τ^{ab} of the first-order Eckart theory constructed here. Gravitational interactions may be taken into account (if desired) by including the Einstein equations,

(27) $$G_{ab} = 8\pi\, T_{ab}$$

for the spacetime metric, g_{ab}, coupled to the stress-energy tensor of the fluid.

The second-order Israel-Stewart theory results from adopting an expression for the entropy current [Eq. (19)] which includes all possible terms through second order in the deviations from equilibrium. Specifically,

$$(28)\qquad s^a = snu^a + \frac{q^a}{T} - \frac{1}{2}(\beta_0\tau^2 + \beta_1 q^b q_b + \beta_2 \tau^{bc}\tau_{bc})\frac{u^a}{T} + \alpha_0\tau\frac{q^a}{T} + \alpha_1\tau^a{}_b\frac{q^b}{T}.$$

The three new thermodynamic coefficients β_i model the deviations of the physical entropy density from the thermodynamic entropy density, sn. The other two new coefficients, α_i, represent changes in the entropy current due to possible viscous-heat flux couplings. The divergence of Eq. (28) may be computed and simplified, in analogy to the treatment of Eq. (19), to yield

$$(29)\quad T\nabla_a s^a = -\tau\left[\nabla_a u^a + \beta_0 u^a\nabla_a\tau - \alpha_0\nabla_a q^a - \gamma_0 T q^a\nabla_a\left(\frac{\alpha_0}{T}\right) + \frac{1}{2}\tau T\nabla_a\left(\frac{\beta_0}{T}u^a\right)\right]$$
$$- q^a\left[\frac{1}{T}\nabla_a T + u^b\nabla_b u_a + \beta_1 u^b\nabla_b q_a - \alpha_0\nabla_a\tau - \alpha_1\nabla_b\tau^b_a + \frac{1}{2}Tq_a\nabla_b\left(\frac{\beta_1}{T}u^b\right)\right.$$
$$\left. -(1-\gamma_0)\tau T\nabla_a\left(\frac{\alpha_0}{T}\right) - (1-\gamma_1)T\tau_a{}^b\nabla_b\left(\frac{\alpha_1}{T}\right) + \gamma_2\nabla_{[a}u_{b]}q^b\right]$$
$$-\tau^{ab}\langle\nabla_a u_b + \beta_2 u^c\nabla_c\tau_{ab} - \alpha_1\nabla_a q_b + \frac{1}{2}T\tau_{ab}\nabla_c\left(\frac{\beta_2}{T}u^c\right)$$
$$-\gamma_1 T q_a\nabla_b\left(\frac{\alpha_1}{T}\right) + \gamma_3\nabla_{[a}u_{c]}\tau^c{}_b\rangle.$$

As in the first-order case, τ, q^a, and τ^{ab} are defined in the simplest fashion which will ensure that the second law holds in its divergence form [Eq. (18)]. The simplest definitions which will guarantee the quadratic form [Eq. (26)] of the divergence of the entropy current are

$$(30)\quad \tau = -\zeta\left[\nabla_a u^a + \beta_0 u^a\nabla_a\tau - \alpha_0\nabla_a q^a - \gamma_0 T q^a\nabla_a\left(\frac{\alpha_0}{T}\right) + \frac{1}{2}\tau T\nabla_a\left(\frac{\beta_0}{T}u^a\right)\right],$$

$$(31)\quad q^{a} = -\kappa T q^{ab}\left[\frac{1}{T}\nabla_b T + u^c\nabla_c u_b + \beta_1 u^c\nabla_c q_b - \alpha_0\nabla_b \tau - \alpha_1\nabla_c \tau^c{}_b + \frac{1}{2}Tq_b\nabla_c\left(\frac{\beta_1}{T}u^c\right) - (1-\gamma_0)\tau T\nabla_b\left(\frac{\alpha_0}{T}\right) - (1-\gamma_1)T\tau_b{}^c\nabla_c\left(\frac{\alpha_1}{T}\right) + \gamma_2\nabla_{[b}u_{c]}q^c\right],$$

$$(32)\quad \tau^{ab} = -2\eta\langle\nabla^a u^b + \beta_2 u^c\nabla_c\tau^{ab} - \alpha_1\nabla^a q^b + \frac{1}{2}T\tau^{ab}\nabla_c\left(\frac{\beta_2}{T}u^c\right) - \gamma_1 Tq^a\nabla^b\left(\frac{\alpha_1}{T}\right) + \gamma_3\nabla^{[a}u^{c]}\tau_c{}^b\rangle.$$

There are two new coefficients present in Eqs. (30-32), γ_0 and γ_1, that appear because of ambiguity in factoring the cross-terms on the right hand side of Eq. (29) which involve τq^a and $\tau^a{}_b q^b$. There are also two new coefficients, γ_2 and γ_3, which couple the heat flow vector and stress tensor to the vorticity of the fluid. Within the phenomenological fluid theory, the magnitudes of the γ_i are unknown and could in principle be large compared to unity.

Equations (1), (2), and (30-32) form the complete set of equations of motion for the second-order Israel-Stewart theory, written in the so-called "Eckart frame" in which the four-velocity is chosen to be parallel to the particle number current. Israel and Stewart[12-15] have also given the equations of motion for the second-order theory in a frame with arbitrary four-velocity. Their more general theory still adopts Eqs. (9) and (11) to define ρ and n; no one has yet had the temerity to construct the analogue of the Havas-Swenson theory at second order.

(b) *Equilibrium states.* Before the propagation of perturbations on an equilibrium background, or the stability of the equilibrium states can be studied, the nature of the equilibrium states themselves must be determined. Equilibrium is defined by the condition that the entropy of the fluid must not change with time; this implies that the divergence of the entropy current must be zero. Since the divergence of the entropy current is the sum of positive terms [Eq. (26)], each of these terms must vanish in equilibrium. Thus, in both the first- and second-order theories, the heat flow and viscous stresses must vanish in equilibrium:

$$\tau = q^{a} = \tau^{ab} = 0. \tag{33}$$

The vanishing of these deviations from equilibrium, applied to the defining equations for these quantities [Eqs. (23-25) for the first-order theory, Eqs. (30-32) for the second-order theory] imply the following additional conditions on an equilibrium state:

$$\nabla_{a} u^{a} = 0, \tag{34}$$

$$\langle \nabla_{a} u_{b} \rangle = 0, \tag{35}$$

$$q^{ab}(\nabla_{b} T + T u^{c} \nabla_{c} u_{b}) = 0. \tag{36}$$

Applying the conservation laws [Eqs. (1) and (2)] to a fluid which satisfies these constraints yields

$$u^{a} \nabla_{a} n = 0, \tag{37}$$

$$u^{a} \nabla_{a} \rho = 0, \tag{38}$$

$$q^{ab}[\nabla_{b} p + (\rho + p) u^{c} \nabla_{c} u_{b}] = 0. \tag{39}$$

Equations (37) and (38) imply that all of the equilibrium thermodynamic variables (e.g., s, T, p) must be constant along the fluid flow lines (integral curves of u^{a}), since each of these variables depends only on ρ and n through the equation of state. This result and Eqs. (34) - (36) imply that the vector field u^{a}/T is a Killing vector field, i.e.,

$$\nabla_{a}\left(\frac{u_{b}}{T}\right) + \nabla_{b}\left(\frac{u_{a}}{T}\right) = 0. \tag{40}$$

The final equation, Eq. (39), combined with these other results, is equivalent to the requirement that a certain thermodynamic potential,

(41) $$\Theta = \frac{\rho + p}{nT} - s$$

have vanishing gradient.

Notice that the conditions satisfied by the equilibrium states are identical in the simple first- and second-order theories. The more general first-order Havas-Swenson theory allows the possibility of more general "equilibrium" states; e.g., states with nonzero heat flow or viscous stresses which possess unchanging entropy.

3. STABILITY OF FIRST-ORDER DISSIPATIVE RELATIVISTIC FLUIDS

In this section the dynamics of small perturbations about an equilibrium state of the first-order (Eckart) fluid theory will be studied. First the set of equations governing linear perturbations about equilibrium in the Eckart theory are determined. Then the exponential plane wave solutions to these equations of motion are examined. At least one transverse and one longitudinal solution are found to be growing exponentially in time. Finally, physically acceptable perturbations are expressed as Fourier transforms of the exponential plane waves. This shows that in the first-order Eckart theory, equilibrium states are always unstable to physically reasonable perturbations.

The perturbations about an equilibrium state will be analyzed in the Eulerian framework in order to avoid the gauge ambiguities present in the Lagrangian approach[18,24]. The difference between the actual nonequilibrium value of a field Q at a spacetime point and the value of Q in the fiducial background equilibrium state will be denoted by δQ. Any field which does not include the prefix δ (e.g., ρ, n, u^a, ...) will henceforth refer to the fiducial equilibrium state which satisfies the equilibrium conditions outlined above in Eqs. (33) - (41). In order to simplify the analysis, only perturbations which leave the gravitational field fixed, i.e.,

$\delta g_{ab} = 0$, will be considered. This is appropriate for special relativistic fluids or for short-wavelength perturbations of any equilibrium state.

The perturbations δQ are assumed to be small enough that their evolution is adequately described by the equations of motion [Eqs. (1), (2), and (23) - (25)] linearized about the fiducial background equilibrium state. For the purposes of deriving the linearized equations of motion for the perturbations, no special symmetry of the background equilibrium state is assumed; in particular, it could include rapid rotation and/or strong gravitational fields. The linearized equations of motion are then:

(42) $$\nabla_a \delta T^{ab} = 0,$$

(43) $$\nabla_a \delta N^a = 0,$$

(44) $$\delta\tau = -\zeta \nabla_a \delta u^a,$$

(45) $$\delta q^a = -\kappa T q^{ab}\left[\nabla_b\left(\frac{\delta T}{T}\right) + u^c\nabla_c \delta u_b + \delta u^c \nabla_c u_b\right],$$

(46) $$\delta\tau^{ab} = -2\eta\langle \nabla^a \delta u^b + \delta u^a u^c \nabla_c u^b \rangle,$$

where the linearly perturbed stress-energy tensor and particle number current are given by

(47) $$\delta T^{ab} = (\rho+p)(\delta u^a u^b + u^a \delta u^b) + \delta\rho u^a u^b + (\delta p + \delta\tau) q^{ab} + u^a \delta q^b + u^b \delta q^a + \delta\tau^{ab},$$

(48) $$\delta N^a = \delta n u^a + n \delta u^a.$$

The derivatives which appear in the linearized equations of motion are covariant derivatives compatible with the background spacetime metric g_{ab}; indices are also raised and lowered with the background metric. The linear perturbations satisfy the linearized versions of the constraints outlined in Eq. (15):

(49) $$u^a \delta q_a = \delta\tau_{ab} - \langle \delta\tau_{ab} \rangle = u^a \delta u_a = 0.$$

It will be sufficient to look for solutions to these equations possessing the following properties:

(1) The background spacetime is assumed to be flat Minkowski space,

(2) The fiducial background equilibrium state is assumed to be homogeneous, so that all background field variables have vanishing gradients,

(3) The solutions represent exponential plane waves, i.e., they have the form

$$\delta Q = \delta Q_0 \exp(\Gamma t + ikx), \tag{50}$$

where δQ_0 is constant, t and x are two of the coordinates of the background Minkowski space. The case where the background equilibrium state is at rest in this coordinate system, i.e.,

$$u^{\mathrm{a}}\nabla_{\mathrm{a}} = \partial_{\mathrm{t}}, \tag{51}$$

will be considered first. After making these simplifying assumptions, the equations of motion may be put into matrix form,

$$M^{\mathrm{A}}{}_{\mathrm{B}}\delta Y^{\mathrm{B}} = 0, \tag{52}$$

where δY^{B} represents the list of linear perturbation fields, and $M^{\mathrm{A}}{}_{\mathrm{B}}$ is a 14×14 complex valued matrix which describes the linearized equations of motion, specialized to plane wave solutions with the restrictions outlined above. The matrix takes on a particularly nice block-diagonal form when the following 14 fields are chosen as the perturbation variables (note that the order of the fields in this equation defines the columns of the matrix $M^{\mathrm{A}}{}_{\mathrm{B}}$);

$$\delta Y^{\mathrm{B}} = \{\delta\rho,\ \delta n,\ \delta u^{\mathrm{x}},\ \delta\tau,\ \delta q^{\mathrm{x}},\ \delta\tau^{\mathrm{xx}},\ \delta u^{\mathrm{y}},\ \delta q^{\mathrm{y}},\ \delta\tau^{\mathrm{xy}},\ \delta u^{\mathrm{z}},\ \delta q^{\mathrm{z}},\ \delta\tau^{\mathrm{xz}},\ \delta\tau^{\mathrm{yz}},\ \delta\tau^{\mathrm{yy}} - \delta\tau^{\mathrm{zz}}\}. \tag{53}$$

The matrix $M^{\mathrm{A}}{}_{\mathrm{B}}$ then block diagonalizes as follows:

(54)
$$\mathbf{M} = \begin{bmatrix} \mathbf{Q} & 0 & 0 & 0 \\ 0 & \mathbf{R} & 0 & 0 \\ 0 & 0 & \mathbf{R} & 0 \\ 0 & 0 & 0 & \mathbf{I} \end{bmatrix},$$

where the submatrices **Q**, **R**, and **I** are defined as follows:

(55)
$$\mathbf{Q} = \begin{bmatrix} 0 & \Gamma & ink & 0 & 0 & 0 \\ \Gamma & 0 & i(\rho+p)k & 0 & ik & 0 \\ i(\partial p/\partial\rho)_n k & i(\partial p/\partial n)_\rho k & (\rho+p)\Gamma & ik & \Gamma & ik \\ 0 & 0 & ik & 1/\zeta & 0 & 0 \\ \frac{i}{T}(\partial T/\partial\rho)_n k & \frac{i}{T}(\partial T/\partial n)_\rho k & \Gamma & 0 & 1/\kappa T & 0 \\ 0 & 0 & ik & 0 & 0 & 3/4\eta \end{bmatrix},$$

(56)
$$\mathbf{R} = \begin{bmatrix} (\rho+p)\Gamma & \Gamma & ik \\ \Gamma & 1/\kappa T & 0 \\ ik & 0 & 1/\eta \end{bmatrix},$$

and **I** is the 2×2 unit matrix. It is immediately obvious that the two components of the shear stress, $\delta\tau^{yy} - \delta\tau^{zz}$ and $\delta\tau^{yz}$, vanish identically at linear order.

There exist exponential plane-wave solutions of Eq. (52) whenever Γ and k have values which satisfy the dispersion relation,

(57)
$$\det \mathbf{M} = 0.$$

The determinant of **M** is simply the product of the determinants of its diagonal blocks, and thus the roots of Eq. (57) are simply the collection of roots obtained by separately setting the determinants of **Q** and **R** equal to zero. The roots obtained by setting the determinant of **Q** equal to zero are referred to as longitudinal modes since the matrix **Q** involves only scalars and the components of the perturbation fields which are parallel to the direction of spatial variation (x). The roots obtained by setting the determinant of $\mathbf{R}$ equal to zero describe the propagation of transverse modes of the fluid, since the matrix $\mathbf{R}$ involves only the components of the perturbation fields which are orthogonal to the direction of spatial variation (x) of a perturbation.

The determinant of the matrix $\mathbf{R}$ is

$$(58) \qquad -\eta\kappa T \det \mathbf{R} = \kappa T\Gamma^2 - (\rho + p)\Gamma - \eta k^2 = 0,$$

which can be solved for the frequency, Γ,

$$(59) \qquad \Gamma_{\pm} = \frac{1}{2\kappa T}\{(\rho+p) \pm [(\rho+p)^2 + 4\eta\kappa T k^2]^{1/2}\}.$$

The frequencies of these transverse modes, given by Eq. (59), are purely real for real wave numbers k, and hence these modes do not propagate. An observer at fixed coordinate x would observe only a monotonically growing or decaying perturbation; there would be no superimposed oscillation. The existence of a positive real root (Γ_+) implies the existence of a growing mode, and hence of an instability in the fluid (unless the thermal conductivity is zero). Since the frequency Γ_+ is positive for all real wave numbers k, the fluid is unstable to a growing transverse mode at all wavelengths. Note that as the thermal conductivity approaches zero, the growth timescale associated with the growing mode also approaches zero; the dissipative fluid thus becomes more unstable the less dissipative it is.

The dispersion relation for the longitudinal modes, obtained by setting $\det \mathbf{Q} = 0$, is a quartic polynomial in Γ:

$$(60) \quad F(\Gamma) = \frac{4}{3}\eta\zeta\kappa T \det \mathbf{Q}$$
$$= \kappa T\Gamma^4 - (\rho+p)\Gamma^3 + \left\{\kappa T\left[\left(\frac{\partial p}{\partial \rho}\right)_s + \left(\frac{\partial \Theta}{\partial s}\right)_\rho\right] - \left(\zeta + \frac{4}{3}\eta\right)\right\}k^2\Gamma^2$$
$$- \left\{(\rho+p)\left(\frac{\partial p}{\partial \rho}\right)_s + \kappa\left(\zeta + \frac{4}{3}\eta\right)k^2\left(\frac{\partial T}{\partial \rho}\right)_n\right\}k^2\Gamma$$
$$+ \kappa T\left(\frac{\partial p}{\partial \rho}\right)_s\left(\frac{\partial \Theta}{\partial s}\right)_p k^4.$$

In the special case of a spatially homogeneous perturbation ($k = 0$), the only nonzero root of Eq. (60) is

$$\Gamma = \frac{\rho + p}{\kappa T}, \tag{61}$$

a growing spatially homogeneous perturbation. Note that as the limit of zero thermal conductivity is taken, $\kappa T \to 0$, the growth timescale of this mode ($1/\Gamma$) approaches zero, as in the transverse case. It is therefore clear that there exist growing longitudinal modes for at least some neighborhood in k about $k = 0$.

In the exceptional case where the thermal conductivity is zero, it is possible to show that the transverse modes are still unstable as seen by an observer moving with a nonzero constant velocity relative to the background equilibrium state[21], as long as the shear viscosity coefficient is not also equal to zero. In the even more special case when both the thermal conductivity and shear viscosity are zero (leaving the bulk viscosity as the only source of dissipation), it is possible to show that at least long wavelength longitudinal modes will be unstable as seen by an observer moving relative to the background state. A weaker version of this result, which assumed the positivity of certain thermodynamic derivatives, can be found in Ref. (21).

This analysis has shown that the equilibrium states of the first-order Eckart theory of dissipative relativistic fluids always possess linear plane-wave perturbations which grow exponentially in time; does this necessarily imply that physically reasonable perturbations will grow exponentially in time also? The answer to this question clearly depends to some extent on how one defines "physically reasonable" perturbations. If square-integrable initial data (i.e., in L^2) are defined as being physically reasonable, then these perturbations do grow exponentially in time. Since the initial data is taken to be in L^2, it has a well-defined spatial Fourier transform; if it is a generic perturbation then its spatial Fourier transform will have non-zero support on the set of growing plane-wave modes. It is then possible to show that the L^2 norm of the perturbation will in fact diverge exponentially in time[21].

Finally, it is important to estimate the growth timescales associated with these instabilities to determine their relevance to physics. If, for all imaginable conditions, the growth timescales were absurdly large (e.g., greater than the age of the universe), then one could claim that the instabilities were only of academic interest; in any conceivable application the instabilities would be undetectable in the dynamics of the fluid. In fact, the growth timescales for these instabilities are absurdly *short* for nearly Newtonian, weakly dissipative systems. For example, the frequency for the transverse modes described by Eq. (59) is bounded below by

$$\Gamma_+ \geqslant \frac{(\rho c^2 + p)c^2}{\kappa T}, \tag{62}$$

where the speed of light (c) has been explicitly inserted into this equation. The characteristic growth time is given by $\tau = \Gamma_+^{-1}$, so

$$\tau \leqslant \frac{\kappa T}{(\rho c^2 + p)c^2}. \tag{63}$$

This timescale is ridiculously short for everyday fluids; for example, water at room temperature ($\approx$ 300 K) and pressure ($\approx$ 1 bar) has a growth timescale given by

$$\tau \lesssim 2 \times 10^{-34} \text{ sec.} \tag{64}$$

4. STABILITY AND CAUSALITY IN SECOND-ORDER DISSIPATIVE RELATIVISTIC FLUIDS

In this section the stability of equilibrium states in the second-order Israel-Stewart theory of dissipative relativistic fluids is examined. This analysis reveals that it is possible for all equilibria in this theory to be stable, provided certain constraints on the thermodynamic derivatives and second-order coefficients (the α_i and β_i) are satisfied. These stability conditions are shown to imply that all linear perturbations propagate causally (i.e., subluminally) and obey a symmetric hyperbolic system of equations. The converse theorem is also shown to be true: if the linear perturbations propagate causally and obey a symmetric hyperbolic (in a particular sense) system of equations, then all equilibria are stable. The definition of hyperbolicity needed to establish the converse theorem is rather strict and cannot be replaced by a more naive definition. A new result presented in this section is an example of a fluid system in which all the characteristic velocities are real and less than the speed of light, but which violates the stability conditions, so that no equilibrium state is stable. This shows that simple, naive ideas about hyperbolicity are insufficient to establish stability.

(a) *Stability.* As in the previous section, the perturbations are analyzed here within the Eulerian framework, and only linear departures from equilibrium are considered. The equations of motion for the perturbations are obtained by linearizing the fluid's full set of equations of motion [Eqs. (1), (2), and (30)-(32)] about a fiducial equilibrium background state; the resulting equations of motion for the linearized perturbations are given by Eq. (42), (43), and

$$(65) \qquad \delta\tau = -\zeta\left[\nabla_a \delta u^a + \beta_0 u^a \nabla_a \delta\tau - \alpha_0 \nabla_a \delta q^a - \gamma_0 T \delta q^a \nabla_a\left(\frac{\alpha_0}{T}\right)\right],$$

$$(66) \qquad \delta q^a = -\kappa T q^{ab}\left[\nabla_b\left(\frac{\delta T}{T}\right) + u^c\nabla_c \delta u_b + \delta u^c \nabla_c u_b + \beta_1 u^c \nabla_c \delta q_b - \alpha_0 \nabla_b \delta\tau - \alpha_1 \nabla_c \delta\tau^c_{\ b} - (1-\gamma_0) T \delta\tau \nabla_b\left(\frac{\alpha_0}{T}\right) - (1-\gamma_1) T \delta\tau^c_{\ b} \nabla_c\left(\frac{\alpha_1}{T}\right) + \gamma_2 \nabla_{[b} u_{c]} \delta q^c\right],$$

$$(67) \qquad \delta\tau^{ab} = -2\eta\langle \nabla^a \delta u^b + \delta u^a u^c \nabla_c u^b + \beta_2 u^c \nabla_c \delta\tau^{ab} - \alpha_1 \nabla^a \delta q^b - \gamma_1 T \delta q^a \nabla^b\left(\frac{\alpha_1}{T}\right) + \gamma_3 \nabla^{[a} u^{c]}\, \delta\tau_c^{\ b}\rangle,$$

The approach to analyzing the stability of equilibrium states in the second-order theory is rather different than that used in the previous section. In that case, the equilibrium states were unstable and the aim was to identify some particular unstable perturbations.

In this case, the aim is to find the necessary and sufficient set of conditions which will imply the stability of all equilibrium states. This will be done by constructing an energy functional for the perturbations, i.e., a monotonically decreasing function of time which depends quadratically on the perturbation variables. Such a functional can be constructed for Israel-Stewart fluids in terms of an energy current vector defined by

$$(68)\quad TE^a = \delta T^a{}_b\delta u^b - \frac{1}{2}(\rho+p)u^a\delta u^b\delta u_b - \alpha_0\delta\tau\delta q^a - \alpha_1\delta\tau^a{}_b\delta q^b + \frac{\delta T}{T}\delta q^a$$
$$+ \frac{1}{2}(\rho+p)^{-1}\left[\left(\frac{\partial\rho}{\partial p}\right)_s(\delta p)^2 + \left(\frac{\partial\rho}{\partial s}\right)_p\left(\frac{\partial p}{\partial s}\right)_\theta(\delta s)^2\right]u^a$$
$$+ \frac{1}{2}\,[\beta_0(\delta\tau)^2 + \beta_1\delta q^b\delta q_b + \beta_2\delta\tau^{bc}\delta\tau_{bc}]u^a\,.$$

The total energy associated with a spacelike surface Σ (i.e., the energy at one instant of time) is given by the integral of the energy current over the surface:

$$(69)\qquad E(\Sigma) = \int_\Sigma E^a d\Sigma_a.$$

From an argument analogous to that given for the entropy of the fluid in Section 2, this energy will be a decreasing function of time (for fluids with compact spatial support) as long as the divergence of the energy current is negative. The divergence of this energy current can be computed, and the resulting expression simplified using the perturbation equations to yield:

$$(70)\qquad \nabla_a E^a = -\left[\frac{(\delta\tau)^2}{T\zeta} + \frac{\delta q^a\delta q_a}{\kappa T^2} + \frac{\delta\tau^{ab}\delta\tau_{ab}}{2\eta T}\right].$$

The energy functional is thus a monotonically decreasing function of time.

The motivation for constructing this energy functional for Israel-Stewart fluids was to study their stability. Such a functional is expected to provide a useful indication of stability because of the following qualitative argument. If the energy functional were non-negative for all possible values of the perturbation variables, then it would suggest that the equilibrium state is stable. The energy of any perturbation would be bounded above by its initial value and below by zero in this case. If, on the other hand, perturbations exist having negative energy, then the energy is unbounded below and could evolve towards negative infinity, suggesting the presence of an instability. To establish a rigorous connection between the sign of the energy functional and stability, a detailed analysis of the equations of motion for the perturbations is needed; such an analysis is supplied in the Appendix of Ref. (23), where we prove the following two propositions:

PROPOSITION A. *The perturbations of an Israel-Stewart fluid will not grow without bound (as measured by a square integral norm) if the energy functional is non-negative for all perturbations.*

PROPOSITION B. *Consider an equilibrium state of an Israel-Stewart fluid in which the thermodynamic inequalities,*

$$0 < \left(\frac{\partial p}{\partial \rho}\right)_s < 1, \tag{71}$$

and

$$\left(\frac{\partial \rho}{\partial s}\right)_p \left(\frac{\partial p}{\partial s}\right)_\theta > 0. \tag{72}$$

are satisfied. If there exist perturbations having negative energy, then they will grow without bound as they evolve in time.

These two propositions demonstrate that a positive energy functional is a sufficient condition for the stability of Israel-Stewart fluids, and that it is a necessary condition for the stability of those equilibrium configurations that satisfy Eqs. (71) and (72). It has not yet been proven that a positive energy functional is strictly necessary for the stability of these fluids; however, another result which indicates that this is probably the case has been established. In the appendix the following proposition is proven:

PROPOSITION C. *Any solution of the linearized Israel-Stewart fluid perturbation equations which has vanishing variation in the conserved particle number,* $\delta N = 0$, *and vanishing variation of the conserved momenta,* $\delta P(k^a) = 0$, *for each Killing vector field* k^a *of the background spacetime, will grow without bound if the energy functional for this solution is negative on some spacelike surface.*

Proposition C applies to the stability of any equilibrium configuration, including those which violate Eqs. (71) and (72). It is limited, however, to the class of perturbations that leave the conserved particle number and conserved momenta of the background equilibrium configuration unchanged. Thus, Propositions B and C are complementary; they establish that a positive energy functional is a necessary condition for stability under different circumstances.

The energy functional defined by Eqs. (68) and (69) is a complicated function of the perturbation variables. Since the sign of the energy determines which equilibrium states are stable, it is desirable to factor the energy into a form which makes the conditions necessary for it to be positive self-evident.

Let the vector t^a be the future-directed unit normal to the spacelike surface Σ upon which E is defined. Associated with t^a are several useful tensors. The vector λ^a is defined to be the velocity of observers moving along t^a relative to the fluid

(73) $$\lambda^a = q^a{}_b t^b / u^c t_c.$$

It is easy to see that the norm of λ^a is bounded between zero and one. The projection tensor associated with λ^a will also be useful:

(74) $$\gamma^a{}_b = q^a{}_b - \lambda^{-2}\lambda^a\lambda_b.$$

The expression for the energy can now be written in terms of an energy density e, defined by

(75) $$e = TE^a t_a / u^b t_b,$$

so that the energy is given by

$$E(\Sigma) = \int_\Sigma e \frac{u^a}{T} d\Sigma_a. \tag{76}$$

The total energy E will then be positive for all possible perturbations if and only if the energy density is also positive for all perturbations at every point in the fluid. Fortunately, it is possible to factor the energy density functional into the following form:

$$e = \frac{1}{2} \sum_A \Omega_A (\delta Z_A)^2, \tag{77}$$

where the Ω_A are the following functions of the thermodynamic variables:

$$\Omega_1 = (\rho + p)^{-1} \left(\frac{\partial \rho}{\partial p}\right)_s, \tag{78}$$

$$\Omega_2 = (\rho + p)^{-1} \left(\frac{\partial \rho}{\partial s}\right)_p \left(\frac{\partial p}{\partial s}\right)_\Theta, \tag{79}$$

$$\Omega_3 = (\rho + p)\left[1 - \lambda^2 \left(\frac{\partial p}{\partial \rho}\right)_s\right] - \left[\frac{1}{\beta_0} + \frac{2}{3\beta_2} + \frac{K^2}{\Omega_6}\right]\lambda^2, \tag{80}$$

$$\Omega_4 = (\rho + p) - \frac{2\beta_2 + (\beta_1 + 2\alpha_1)\lambda^2}{2\beta_1\beta_2 - \alpha_1^2\lambda^2}, \tag{81}$$

$$\Omega_5 = \beta_0, \tag{82}$$

$$\Omega_6 = \frac{\beta_1}{\lambda^2} - \left[\frac{\alpha_0^2}{\beta_0} + \frac{2\alpha_1^2}{3\beta_2} + \frac{1}{nT^2}\left(\frac{\partial T}{\partial s}\right)_n\right], \tag{83}$$

$$\Omega_7 = \beta_1 - \frac{\alpha_1^2}{2\beta_2}\lambda^2, \tag{84}$$

$$\Omega_8 = \beta_2, \tag{85}$$

and the δZ_A represent certain linearly independent combinations of the perturbation functions, given by

(86) $$\delta Z_1 = \delta p + (\rho + p)\left(\frac{\partial p}{\partial \rho}\right)_s\left[\lambda_a \delta u^a + \frac{1}{T}\left(\frac{\partial T}{\partial p}\right)_s \lambda_a \delta q^a\right],$$

(87) $$\delta Z_2 = \delta s + \frac{1}{T}(\rho + p)\left(\frac{\partial T}{\partial \rho}\right)_p\left(\frac{\partial s}{\partial p}\right)_\Theta \lambda_a \delta q^a,$$

(88) $$\delta Z_3 = \lambda^{-1}\lambda_a \delta u^a,$$

(89) $$\delta Z^a{}_4 = \gamma^a{}_b \delta u^b,$$

(90) $$\delta Z_5 = \delta \tau + \frac{1}{\beta_0}\lambda_a \delta u^a - \frac{\alpha_0}{\beta_0}\lambda_a \delta q^a,$$

(91) $$\delta Z_6 = \lambda_a \delta q^a + \frac{K}{\Omega_6}\lambda_a \delta u^a,$$

(92) $$\delta Z_7{}^a = \gamma^a{}_b \delta q^b + \frac{2\beta_2 + \alpha_1 \lambda^2}{2\beta_1\beta_2 - \alpha_1^2\lambda^2}\gamma^a{}_b \delta u^b,$$

(93) $$\delta Z_8{}^{ab} = \delta \tau^{ab} + \frac{1}{\beta_2}\langle \lambda^a \delta u^b \rangle - \frac{\alpha_1}{\beta_2}\langle \lambda^a \delta q^b \rangle,$$

where

(94) $$K = \frac{1}{\lambda^2} + \frac{\alpha_0}{\beta_0} + \frac{2\alpha_1}{3\beta_2} - \frac{n}{T}\left(\frac{\partial T}{\partial n}\right)_s.$$

The conditions necessary for E to be positive are then simply that the Ω_A must all be positive. A number of the Ω_A depend on the choice of spacelike surface because of their dependence on the parameter λ^2. There is enough freedom in the choice of spacelike surface so that λ^2 can take on any value between zero and one at any point in the star. The positivity of the Ω_A must therefore be imposed for all values of λ^2 in order to assure the positivity of the energy for all perturbations and all choices of spacelike surface. It is possible to show[23] that the most restrictive case is when $\lambda^2 = 1$, so that the conditions $\Omega_A\,(\lambda^2 = 1) > 0$ imply $\Omega_A > 0$ for all $0 \leqslant \lambda^2 \leqslant 1$.

Before proceeding further, it is worthwhile to note several implications of the stability conditions $\Omega_A > 0$. The first two conditions are the usual stability conditions for a relativistic perfect fluid; the positivity of Ω_1 guarantees that the square of the "adiabatic sound speed" will be positive (note that there is no *a priori* theoretical reason to believe that any perturbation will propagate at the adiabatic sound speed in the *dissipative* fluid theory); the positivity of Ω_2 is just the relativistic Schwarzchild criterion for stability against convection[25]. The three new second-order thermodynamic coefficients, β_i are required to be positive by the conditions on Ω_5, Ω_7, and Ω_8; and are further bounded from below by $\beta_0 > (\rho + p)^{-1}$, $\beta_1 > (\rho + p)^{-1}$, and $\beta_2 > \frac{2}{3}(\rho + p)^{-1}$ as a result of the positivity of Ω_3. That the β_i must be positive confirms the expectation that the nonequilibrium entropy density will be smaller in magnitude than the equilibrium value, sn.

(b) *Causality and Hyperbolicity.* The original motivation for constructing the second-order theory of relativistic dissipative fluids[12] was the desire to obtain a theory in which all perturbations propagated causally. Stewart and Israel[13,14] have investigated the extent to which the Israel theory succeeds in this respect. They derived expressions for the characteristic velocities for the system of perturbation equations [essentially, Eqs. (42), (43), and (65) - (67)]. These expressions are so complicated in form that it is not possible to determine by inspection whether the velocities are necessarily less than the speed of light, or whether they are even all real. In order to proceed, Israel and Stewart then specialized to the dilute gas limit where relativistic kinetic theory can be used to obtain explicit expressions for the α_i and β_i. In this limit, they were able to conclude that the characteristic velocities for this system of equations were less than the speed of light. The following analysis shows that perturbations propagate causally and obey a set of hyperbolic equations in a far wider range of circumstances, namely whenever the equilibrium states are stable.

The system of equations for the perturbations of an equilibrium state of an Israel second-order fluid, Eqs. (42), (43), and (65) - (67), have the following general form:

$$A^{A}{}_{B}{}^{a}\nabla_a \delta Y^B + B^{A}{}_{B}\delta Y^B = 0, \tag{95}$$

where δY^B represents the list of the fourteen perturbation variables, i.e., $\delta\rho$, δn, δu^a, δT, δq^a, $\delta\tau^{ab}$. The index B runs over these fourteen fields, while the index A runs over the fourteen equations of motion for the perturbations. The matrices $\mathcal{A}^A{}_B{}^a$ and $\mathcal{B}^A{}_B$ are functions of the unperturbed equilibrium fluid configuration.

A three dimensional surface is a characteristic surface for these equations if the initial values of the fields δY^B cannot be freely specified on that surface. The characteristic surfaces coincide with the level surfaces of a scalar function φ which satisfies the equation (see, e.g., ref. (19), page 170):

$$\det(\mathcal{A}^A{}_B{}^a\nabla_a\varphi) = 0. \tag{96}$$

The characteristic velocities are the slopes of these surfaces. To solve the characteristic equation [Eq. (96)], a Cartesian coordinate system is chosen which is at some point in the fluid momentarily co-moving with the fluid; further, the scalar field φ is chosen to vary spatially only in the (arbitrarily chosen) x^1 direction. These conditions may be summarized as follows:

$$g^{ab}\partial_a\partial_b = -(\partial_0)^2 + (\partial_1)^2 + (\partial_2)^2 + (\partial_3)^2, \tag{97}$$

$$u^a\partial_a = \partial_0, \tag{98}$$

and

$$\varphi = \varphi(x^0,x^1). \tag{99}$$

These conditions do not restrict the background equilibrium state in any way; it can be rotating, inhomogeneous, and in a curved spacetime.

In this coordinate system the characteristic equation takes the form

$$\det(v\,\mathcal{A}^A{}_B{}^0 - \mathcal{A}^A{}_B{}^1) = 0, \tag{100}$$

where the characteristic velocity, v, is defined by

(101) $$v = -\partial_0\varphi/\partial_1\varphi.$$

In analogy with the calculation in Section 3, the characteristic equation matrices may be simplified by choosing the following set of perturbation variables:

(102) $$\delta Y^{\mathrm{B}} = \left\{T\delta\Theta, \frac{\delta T}{T}, \delta\tau, \delta u^1, \delta q^1, \delta\tau^{11}, \delta u^2, \delta q^2, \delta\tau^{21}, \delta u^3, \delta q^3, \delta\tau^{31}, \delta\tau^{22} - \delta\tau^{33}, \delta\tau^{23}\right\}.$$

The characteristic matrix then block diagonalizes:

(103) $$v\,A^0 - A^1 = \begin{bmatrix} \mathbf{Q} & 0 & 0 & 0 \\ 0 & \mathbf{R} & 0 & 0 \\ 0 & 0 & \mathbf{R} & 0 \\ 0 & 0 & 0 & \mathbf{S} \end{bmatrix}.$$

where the submatrices **Q**, **R**, and **S** are defined as follows:

(104) $$\mathbf{Q} = \begin{bmatrix} \frac{v}{T}\left(\frac{\partial n}{\partial\Theta}\right)_T & \frac{v}{T}\left(\frac{\partial\rho}{\partial\Theta}\right)_T & 0 & -n & 0 & 0 \\ \frac{v}{T}\left(\frac{\partial\rho}{\partial\Theta}\right)_T & vT\left(\frac{\partial\rho}{\partial T}\right)_\Theta & 0 & -(\rho+p) & -1 & 0 \\ 0 & 0 & \beta_0 v & -1 & \alpha_0 & 0 \\ -n & -(\rho+p) & -1 & v(\rho+p) & v & -1 \\ 0 & -1 & \alpha_0 & v & \beta_1 v & \alpha_1 \\ 0 & 0 & 0 & -1 & \alpha_1 & \frac{3}{2}\beta_2 v \end{bmatrix},$$

(105) $$\mathbf{R} = \begin{bmatrix} v(\rho+p) & v & -1 \\ v & \beta_1 v & \alpha_1 \\ -1 & \alpha_1 & 2\beta_2 v \end{bmatrix},$$

and

(106) $$\mathbf{S} = \begin{bmatrix} 2\beta_2 v & 0 \\ 0 & \beta_2 v \end{bmatrix}.$$

The determinants of the submatrices are given by

(107) $$\det \mathbf{Q} = \frac{3}{2} v^2[Av^4 + Bv^2 + C]\left[\left(\frac{\partial n}{\partial \theta}\right)_T\left(\frac{\partial \rho}{\partial T}\right)_\theta - \left(\frac{\partial n}{\partial T}\right)_\theta\left(\frac{\partial \rho}{\partial \theta}\right)_T\right],$$

(108) $$\det \mathbf{R} = v\{2\beta_2[\beta_1(\rho + p) - 1]v^2 - [(\rho + p)\alpha_1{}^2 + 2\alpha_1 + \beta_1]\},$$

(109) $$\det \mathbf{S} = 2(\beta_2 v)^2.$$

The functions A, B, and C which appear in Eq. (107) are defined by

(110) $$A = \beta_0\beta_2[\beta_1(\rho + p) - 1],$$

(111) $$B = -(\rho + p)D - \beta_1 E - 2F,$$

(112) $$C = (DE - F^2)/\beta_0\beta_2,$$

where D, E, and F are given by

(113) $$D = \beta_0\beta_2\left[\frac{\alpha_0{}^2}{\beta_0} + \frac{2\alpha_1{}^2}{3\beta_2} + \frac{1}{nT^2}\left(\frac{\partial T}{\partial s}\right)_n\right],$$

(114) $$E = \beta_0\beta_2\left[(\rho + p)\left(\frac{\partial p}{\partial \rho}\right)_s + \frac{1}{\beta_0} + \frac{2}{3\beta_2}\right],$$

(115) $$F = \beta_0\beta_2\left[\frac{\alpha_0}{\beta_0} + \frac{2\alpha_1}{3\beta_2} - \frac{n}{T}\left(\frac{\partial T}{\partial n}\right)_s\right].$$

The characteristic velocities are obtained by setting the determinants of the submatrices separately to zero and solving for v. The two characteristic velocities corresponding to the zeros of det(**S**) are both zero. The matrix **R**, which describes transverse perturbations of the fluid, has one characteristic velocity which is zero, and two non-zero characteristic velocities, given by

(116) $$v_T^2 = \frac{(\rho + p)\alpha_1^2 + 2\alpha_1 + \beta_1}{2\beta_2[\beta_1(\rho + p) - 1]}.$$

The longitudinal matrix, **Q**, has two characteristic velocities which are zero, and four nonzero characteristic velocities which are the roots of the quartic polynomial

(117) $$P(v^2) = Av^4 + Bv^2 + C = 0.$$

One of the pairs of roots of Eq. (117) should correspond in some sense to the propagation of sound in an Israel-Stewart fluid, and the other should correspond to the propagation of temperature fluctuations (second sound).

The system of perturbation equations for the Israel-Stewart fluid is called a symmetric system because the matrices $\mathbf{A}^a$ are all symmetric. The system would also be *symmetric hyperbolic* (see Ref. (19), p. 593) if some linear combination of the $\mathbf{A}^a$ is (positive) definite. Other definitions of hyperbolicity, which impose conditions on the characteristic velocities (such as reality) fail when there are multiple characteristics having the same velocity. Since this is always the case for the Israel-Stewart fluids, the matrix condition given above is used to determine the hyperbolicity of the system.

The following may be shown to be the necessary and sufficient conditions for the matrix $\mathbf{A}^0$ to be positive definite:

(118) $$\left(\frac{\partial n}{\partial \theta}\right)_T > 0,$$

(119) $$\left(\frac{\partial \rho}{\partial T}\right)_\theta > 0,$$

(120) $$\beta_i > 0,$$

for $i = 1,2,3$,

(121) $$\beta_1(\rho + p) - 1 > 0,$$

and

$$(122)\qquad \left(\frac{\partial \rho}{\partial T}\right)_{\mathrm{n}} > 0.$$

The perturbation equations for an Israel fluid will then form a symmetric hyperbolic system if Eqs. (118) - (122) are satisfied. It is not known whether these are the weakest conditions that imply that the system is symmetric hyperbolic or not.

The stability conditions, $\Omega_A > 0$ [Eqs. (78) - (85)] and the conditions for the system of perturbation equations to be symmetric hyperbolic [Eqs. (118) - (122)] are clearly related to one another. Both sets of conditions imply that the β_i must be positive, for example. The final portions of this section establish the relationships that exist between these conditions together with the conditions that guarantee that the characteristic velocities are subluminal.

(c) *Stability implies causality and hyperbolicity.* One relationship that exists between the stability and causality conditions is that perturbations will propagate causally according to a symmetric hyperbolic system of equations in any Israel-Stewart fluid that satisfies the stability conditions. The proof of this relationship follows.

Using the expression for the $\Omega_A(\lambda^2)$ [Eqs. (78) - (85)], it is easy to verify that the transverse characteristic velocities [given by Eq. (116)] are constrained by

$$(123)\qquad 1 - v_T^2 = \frac{\Omega_4(1)\Omega_7(1)}{\Omega_4(0)\Omega_7(0)} > 0,$$

and

$$(124)\qquad v_T^2 = \frac{1}{2\beta_2}\left[\frac{\rho + p}{\Omega_4(0)\Omega_7(0)}\left(\alpha_1 + \frac{1}{\rho + p}\right)^2 + \frac{1}{\rho + p}\right] > 0,$$

which together guarantee that $0 < v_T^2 < 1$.

The longitudinal velocities are slightly more complicated to handle. The longitudinal velocities will be real only if their squares are real; their squares can be real only if the discriminant of the quartic polynomial in Eq. (117) is nonnegative. The discriminant can be put into the following form using Eqs. (110) - (115):

$$(125)\quad B^2 - 4AC = \frac{1}{\beta_1(\rho + p)}\{[(\rho + p)D + \beta_1 E + 2\beta_1(\rho + p)F]^2 + [\beta_1(\rho + p) - 1][(\rho + p)D - \beta_1 E]^2\} > 0.$$

Thus, the squares of the longitudinal velocities are real if the stability conditions are satisfied. The longitudinal velocities will then be real and less than the speed of light if the zeroes of the quartic polynomial lie between zero and one. The zeroes of the quartic polynomial can be located by using a geometrical argument. The coefficient of the quartic term, A, is positive if the stability conditions are satisfied; therefore $P(v^2)$ will be positive for large v^2. It is possible, after a fair amount of algebra, to show that the stability conditions [Eqs. (78) - (85)] imply the following conditions on P:

$$(126)\quad P(0) = C > 0,$$

$$(127)\quad \frac{dP}{dv^2}(0) = B < 0,$$

$$(128)\quad P(1) = A + B + C > 0,$$

$$(129)\quad \frac{dP}{dv^2}(1) = 2A + B > 0.$$

These four conditions imply that the zeroes of P lie between zero and one; i.e.,

$$(130)\quad 0 < {v_L}^2 < 1.$$

The conditions under which the perturbation equations form a symmetric hyperbolic system [Eqs. (118) - (122)] are easily shown to be consequences of the stability conditions. Thus, we have shown that the perturbations of a second-order Israel-Stewart fluid propagate causally according to a symmetric hyperbolic set of equations if the stability criteria ($\Omega_A > 0$) are satisfied.

(d) *Causality and hyperbolicity imply stability.* A fairly tedious and complicated argument can also be made to prove the converse theorem, namely that an Israel-Stewart fluid must be stable if perturbations propagate causally and the perturbation equations form a symmetric hyperbolic system [in the sense of Eqs. (118) - (122)]. The details of the proof of this theorem are given in Ref. (23). The discussion here will be confined to an examination of the necessity of making such a strong assumption about hyperbolicity in this theorem.

It is interesting that it appears to be necessary to use the definition of a hyperbolic system given in Eqs. (118) - (122) in order to prove that causality plus hyperbolicity implies stability. One might guess that causality plus the weaker (and more familiar) requirement that the characteristic velocities all be real might be enough to guarantee stability. A counterexample, however, exists to that conjecture, in which the characteristic velocities are all real and bounded between zero and one, so causality is assured, yet the stability conditions are *not* satisfied.

Consider a fluid which has an equation of state such that

(131) $$0 < \left(\frac{\partial p}{\partial \rho}\right)_s < 1$$

and

(132) $$0 < \left(\frac{\partial \rho}{\partial s}\right)_p \left(\frac{\partial p}{\partial s}\right)_\Theta .$$

These two thermodynamic constraints then imply that $(\partial p/\partial n)_s > 0$, $(\partial T/\partial s)_p > 0$, and $(\partial \Theta/\partial s)_\rho < 0$, as can be shown using the thermodynamic identities derived in Ref. (23), Sec. III(c). Now choose the second order coefficients to have the following values:

(133) $$\beta_0 = -\frac{3}{2}\beta_2 = -1/(\rho + p),$$

(134) $$\alpha_0 = \alpha_1 = -1/(\rho + p),$$

(135) $$\beta_1 = \left\{1 + \left[1 - \left(\frac{\partial \Theta}{\partial s}\right)_p\right]\left[1 - \left(\frac{\partial p}{\partial \rho}\right)_s\right]^{-1}\right\}(\rho + p)^{-1} > (\rho + p)^{-1}.$$

It is clear that the stability criteria are not all satisfied with these choices; in particular, $\Omega_5 = \beta_0$ is negative. Nevertheless, it is easy to show that the transverse characteristic velocity, defined in Eq. (116), is given by $v_T^2 = 3/4$, so that $0 < v_T^2 < 1$. With the above choices for the α_i, β_i, and thermodynamic derivative signs, it is also possible to show that the coefficients A, B, and C, defined in Eqs. (110) - (113), which determine the longitudinal velocities through Eq. (117), satisfy the inequalities given in Eqs. (125) - (129), so that the longitudinal velocities also all satisfy $0 < v_L^2 < 1$.

This then is a specific example demonstrating that it is possible to have a second order Israel-Stewart fluid whose characteristic velocities are all real and less than the speed of light, but which nonetheless contains unstable equilibria. It does seem necessary, therefore, to make the stronger assumption that the perturbation equations form a symmetric hyperbolic system in the specific sense of Eqs. (118) - (122) in order to conclude that causality plus hyerbolicity implies stability.

APPENDIX

This appendix presents the proof of the following proposition [see Ref. (26) for an analogous Newtonian result]:

PROPOSITION. *Any solution of the linearized Israel fluid perturbation equations which has vanishing variation in the conserved particle number,* $\delta N = 0$, *and vanishing variation of the conserved momenta,* $\delta P(k^a) = 0$, *for each Killing vector field* k^a *of the background spacetime, will grow without bound if the energy functional for this solution is negative on some spacelike surface.*

Proof. Consider an equilibrium solution of the Israel-Stewart fluid equations on a spacetime manifold $\mathcal{M}$. Assume that $\mathcal{M}$ can be foliated by spacelike surfaces Σ: $\mathcal{M} = \{t\} \times \Sigma$. Define the energy $E(t)$ associated with the solutions to the linearized perturbation equations by

(A1) $$E(t) = -\int_{\Sigma(t)} E^a t_a d\Sigma\ ,$$

where E^a is the energy current defined in Eq. (68), and t^a is the unit normal vector to the surface $\Sigma(t)$. The derivative of this energy is determined by Eq. (70) to be:

(A2) $$\frac{dE}{dt} = -\int_{\Sigma(t)} \left[\frac{(\delta\tau)^2}{\zeta T} + \frac{\delta q^a \delta q_a}{\kappa T^2} + \frac{\delta\tau^{ab}\delta\tau_{ab}}{2\eta T}\right](-\nabla^c t \nabla_c t)^{-1/2} d\Sigma.$$

First show that the set of perturbations for which dE/dt vanishes is a subset of the perturbations for which E itself vanishes. From Eq. (A2) it follows that $dE/dt = 0$ implies that

(A3) $$\delta\tau = \delta q^a = \delta\tau^{ab} = 0.$$

Using these conditions, and the linearized perturbation equations [Eqs. (42) - (43), and (65) - (67)], it is straightforward to show that any such solution also has

(A4) $$\nabla_a \delta\Theta = 0,$$

and the vector

(A5) $$\xi^a = \delta u^a/T - \delta T u^a/T^2$$

must be a Killing vector field. The energy functional, Eq. (A1), for this set of perturbations can now be evaluated. Using Eq. (81) it follows that

(A6) $$E^a = \tfrac{1}{2}\delta T^a{}_b \xi^b + \tfrac{1}{2}\delta\Theta[n\delta u^a + \delta n u^a],$$

so that

(A7) $$E(t) = \tfrac{1}{2}\,\delta P(\xi^a) + \tfrac{1}{2}\,\delta\Theta\delta N$$

where

(A8) $$\delta N = -\int_{\Sigma(t)} [n\delta u^a + \delta n u^a] t_a d\Sigma$$

is the variation in the conserved particle number, and

$$\delta P(\xi^a) = -\int_{\Sigma(t)} \delta T^a{}_b \xi^b t_a d\Sigma \tag{A9}$$

is the variation in the conserved momentum associated with the Killing vector field ξ^a. For those perturbations where the variations in these conserved quantities vanish, it follows that $E(t) = 0$ from Eq. (A7). Thus, perturbations that are constrained by $\delta N = \delta P(\xi^a) = 0$, which are elements of the kernel of dE/dt (i.e., $dE/dt = 0$) are also elements of the kernel of E (i.e., $E = 0$).

Next consider initial values of the perturbation functions on some surface $\Sigma(t)$ for which the energy $E < 0$. This energy is monotonically decreasing from Eq. (A2). Since the functionals E and dE/dt are continuous (in an appropriate square integrable norm) functionals of the perturbation solutions, it follows that any perturbation having initially negative E must remain outside an open set containing the kernel of E. Since the kernel of dE/dt is a subset of the kernel of E, it follows that dE/dt will also be bounded away from zero as the fluid perturbations evolve. It follows that E will decrease without bound. Since E can be diagonalized as in Eq. (88), it follows that one of the perturbation functions δZ_A associated with a negative eigenvalue, $\Omega_A < 0$, of the energy must grow without bound in this case.

BIBLIOGRAPHY

1. J. C. Maxwell, "On the Dynamical Theory of Gases", Philos. Trans. R. Soc. London, 157 (1867), 49-88.

2. C. Eckart, "The Thermodynamics of Irreversible Processes III. Relativistic Theory of the Simple Fluid", Phys. Rev., 58 (1940), 919-924.

3. L. Landau and E. M. Lifshitz, Fluid Mechanics, Addison-Wesley, Reading, Mass., 1958. Sec. 127.

4. C. W. Misner, K. S. Thorne, and J. A. Wheeler, Gravitation, Freeman, San Francisco, 1973, pp. 567 and 568.

5. S. Weinberg, Gravitation and Cosmology, Wiley, New York, 1972, pp. 53-57.

6. P. Havas and R. J. Swenson, "Relativistic Thermodynamics of Fluids. I", Ann. Phys. (N.Y.), 118 (1979), 259-306.

7. H. Grad, "On the Kinetic Theory of Rarefied Gases", Comm. Pure Appl. Math., 2 (1949), 331-407.

8. I. Müller, "Zum Paradoxen der Wärmeleitungstheorie", Z. Physik, 198 (1967), 329-344.

9. J. M. Stewart, Non-equilibrium Relativistic Kinetic Theory, Springer-Verlag, Berlin, 1971.

10. J. L. Anderson, in Relativity (M. Carmeli, S. I. Fickler, and L. Witten, editors), Plenum, New York, 1970, pp. 109-124.

11. M. Kranys, "Nontruncated relativistic Chernikov-Grad 13-moment equations as an approach to nonstationary thermodynamics", Phys. Lett., A33 (1970), 77-78.

12. W. Israel, "Nonstationary Irreversible Thermodynamics: A Causal Relativistic Theory", Ann. Phys. (N.Y.), 100 (1976), 310-331.

13. J. M. Stewart, "On transient relativistic thermodynamics and kinetic theory", Proc. R. Soc. London, A357 (1977), 59-75.

14. W. Israel and J. M. Stewart, "On transient relativistic thermodynamics and kinetic theory. II", Proc. R. Soc. London, A365 (1979), 43-52.

15. W. Israel and J. M. Stewart, "Transient Relativistic Thermodynamics and Kinetic Theory", Ann. Phys. (N.Y.), 118 (1979), 341-372.

16. W. Israel and J. M. Stewart, in General Relativity and Gravitation, edited by A. Held, Plenum, New York, 1980, Vol. II.

17. R. J. Swenson, "Heat Conduction - Finite or Infinite Propagation", J. Non-Equilib. Thermodyn., 3 (1978), 39-48.

18. L. Lindblom and W. A. Hiscock, "On the stability of rotating stellar models in general relativity theory", Astrophys. Jour., 267 (1983), 384-401.

19. R. Courant and D. Hilbert, Methods of Mathematical Physics, Interscience, New York, 1962, Vol. II.

20. L. Brillouin, Wave Propagation and Group Velocity, Academic Press, New York, 1960.

21. W. A. Hiscock and L. Lindblom, "Generic instabilities in first-order dissipative relativistic fluid theories", Physical Review, D31 (1985), 725-733.

22. W. A. Hiscock, "Generic instabilities in first-order dissipative relativistic fluid theories. II. Havas-Swenson-type theories", Physical Review, D33 (1986), 1527-1532.

23. W. A. Hiscock and L. Lindblom, "Stability and Causality in Dissipative Relativistic Fluids", Ann. Phys. (N.Y.), 151 (1983), 466-496.

24. J. L. Friedman, "Generic Instability of Rotating Relativistic Stars", Comm. Math. Phys., 62 (1978), 247-278.

25. K. S. Thorne, in High Energy Astrophysics, ed. by C. DeWitt, E. Schutzman, and P. Veron, Gordon and Breach, New York, 1967, Vol. 3, p. 307.

26. L. Lindblom, "Necessary Conditions for the Stability of Rotating Newtonian Stellar Models", Astrophys. Jour., 267 (1983), 402-408.

W. A. HISCOCK
L. LINDBLOM
DEPARTMENT OF PHYSICS
MONTANA STATE UNIVERSITY
BOZEMAN, MONTANA 59717

Contemporary Mathematics
Volume 71, 1988

THE HAMILTONIAN FORMULATION OF CLASSICAL FIELD THEORY

Jerrold E. Marsden*

1. INTRODUCTION

I shall present some results from the theory of classical non-relativistic field theory and discuss how they might be useful in the general relativistic context. Some of the Hamiltonian formalism has already been successfully employed in the general relativistic context, but much more remains to be done in the area of dynamic stability, linearization stability, bifurcation, symmetry breaking, and covariant reduction.

2. POISSON MANIFOLDS

Let us begin with some terminology. A *Poisson manifold* is a manifold P together with a bracket $\{\ ,\ \}$ on $\mathcal{F}(P) := C^\infty(P,\mathbf{R})$ satisfying

(PB1) $\{\ ,\ \}$ makes $\mathcal{F}(P)$ into a Lie algebra

and

(PB2) $\{FG,H\} = F\{G,H\} + G\{F,H\}$.

We call $C \in \mathcal{F}(P)$ a *Casimir* if

$$\{C,F\} = 0$$

*Research partially supported by NSF grant DMS 84-04506.

for all $F \in \mathcal{F}(P)$. If $H \in \mathcal{F}(P)$, the *Hamiltonian vector field* X_H associated to H is defined by

$$X_H[F] = \{F,H\}$$

for all $F \in \mathcal{F}(P)$, where $X_H[F] = dF \cdot X_H$ denotes the derivative of F in the direction X_H. Thus, C is a Casimir when $X_C = 0$.

Besides the standard canonical, or symplectic examples, one has the following non-canonical examples:

EXAMPLES. A. Let $P = \mathbf{R}^3$ with elements denoted $\boldsymbol{\ell}$. Let

$$\{F,H\}(\boldsymbol{\ell}) = -\boldsymbol{\ell} \cdot (\nabla F \times \nabla H).$$

One checks that this makes P into a Poisson manifold and that the functions

$$C(\boldsymbol{\ell}) = \Phi(\|\boldsymbol{\ell}\|^2)$$

for any $\Phi \in \mathcal{F}(\mathbf{R})$, are Casimirs. For $H \in \mathcal{F}(P)$, one has

$$X_H(\boldsymbol{\ell}) = \nabla H \times \boldsymbol{\ell}.$$

In particular, if

$$H = \tfrac{1}{2}\langle \mathbf{I}^{-1}\boldsymbol{\ell}, \boldsymbol{\ell}\rangle$$

for a positive definite symmetric matrix $\mathbf{I}$ (the inertia tensor), then

$$X_H(\boldsymbol{\ell}) = \omega \times \boldsymbol{\ell}, \qquad \omega = \mathbf{I}^{-1}\boldsymbol{\ell}$$

gives Euler's rigid body equations.

B. Let P be the space of triples $(\mathbf{M},\rho,\eta)$ where $\mathbf{M}$ = a one form density, ρ = a density, and η = a function on a domain D in $\mathbf{R}^3$, with appropriate boundary conditions. Let

$$\{F,H\} = \int_D \mathbf{M}\left[\left(\frac{\delta H}{\delta \mathbf{M}}\cdot\nabla\right)\frac{\delta F}{\delta \mathbf{M}} - \left(\frac{\delta F}{\delta \mathbf{M}}\cdot\nabla\right)\frac{\delta H}{\Delta \mathbf{M}}\right]$$
$$+ \int_D \rho\left[\left(\frac{\delta H}{\delta \mathbf{M}}\cdot\nabla\right)\frac{\delta F}{\delta \rho} - \left(\frac{\delta F}{\delta \mathbf{M}}\cdot\nabla\right)\frac{\delta H}{\delta \rho}\right]$$
$$+ \int_D \eta\nabla\cdot\left[\frac{\delta H}{\delta \mathbf{M}}\frac{\delta F}{\delta \eta} - \frac{\delta F}{\delta \mathbf{M}}\frac{\delta H}{\delta \eta}\right]$$

which is a Poisson structure on P. Here $\delta F/\delta \mathbf{M}$ is the functional derivative, a vector field on D. The function

$$C(\mathbf{M},\rho,\eta) = \int_D \rho\Phi(\eta,\Omega)$$

where $\Omega = \frac{1}{\rho}[\mathbf{d}\eta \wedge \mathbf{d}(\mathbf{M}/\rho)]$ (the potential vorticity) and $\Phi \in \mathcal{F}(\mathbf{R}^2)$, is a Casimir.

The equations for adiabatic compressible flow are

$$\rho\frac{D\mathbf{v}}{dt} = -\nabla p, \quad \frac{D\rho}{dt} + p \operatorname{div} \mathbf{v} = 0, \quad \frac{D\eta}{dt} = 0$$

where $D/dt = \partial/\partial t + \mathbf{v}\cdot\nabla$, $\mathbf{v}$ = velocity field, ρ = density, η = entropy, p = pressure. Here $p = \rho^2(\partial w/\partial\rho)$ where $w(\rho,\eta)$ is the energy density per unit mass, a given function of ρ, η, and we assume $c^2 := \partial p/\partial\rho > 0$. With $M = \rho\mathbf{v}$ and

$$H = \int_D \rho[\|\mathbf{v}\|^2/2 + w],$$

the equations are Hamiltonian relative to the above bracket.

C. The two preceding brackets are special cases of Lie-Poisson brackets. These are defined on the dual $\mathfrak{g}^*$ of a Lie algebra $\mathfrak{g}$ by

$$\{F,H\}_\pm(\mu) = \pm\left\langle\mu, \left[\frac{\delta F}{\delta\mu}, \frac{\delta H}{\delta\mu}\right]\right\rangle$$

where $\mu \in \mathfrak{g}^*$, $\langle\ ,\ \rangle$ is the pairing between $\mathfrak{g}^*$ and $\mathfrak{g}$, $\delta F/\delta\mu \in \mathfrak{g}$ is the functional derivative:

$$DF(\mu)\cdot\nu := \frac{d}{d\epsilon}F(\mu + \epsilon\nu)\Big|_{\epsilon=0} = \left\langle\nu, \frac{\delta F}{\delta\mu}\right\rangle$$

and [,] is the Lie algebra bracket. For example A, $\mathfrak{g} = so(3) = (\mathbb{R}^3, \times)$ (with the '−' sign), and for example *B*, $\mathfrak{g}$ is the semi-direct product (vector fields) *s* (functions × densities) (with the '+' sign). ▲

These brackets are not conjured out of thin air, but are produced by the methods of reduction (see Marsden, Weinstein, et. al. [1983] for a review). For example, for the rigid body, one starts with phase space $T^*SO(3)$ with the canonical bracket and reduces by SO(3), an intrinsic symmetry group. This produces a map

$$T^*SO(3) \to so(3)^* \quad (6 \text{ dimensions} \to 3)$$

given by left translation to the identity. This map is a *Poisson map*; i.e., is bracket preserving. This map is in fact a momentum map (Noether conserved quantity)

$$J_G: P \to \mathfrak{g}^*$$

which is defined for a Poisson action of a Lie group *G* with associated Lie algebra $\mathfrak{g}$, on *P* by

$$\xi_P[F] = \{F, \langle J, \xi \rangle\}$$

where $\xi \in \mathfrak{g}$ and ξ_P is the corresponding infinitesimal generator of the action. (For SO(3) use the action by SO(3) on the *right.*) It is a general fact that equivariant momentum maps are Poisson maps (use $\mathfrak{g}^*_+$ for left actions, $\mathfrak{g}^*_-$ for right actions). For fluids, the relevant group one reduces by is the particle relabelling (or rearrangement) group.

Obviously, the Casimirs are conserved quantities for any Hamiltonian system. If *H* is *G*-invariant, a momentum map *J* is conserved as well. For Lie-Poisson brackets, conservation of the Casimirs corresponds to the fact that any Hamiltonian system leaves the coadjoint orbits invariant. For the rigid body these are the body angular momentum spheres $\|\boldsymbol{\ell}\|^2$ = constant and for adiabatic flow, they are the states that are "kinematically connected"; i.e., there is a diffeomorphism (representing a conceivable particle motion) mapping one to another. These coadjoint orbits are also the *symplectic leaves*;

3. THE ENERGY-CASIMIR METHOD

Two important applications of this formalism are to stability and to bifurcation. The first is based on the *energy-Casimir* method, developed and used by Arnold [1969] and Holm et. al. [1985]. To motivate it, first consider the classical canonical case.

For the Hamiltonian systems in canonical form

$$\dot{q}^i = \frac{\partial H}{\partial p_i}, \quad \dot{p}_i = -\frac{\partial H}{\partial q_i},$$

there is a classical stability criterion due to Lagrange and Dirichlet. Clearly, an equilibrium point in this case is a critical point of H. If the $2n \times 2n$ matrix D^2H of second partial derivatives evaluated at (q_e,p_e) is positive or negative definite (i.e. all the eigenvalues have the same sign), then (q_e,p_e) is stable. This follows from conservation of energy and the fact, proven in advanced calculus, that the level sets of H near (q_e,p_e) are approximately ellipsoids. Apart from KAM theory, which gives stability of periodic solutions for two degree of freedom systems, the Lagrange-Dirichlet theorem is the only known general stability theorem for canonical systems.

The energy-Casimir method is a generalization of the Lagrange-Dirichlet method. Given an equilibrium u_e for evolution equations $\dot{u} = X(u)$, it proceeds in the following steps:

Energy-Casimir Method

Step A. Write the equations in Hamiltonian form $\dot{F} = \{F,H\}$.

Step B. Find a family of conserved quantities C, such as a family of Casimirs.

Step C. Select C such that $H + C$ has a critical point at u_e.

Step D. Check to see if $D^2(H + C)(u_e)$, the matrix of second partial derivatives of $H + C$ at u_e, is positive or negative definite.

With regard to step C, we point out that an equilibrium solution need not be a critical point of H alone; in general $DH(u_e) \neq 0$. An example where this occurs is a rigid body spinning about one of its principal axes of inertia. In this case, a critical point of H alone would have zero angular velocity; but a critical point of $H + C$ is a (nontrivial) stationary rotation about one of the principal axes.

Formally, the same argument used to establish the Lagrange-Dirichlet test also works here. Unfortunately, for systems with infinitely many degrees of freedom (like fluids and plasmas), there is a snag. The calculus argument used before simply runs into problems; one might think these are just technical and that we just need to improve the methods. In fact there is widespread belief in this "energy criterion" (see, for instance, the discussion and references in Marsden and Hughes [1983], Chapter 6). However, Ball and Marsden [1984] have shown by means of a "realistic" example from elasticity theory that the difficulty is genuine. One way to overcome this difficulty is to modify step D using a convexity argument of Arnold [1969].

Convexity Analysis

Modified Step D

(a) Let $\Delta u = u - u_e$ denote a finite variation in phase space.

(b) Find quadratic functions Q_1 and Q_2 such that

$$Q_1(\Delta u) \leqslant H(u_e + \Delta u) - H(u_e) - DH(u_e)\cdot\Delta u$$

$$Q_2(\Delta u) \leqslant C(u_e + \Delta u) - C(u_e) - DC(u_e)\cdot\Delta u.$$

(c) Require that $Q_1(\Delta u) + Q_2(\Delta u) > 0$ for all $\Delta u \neq 0$.

(d) Introduce the *norm* $\|\Delta u\|$ by

$$\|\Delta u\|^2 = Q_1(\Delta u) + Q_2(\Delta u),$$

so $\|\Delta u\|$ as a measure of the distance from u to u_e; $d(u,u_e) = \|\Delta u\|$.

(e) Require that

$$|H(u_e + \Delta u) - H(u_e)| \leqslant C_1\|\Delta u\|$$

and

$$|C(u_e + \Delta u) - C(u_e)| \leqslant C_2\|\Delta u\|$$

for constants C_1 and C_2, and $\|\Delta u\|$ sufficiently small.

These conditions guarantee stability of u_e and provide the distance measure relative to which stability is defined. The key part of the proof is simply the observation that if we add the two inequalities in (b), we get

$$\|\Delta u\|^2 \leqslant H(u_e + \Delta u) + C(u_e + \Delta u) - H(u_e) - C(u_e);$$

here, $DH(u_e)\cdot\Delta u$ and $DC(u_e)\cdot\Delta u$ have added up to zero by step C. But H and C are constant in time so

$$\|(\Delta u)_{\text{time } t}\|^2 \leqslant [H(u_e + \Delta u) + C(u_e + \Delta u) - H(u_e) - C(u_e)]|_{\text{time } 0}\,.$$

Now employ the inequalities in (e) to get

$$\|(\Delta u)_{\text{time } t}\|^2 \leqslant (C_1 + C_2)\|(\Delta u)_{\text{time } 0}\|\,.$$

This estimate bounds the temporal growth of finite perturbations in terms of initial perturbations, which is exactly what is needed for stability.

In the next section we give an example of how this technique applies in a concrete example. See Holm, Marsden, Ratiu and Weinstein [1985] and Abarbanel, Holm, Marsden and Ratin [1986] for a more extensive analysis.

4. LIQUID DROPS WITH SURFACE TENSION

Because of the historic and sustained interest in the dynamics of gravitating masses, I shall present a somewhat related example, the rotating liquid drop. Consider a planar liquid drop consisting of an incompressible, inviscid fluid with a free boundary and forces of surface tension on the boundary. The dynamic variables are the free boundary Σ and the spatial velocity field $\mathbf{v}$, a divergence free vector field on the region D_Σ bounded by Σ. The surface Σ is an element of the set S of closed curves (respectively surfaces) in $\mathbf{R}^2$ (respectively $\mathbf{R}^3$) diffeomorphic to the boundary of a reference region D and enclosing the same area (respectively volume) as D. We let N denote the space of all such pairs $(\Sigma,\mathbf{v})$. The Hamiltonian approach to hydrodynamic problems was introduced in the fixed boundary case by Arnold [1966] and developed by Marsden and Weinstein [1974,1982,1983]. The free boundary case has also been studied by Sedenko and Iudovich [1978].

The equations of motion for an ideal fluid with a free boundary Σ with surface tension τ are

$$\frac{\partial \mathbf{v}}{\partial t} + (\mathbf{v} \cdot \nabla)\mathbf{v} = -\nabla p,$$

$$\frac{\partial \Sigma}{\partial t} = \langle \mathbf{v},\nu \rangle,$$

$$\operatorname{div} \mathbf{v} = 0 \qquad \text{and} \qquad p|\Sigma = \tau\kappa,$$

where ν is the unit normal to the surface Σ, κ is the mean curvature of Σ and τ is the surface tension coefficient, a numerical constant.

The Poisson bracket will be defined for functions F, G: $N \to \mathbf{R}$ which possess *functional derivatives*, defined as follows:

i) $\delta F/\delta \mathbf{v}$ is a divergence free vector field on D_Σ such that

$$D_{\mathbf{v}}F(\Sigma,\mathbf{v})\cdot\delta\mathbf{v} = \int_{D_\Sigma} \langle \frac{\delta F}{\delta \mathbf{v}}, \delta\mathbf{v} \rangle dA$$

where the partial (Fréchet) derivative $D_{\mathbf{v}}F$ is computed with Σ fixed.

ii) $\delta F/\delta\varphi$ is the function on Σ with zero integral given by

$$\frac{\delta F}{\delta \varphi} = \langle \frac{\delta F}{\delta \mathbf{v}}, \nu \rangle.$$

(The symbol φ represents the potential for the gradient part of $\mathbf{v}$ in the Helmholtz, or Hodge, decomposition.)

iii) $\delta F/\delta\Sigma$ is a function on Σ determined up to an additive constant as follows. A variation $\delta\Sigma$ of Σ is identified with a function on Σ representing the infinitesimal variation of Σ in its normal direction. It follows from the incompressibility assumption that $\delta\Sigma$ has zero integral. Let $\delta F/\delta\Sigma$ be the function determined up to an additive constant by

$$\int_\Sigma \frac{\delta F}{\delta\Sigma}\,\delta\Sigma\,ds = D_\Sigma F(\Sigma,\mathbf{v})\cdot\delta\Sigma.$$

We now define a Poisson bracket on N as follows. For functions F and G mapping N to $\mathbf{R}$ and possessing functional derivatives as defined above, set

$$\{F,G\} = \int_{D_\Sigma} \langle\omega, \frac{\delta F}{\delta\mathbf{v}}\times\frac{\delta G}{\delta\mathbf{v}}\rangle dA + \int_\Sigma \left[\frac{\delta F}{\delta\Sigma}\frac{\delta G}{\delta\varphi} - \frac{\delta G}{\delta\Sigma}\frac{\delta F}{\delta\varphi}\right]ds,$$

where $\omega = \operatorname{curl}\mathbf{v}$. This Poisson bracket on N is derived from the canonical cotangent bracket on T^* , where, in the two-dimensional case, $C = \mathrm{Emb}_{\mathrm{vol}}(D,\mathbf{R}^2)$ is the manifold of volume-preserving embeddings of a two-dimensional reference manifold D into $\mathbf{R}^2$, by reduction by the group $G = \mathrm{Diff}_{\mathrm{vol}}(D)$, the group of volume-preserving diffeomorphisms of D (i.e. the group of particle relabelling transformations). (See Lewis, Marsden, Montgomery and Ratiu [1986a] for details.)

We take our Hamiltonian to be

$$H(\Sigma,\mathbf{v}) = \int_{D_\Sigma} \tfrac{1}{2}|\mathbf{v}|^2 dA + \tau\int_\Sigma ds.$$

The functional derivatives of H are computed to be

$$\frac{\delta H}{\delta\mathbf{v}} = \mathbf{v},$$

$$\frac{\delta H}{\delta\varphi} = \langle\frac{\delta H}{\delta\mathbf{v}},\nu\rangle = \langle\mathbf{v},\nu\rangle,$$

and

$$\frac{\delta H}{\delta\Sigma} = \tfrac{1}{2}|\mathbf{v}|^2 + \tau\kappa,$$

where $\delta H/\delta\Sigma$ is taken modulo constants. For this H and the Poisson bracket defined above, the equations of motion for the free boundary fluid with surface tension are equivalent to the relation $\dot{F} = \{F,H\}$ for all functions F on N possessing functional derivatives.

This explains the Hamiltonian structure for the free boundary problem. It is used in the following two ways:

1. *Stability.* The circular planar drop is nonlinearly stable provided

$$\frac{3\tau}{r^3} > (\Omega/2)^2$$

where Ω is the rotation rate of the circular drop. Formal stability is proved using the energy-Casimir method, taking $H + C$ to be the energy plus a multiple of the angular momentum. (See Lewis, Marsden and Ratiu [1986b].) To prove nonlinear stability rigorously requires some more work, analogous to using the convexity estimates. Here it is the Weierstrass theory in a version due to Hestenes that does the job. See Lewis [1987] for details.

2. *Bifurcation.* If the parameters τ, r or Ω are varied to violate the stability condition, one can prove that a bifurcation occurs, so in this sense the stability condition is sharp. The result, due to Lewis, Marsden and Ratiu [1986c] uses the bifurcation theory for Hamiltonian systems with symmetry (see Golubitsky and Stewart [1986]).

5. COMMENTS ON GENERAL RELATIVISTIC FLUIDS

The Poisson structure given in Section 2 has been shown to be relevant for general relativistic fluids by BMW (Bao, Marsden and Walton) [1985]. They show that the evolution equations in a general lapse and shift (not necessarily comoving with the fluid) have the form $\dot{F} = \{F,H\}$ where $H = N\mathcal{H} + X\cdot\mathcal{J}$ (N = lapse, X = shift, $\mathcal{H}$ = superhamiltonian, $\mathcal{J}$ = supermomentum) and where $\{\ \}$ is the canonical bracket for the ADM variables (g,π) plus the Lie-Poisson bracket for the fluid variables. This is the Poisson bracket form which corresponds to the adjoint form of the ADM equations (Fischer and Marsden [1979]).

Remarks and Open Questions:

1. Does the free boundary bracket of §4 also carry over to the general relativistic case?

2. The energy-Casimir method has been used for fluids in a fixed background by Holm and Kuperschmidt [1984, 1986]; is it useful in the coupled case as well?

3. Is the bifurcation theory with symmetry useful for studying general relativistic gravitating masses, fission, etc? We point out, as a curiosity, that the bifurcation that occurs in the liquid drop example is to a shape with "galactic symmetry" or "propellor symmetry", i.e. symmetry under rotation by π, but not under reflections.

4. The Hamiltonian structure can surely be generalized to include electromagnetic effects, following Marsden and Weinstein [1982] and Marsden, Weinstein, et. al. [1983], for both fluids and plasmas.

5. The Hamiltonian structure of Bao et. al. is useful for *linearization stability.* Recall that the main result of the linearization stability program is that a spacetime (with a compact Cauchy surface of constant mean curvature) is linearization stable if and only if it has no Killing fields; if it does, then obstructions to perturbation expansions are the second order Taub conditions. (See Fischer, Marsden and Moncrief [1980], Arms, Marsden and Moncrief [1982], Isenberg and Marsden [1982].) For fluids, this result appears to be true as well, provided one imposes constraints on the perturbations corresponding to, for example, preserving baryon number - these constraints are, mathematically, preserving the coadjoint orbit structure and are closely related to "Lin constraints" (see Cendra and Marsden [1986]).

6. COVARIANT REDUCTION

Finally, I wish to point out some interesting open issues in classical relativistic field theory which are suggested by the preceding sections.

First of all, one should note that there is a beautiful covariant version of the multisymplectic formalism adapted especially for the linearization stability program, and which incorporates momentum maps, called the "GIMMSY project"; see Gotay et. al. [1987]. This is a covariant analogue of the formalism linking Lagrangians on TQ (Q = configuration space) with Hamiltonians on T^*Q via the Legendre transformation.

PROBLEM 1. *Develop a covariant theory of reduction, starting with the Gimmsy setup and producing, after* 3 + 1*ing, the* BMW *bracket for* GR *fluids.* The work of Kunzle and Nester [1984] and Holm [1985] suggests this is reasonable.

On the other hand, it seems that there is also a *covariant* version of Poisson brackets (where the bracket does *not* depend on any hypersurface and involves integration over *spacetime*). In this formalism, due to Marsden, Montgomery, Morrison and Thompson [1986] (M^3T), the Euler-Lagrange field equations take the form $\{F,S\} = 0$ where S is an appropriate action integral. It does work for general relativistic fluids and plasmas, including electromagnetic effects. One obtains covariant analogues of Lie-Poisson brackets.

A more grandiose project is:

PROBLEM 2. *Make a commutative diagram of the following sort*:

Covariant Multisymplectic Formalism for GR fluids (GIMMSY)	Covariant Reduction ⟶	Covariant Lie Poisson Formalism for GR Fluids (M^3T)
3 + 1 procedure ↓		↓ 3 + 1 procedure
3+1 GR fluids in material (Lagrangian) or inverse material representation (Kunzle and Nester)	Standard material to spatial reduction (Holm [1985], Holm, Marsden and Ratiu [1986] ⟶	3+1 GR fluids in spatial representation (BMW)

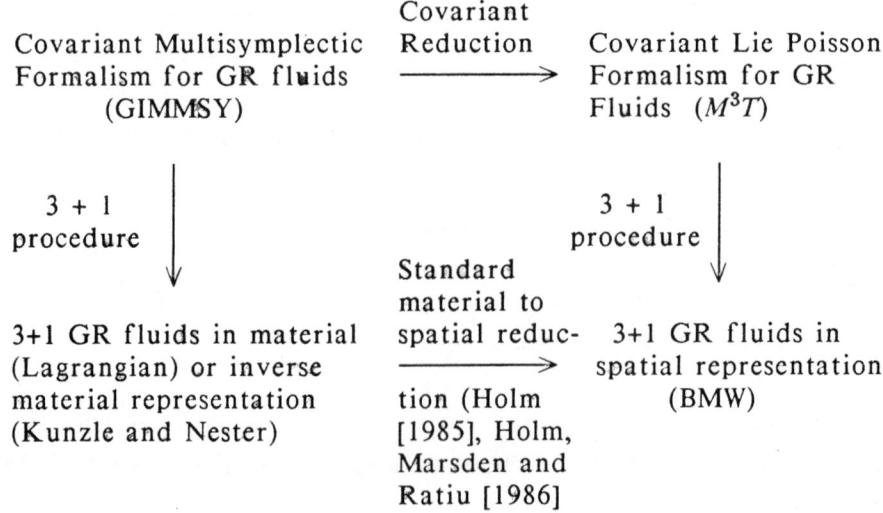

Bits of this scheme are known (see the quoted references); it would greatly clarify classical field theory if the whole picture were known in detail for fluids in particular and for GR fluids more generally.

REFERENCES

H. Abarbanel, D. Holm, J. Marsden and T. Ratiu [1986]. Nonlinear Stability of Stratified Flow, *Phys. Rev. Lett.* **52**, 2352-2355 and *Phil. Trans. Roy. Soc. Lond.* **A318**, 349-409.

J. M. Arms, J. E. Marsden and V. Moncrief [1982]. The structure of the space of solutions of Einstein's equations, II: Several Killing fields and the Einstein-Yang Mills equations, *Ann. Phys.* **144**, 81-106.

V. I. Arnold [1966]. Sur la geometrie differentielle des groupes de Lie de dimension infinie et ses aplications a l'hydrodynamique des fluids parfaits, *Ann. Inst. Fourier*, Grenoble **16**, 319-361.

V. I. Arnold [1969]. On an a priori estimate in the theory of hydrodynamical stability, *American Math. Soc. Transl.* **79**, 267-269.

J. M. Ball and J. E. Marsden [1984]. Quasiconvexity, second variations and the energy criterion in nonlinear elasticity, *Arch. Rat. Mech. Analysis* **86**, 251-277.

D. Bao, J. E. Marsden and R. Walton [1985]. The Hamiltonian Structure of General Relativistic Perfect Fluids, *Comm. Math. Phys.* **99**, 319-345.

H. Cendra and J. E. Marsden [1986]. Lin Constraints, Clebsch Potentials and Variational Principles, *Physica D*; to appear.

A. E. Fischer and J. E. Marsden [1979]. Topics in the dynamics of general relativity, in *Isolated Gravitating Systems in General Relativity*, J. Ehlers, ed., Italian Physical Society, 322-395.

A. E. Fischer, J. E. Marsden and V. Moncrief [1980]. The structure of the space of solutions of Einstein's equations, I: One Killing field, *Ann. Inst. H. Poincare* **33**, 147-194.

M. Golubitsky and I. Stewart [1986]. Generic Bifurcation of Hamiltonian Systems with Symmetry, *Physica D* (to appear).

M. Gotay, J. Isenberg, J. E. Marsden, R. Montgomery, J. Sniatycki and P. Yasskin [1987]. Momentum Maps and the Hamiltonian Treatment of Classical Field Theories with Constraints (preprint).

D. Holm [1985]. Hamiltonian Formalism of General Relativistic Adiabatic Fluids, *Physica* **17D**, 1-36.

D. D. Holm and B. A. Kuperschmidt [1984]. Relativistic fluid dynamics as a Hamiltonian system, *Phys. Lett.* **101A**, 23-26.

D. D. Holm and B. A. Kuperschmidt [1986]. Lyapunov stability conditions for relativistic multifluid plasma, *Physica* **18D**, 405-409.

D. D. Holm, J. E. Marsden and T. Ratiu [1986]. The Hamiltonian structure of continuum mechanics in material, inverse material, spatial and convective representations, *Seminaire de Mathêmatiques superieurs, Les Presses de L'Université de Montrèal*, **100**, 11-122.

D. D. Holm, J. E. Marsden, T. Ratiu and A. Weinstein [1985]. Nonlinear stability of fluid and plasma equilibria, *Physics Reports* **123**, 1-116.

J. Isenberg and J. E. Marsden [1982]. A slice theorem for the space of solutions of Einstein's equations, *Physics Reports* **89**, 179-222.

H. P. Kunzle and J. M. Nester [1984]. Hamiltonian formulation of gravitating perfect fluids and the Newtonian limit, *J. Math. Phys.* **25**, 1009-1018.

D. Lewis [1987]. *Rotating liquid drops: Hamiltonian structure, stability and bifurcation*, Thesis, University of California, Berkeley.

D. Lewis, J. E. Marsden, R. Montgomery, and T. Ratiu [1986a]. The Hamiltonian Structure for Dynamic Free Boundary Problems, *Physica* **18D**, 391-404.

D. Lewis, J. E. Marsden, and T. Ratiu [1986b]. Formal stability of liquid drops with surface tension, *Perspectives in Nonlinear Dynamics*, (ed. by M. F. Shlesinger, R. Cawley, A. W. Saenz and W. Zachary), World Scientific, 71-83.

D. Lewis, J. E. Marsden, and T. Ratiu [1986c]. Stability and bifurcation for a rotating planar liquid drop, *J. Math. Phys.*, to appear.

J. E. Marsden and T. Hughes [1983]. *Mathematical Foundations of Elasticity*, Prentice-Hall.

J. E. Marsden, R. Montgomery, P. J. Morrison and W. B. Thompson [1986]. Covariant Poisson brackets for classical fields, *Annals of Physics* **169**, 29-48.

J. E. Marsden, T. Ratiu, and A. Weinstein [1984b]. Reduction and Hamiltonian Structures on Duals of Semidirect Product Lie Algebras, *Cont. Math. AMS* **28**, 55-100.

J. E. Marsden and A. Weinstein [1974]. Reduction of Symplectic Manifolds with Symmetry, *Rep. Math. Phys.* **5**, 121-130.

J. E. Marsden and A. Weinstein [1982]. The Hamiltonian Structure of the Maxwell-Vlasov Equations, *Physica* **4D**, 394-406.

J. E. Marsden and A. Weinstein [1983]. Coadjoint Orbits, Vortices and Clebsch Variables for Incompressible Fluids, *Physica* **7D**, 305-323.

J. E. Marsden, A. Weinstein, T. Ratiu, R. Schmid, and R. G. Spencer [1983]. Hamiltonian Systems with Symmetry, Coadjoint Orbits and Plasma Physics, *Proc. IUTAM-ISIMM Symposium on Modern Developments in Analytical Mechanics*, Torino, June 7-11, 1982, *Atti della Academia della Scienze di Torino* **117**, 289-340.

V. I. Sedenko and V. I. Iudovich [1978]. Stability of steady flows of ideal incompressible fluid with free boundary, *Applied Math. and Mechanics* **42**, 1148-1155.

JERROLD E. MARSDEN
DEPARTMENT OF MATHEMATICS
UNIVERSITY OF CALIFORNIA
BERKELEY, CA 94720

Contemporary Mathematics
Volume 71, 1988

THE RICCI FLOW ON SURFACES

Richard S. Hamilton

In this paper we will discuss the evolution of a Riemannian metric g_{ij} on a compact surface M by its curvature R under the equation

$$* \qquad \frac{\partial}{\partial t} g_{ij} = (r - R) g_{ij}$$

where r is the average value of R. We shall prove the following results.

1.1 **Theorem.** *For any initial data, the solution exists for all time. (See also Cao [1]).*

1.2 **Theorem.** *If* $r \leq 0$, *the metric converges to one of constant curvature.*

1.3. **Theorem.** *If* $R > 0$, *the metric converges to one of constant curvature.*

Of course we conjecture that any metric on a compact surface converges to one of constant curvature, but the case of a metric in S^2 with curvature of varying sign is still open.

As consequences we obtain new proofs of many classical results, such as the uniformization theorem for Riemann surfaces, the topological classification of surfaces, and the topological type of the diffeomorphism group of surfaces. The proofs depend on

0271-4132/88 $1.00 + $.25 per page

some new a *priori* estimates on higher derivatives which are interesting in their own right. We hope these results will generalize to higher dimensional Kähler manifolds, and maybe throw light on the problem of S^2 necks pinching off on three manifolds with positive scalar curvature under the Ricci flow.

2. We have previously studied the flow of a metric by its Ricci curvature

$$\frac{\partial}{\partial t} g_{ij} = 2\left[\frac{r}{n} g_{ij} - R_{ij}\right]$$

on three-manifolds [3] and on four-manifolds [4]. In some repects the higher dimensional cases are easier, due to the information contained in the second Bianchi identity. For surfaces with positive curvature, the gradient estimate on the scalar curvature fails for this reason. Therefore a new approach is needed.

On a surface, the Ricci flow equation simplifies, because all of the information about curvature is contained in the scalar function R. In our notation, $R = 2K$ where K is the Gauss curvature, with $K = 1$ on the sphere of radius 1. Thus the Ricci curvature is given by

$$R_{ij} = \frac{1}{2} R\, g_{ij}$$

and the Ricci flow equation simplifies the following to the equation for the metric:

$$\frac{\partial}{\partial t} g_{ij} = (r - R) g_{ij}. \tag{2.1}$$

Notice that the change in the metric is pointwise a multiple of the metric, so the conformal structure is preserved. The term r in the equation is added to keep the area of the surface constant; if $\mu = \sqrt{\det g_{ij}}$ is the area element then

$$\frac{\partial}{\partial t} \mu = (r - R)\mu$$

and as a result, if A is the total area

$$\frac{d}{dt} A = \frac{d}{dt} \int 1 d\mu = \int (r - R) d\mu = 0$$

since r is the mean scalar curvature

$$r = \int R d\mu / \int 1 d\mu .$$

The integral of R over the surface M gives the Euler class $\chi(M)$ by the Gauss-Bonnet formula

$$\int R d\mu = 4\pi \chi(M)$$

and as a consequence, on a surface we see that r is constant; indeed

$$r = 4\pi \chi(M)/A.$$

We could choose to normalize with $A = 4\pi$ and $r = 2$ on the sphere, but we prefer not.

The equation 2.1 makes perfectly good sense in higher dimensions, but differs from the Ricci flow. It is in fact the gradient flow for the Yamabe problem, where we fix the conformal structure and the volume and try to minimize the mean scalar curvature r. Thus in higher dimensions r will decrease. We can prove that the solution exists for all time. When $R < 0$, the solution converges exponentially to a metric with constant scalar curvature. When $R > 0$, we can show the solution exists for all time, and the curvature approaches a constant; however there is some problem with the convergence of the metric. Presumably this problem could be overcome using the positive mass estimate.

3. When the metric g_{ij} evolves, so will its scalar curvature R. The equation for the evolution of the curvature on a surface is particularly simple and elegant. 3.1 The equation for the curvature:

$$\frac{\partial R}{\partial t} = \Delta R + R^2 - rR.$$

It can be found by a straightforward calculation. It has the following interpretation. Let B be any region on the surface with a smooth boundary curve ∂B, and let N be the unit normal to the boundary and λ the arc length measure along the boundary. Then

$$\frac{d}{dt} \iint_B R d\mu = \int_{\partial B} \nabla R \cdot N \, d\lambda.$$

This says that the curvature R flows across any boundary curve with a speed equal to the negative of its gradient. This produces the usual Laplacian term for diffusion in the equation 3.1, while the quadratic self-interaction term $R(R-r)$ is due entirely to the fact that the curvature density R increases or decreases with the change in the area element μ, which changes by the factor $r-R$.

Applying the maximum principle to equation 3.1 gives the following result.

3.2 **Theorem.** *If $R \geq 0$ at the start, it remains so for all time. Likewise if $R \leq 0$ at the start, it remains so for all time. Thus both positive and negative curvature are preserved for surfaces. The same is true in the higher dimensional Yamabe flow, but not in the higher dimensional Ricci flow, where only positive curvature is preserved.*

For negative curvature, the preceeding result can be improved considerably. As an immediate consequence of the maximum principle we get the following result.

3.3 **Theorem.** *If $-C \leq R \leq -\varepsilon < 0$ at the start, then it remains so, and*

$$re^{-\varepsilon t} \leq r - R \leq Ce^{rt}$$

so R approaches r exponentially.

Proof. The maximum of R satisfies the differential inequality

$$\frac{d}{dt} R_{\max} \leq R_{\max}(R_{\max} - r) \leq -\varepsilon(R_{\max} - r)$$

while the minimum of R satisfies

$$\frac{d}{dt} R_{\min} \geq R_{\min}(R_{\min} - r) \geq r(R_{\min} - r).$$

3.4 **Corollary.** *On a compact surface, if $R < 0$ then the solution exists for all time and converges exponentially to a metric of constant negative curvature.*

For positive curvature, the situation is much worse, because $R = r$ is now a repulsive fixed point for the ordinary differential equation

$$\frac{dR}{dt} = R^2 - rR$$

and hence the reaction term in equation 3.1 is fighting the diffusion term. The best we can do is the following.

3.5 **Theorem.** *If* $r > 0$ *and* $R/r \geq c > 0$ *at the start, then* $c < 1$ *and for all time*

$$\frac{R}{r} \geq \frac{1}{1 + (\frac{1}{c} - 1)e^{rt}}.$$

This gives a positive lower bound which deteriorates to zero as $t \to \infty$. Notice that the corresponding upper bound goes to infinity in a finite time.

3.6 **Theorem.** *If* $r > 0$ *and* $R/r \leq C$ *at the start, then* $C > 1$ *and*

$$\frac{R}{r} \leq \frac{1}{1 - (1 - \frac{1}{C})e^{rt}}$$

at least for time

$$t < \frac{1}{r} \log \frac{c}{c - 1}.$$

4. To get results when $R > 0$ somewhere, we need better methods. The first step is to introduce the potential function f.

4.1 **Definition.** The potential f is the solution of the equation

$$\Delta f = R - r$$

with mean value zero.

Note we can always solve the equation since $R - r$ has mean value zero, and the solution is unique up to a constant, so we can make f have mean value zero. The potential function f satisfies a particularly simple equation.

4.2 The equation for the potential:

$$\frac{\partial f}{\partial t} = \Delta f + rf - b$$

where

$$b = \int |Df|^2 d\mu / \int 1\, d\mu$$

is a constant over space and a function only of time.

Proof. Since $\Delta f = R - r$, differentiating in time we compute

$$\Delta \frac{\partial f}{\partial t} = \Delta(\Delta f + rf)$$

which shows that

$$\frac{\partial f}{\partial t} = \Delta f + rf - b$$

for some number b which is a constant over space and a function only of time. It is easy to compute b from the relation

$$\int f\, d\mu = 0.$$

To continue the argument, we introduce a new function h and a tensor M_{ij}.

4.3 **Definition.** We let

$$h = \Delta f + |Df|^2$$

and

$$M_{ij} = D_i D_j f - \frac{1}{2} \Delta f \cdot g_{ij}.$$

Note that M_{ij} is the trace-free part of the second covariant derivative of f.

4.4 The equation for h:

$$\frac{\partial h}{\partial t} = \Delta h - 2|M_{ij}|^2 + rh.$$

4.5 **Corollary.** *If $h \le C$ at the start, then $h \le C\, e^{rt}$ for all time.*

The significance of this estimate is that

$$R = h - |Df|^2 + r$$

so $R \leq C\, e^{rt} + r$. This gives a bound on R from above for all t, which unfortunately deteriorates as $t \to \infty$ if $r > 0$. We also have a lower bound from the maximum principle on equation 3.1. Not using the best possible result, we have the following. If $r \geq 0$ and the minimum of R is negative, it increases. If $r \leq 0$ and the minimum of R is less than r, it increases. This proves the following estimate.

4.6 **Theorem.** For any initial metric on a compact surface, there is a constant C with

$$-C \leq R \leq C\, e^{rt} + r.$$

4.7 **Corollary.** *For any initial metric on a compact surface, the Ricci flow equation has a solution for all time.*

4.8 **Corollary.** *If in addition $r \leq 0$, then the scalar curvature R remains bounded both above and below.*

When $r < 0$, this bound actually shows that $R < 0$ for large time. We can then apply Corollary 3.4. This proves the following.

4.9 **Theorem.** *On a compact surface with $r < 0$, for any initial metric the solution exists for all time and converges to a metric with constant negative curvature.*

5. The case $r = 0$ merits separate attention. We already know the solution exists for all time, and the curvature remains bounded above and below. It remains to see why the solution converges to a flat metric.

Let us write $g_{ij} = e^u \bar{g}_{ij}$ for a conformal change of metric. Then it is easy to compute

$$R = e^{-u}(\bar{R} - \bar{\Delta} u)$$

where $\bar{R}$ is the curvature of $\bar{g}_{ij}$ and $\bar{\Delta}$ is the Laplacian in the metric $\bar{g}_{ij}$. Given $\bar{g}_{ij}$, if we solve $\bar{\Delta}u = \bar{R}$ (which is possible since $\bar{R}$ has mean value zero) then $R = 0$. Hence we can produce a flat metric.

Let us assume now that $\bar{g}_{ij}$ is the flat metric, and study the evolution of g_{ij} by studying its conformal factor u. We easily derive the following equation.

5.1 The equation for the conformal factor

$$\frac{\partial u}{\partial t} = e^{-u}\,\bar{\Delta}u.$$

Applying the maximum principle, we have the following.

5.2 **Corollary.** *There exists a constant C with $-C \le u \le C$. As a result, the metrics $g_{ij}(t)$ are uniformly equivalent for all t. This gives us control of the diameter, the injectivity radius, and the constant in the Sobolev inequality.*

To produce some exponential convergence, we calculate

$$\frac{d}{dt}\int |\bar{D}u|^2\, d\bar{\mu} + 2\int e^{-u}(\bar{\Delta}u)^2\, d\bar{\mu} = 0$$

and use

$$\int (\bar{\Delta}u)^2 d\bar{\mu} \ge c \int |\bar{D}u|^2 d\bar{\mu}$$

and the bounds above and below on u to conclude that for some constant $c > 0$

$$\frac{d}{dt}\int |\bar{D}u|^2\, d\bar{\mu} + c\int |\bar{D}u|^2 d\bar{\mu} \le 0$$

so the integral goes to zero exponentially. Thus

$$\int |\bar{D}u|^2\, d\bar{\mu} \le C\, e^{-ct}$$

for some $c > 0$.

Now if we integrate the previous equation over time, we get

$$2\int_T^\infty \int e^{-u}(\bar{\Delta}u)^2\, d\bar{\mu}\, dt \le \int |\bar{D}u|^2\, d\bar{\mu}(T)$$

which shows that

$$\int_T^\infty \int R^2 d\mu dt \le C\, e^{-cT}.$$

Hence at some point in each interval $T \le t \le T+1$ we will have

$$\int R^2 d\mu \le C\, e^{-ct}.$$

Moreover

$$\frac{d}{dt}\int R^2 d\mu + 2\int |DR|^2 d\mu = \int R^3 d\mu.$$

Since R is bounded, we have

$$\frac{d}{dt}\int R^2 d\mu \le C \int R^2 d\mu.$$

Now since the integral is frequently small, and its growth is controlled, it follows that

$$\int R^2 d\mu \le C\, e^{-ct}$$

for all t. Since R is bounded, any L_p norm of R will go to zero exponentially. Then integrating the previous equation gives

$$\int_0^T \int |DR|^2 d\mu dt \le C\, e^{-cT}$$

for some $c > 0$. This shows

$$\int |DR|^2 d\mu \le C\, e^{-ct}$$

at least once in each interval $T \le t \le T+1$. But we can also bound the growth of this integral.

$$\frac{d}{dt}\int |DR|^2 d\mu + 2\int (\Delta R)^2 d\mu \le -2\int R^2 \Delta R\, d\mu$$

and

$$-2\int R^2 \Delta R\, d\mu \le \int (\Delta R)^2 d\mu + \int R^4 d\mu$$

so

$$\frac{d}{dt}\int |DR|^2 d\mu + \int (\Delta R)^2 d\mu \le \int R^4 d\mu.$$

Since the last term is itself exponentially small, we get

$$\int |DR|^2 d\mu \le C\, e^{-ct}$$

for some $c > 0$ and all t. Then, again integrating over time

$$\int_0^T \int (\Delta R)^2 d\mu dt \le C\, e^{-cT}$$

so

$$\int (\Delta R)^2 d\mu \le C\, e^{-ct}$$

at least once in each interval $T \le t \le T+1$. Now we use the uniform Sobolev inequality to bound the maximum of R by the L^2 norm of R, DR, and D^2R; the bound on D^2R follows from that on ΔR and DR by integrating by parts

$$\int |D^2R|^2 d\mu = \int (\Delta R)^2 d\mu - \frac{1}{2} \int R |DR|^2 d\mu$$

and using the fact that R is bounded. Thus the maximum of R goes to zero exponentially. It now follows by a general result in [3] that the metric converges exponentially to the flat metric.

6. Now we turn our attention to the case where $R > 0$. First we prove a Harnack inequality on R by deriving a maximum principle estimate on the space and time derivatives of $\log R$. This was inspired by a similar proof of the Harnack inequality for the ordinary linear heat equation shown to us by S. T. Yau (see [5]). A similar Harnack estimate can be shown for the curvature of a plane curve moving by its mean curvature vector. We believe that similar Harnack inequalities will play an important role in many geometric problems.

The classical Harnack inequality says the following.

6.1 **Theorem.** *Let M be a compact manifold of dimension n with a fixed metric of non-negative Ricci curvature. Let f be a solution of the ordinary heat equation*

$$\frac{\partial f}{\partial t} = \Delta f$$

for $0 < t < T$ *with* $f > 0$ *everywhere. Then for any two points* (ξ,τ) *and* (X,T) *in space-time with* $0 < \tau < T$ *we have*

$$\tau^{n/2} f(\xi,\tau) \le e^{\Delta/4}\, T^{n/2} f(X,T)$$

where $\Delta = d(\xi,X)^2/(T-\tau)$ *and* $d(\xi,X)$ *is the distance along the shortest geodesic.*

Proof. Let $L = \log f$. Then

$$\frac{\partial L}{\partial t} = \Delta L + |DL|^2.$$

Next let $Q = \dfrac{\partial L}{\partial t} - |DL|^2 = \Delta L$ and compute

$$\frac{\partial Q}{\partial t} \ge \Delta Q + 2DL \cdot DQ + \frac{2}{n} Q^2$$

using

$$|D^2 L|^2 \ge \frac{1}{n}(\Delta L)^2 \text{ and } Rc(DL, DL) \ge 0.$$

It follows from the maximum principle that

$$Q \ge -2/nt$$

regardless of how negative Q might be to start.

Choose a geodesic path from ξ to X parametrized by time t proportionally to the arc length. Along the path

$$\frac{dL}{dt} = \frac{\partial L}{\partial t} + \frac{\partial L}{\partial s}\frac{ds}{dt}.$$

Integrating along the path and using

$$\frac{\partial L}{\partial t} \ge |DL|^2 - \frac{2}{nt} \text{ and } |DL|^2 + \frac{\partial L}{\partial s}\frac{ds}{dt} \ge -\frac{1}{4}\left[\frac{ds}{dt}\right]^2,$$

by Cauchy-Schwartz we get

$$L(X,T) - L(\xi,\tau) = \int_\tau^T \frac{dL}{dt}\,dt$$

$$\geq \int_\tau^T \left\{ |DL|^2 - \frac{2}{nt} + \frac{\partial L}{\partial s}\frac{ds}{dt} \right\} dt$$

$$\geq -\frac{2}{n}\log(T/\tau) - \frac{1}{4}\int_\tau^T \left(\frac{ds}{dt}\right)^2 dt.$$

Now

$$\Delta = d(X,\xi)^2/(T-\tau) = \int_\tau^T \left(\frac{ds}{dt}\right)^2 dt$$

and the result follows by exponentiation.

We imitate the theorem and its proof for the Ricci flow on a surface. Since the metric is changing, we get a more complicated version of Δ.

6.2 **Definition.** On a manifold with a Riemannian metric $g_{ij}(x,t)$ which changes over time t, we define

$$\Delta(\xi,\tau,X,T) = \inf_\gamma \int_\tau^T \left(\frac{ds}{dt}\right)^2 dt$$

taking the infimum over all paths from (ξ,τ) to (X,T) parametrized by time t for $\tau \leq t \leq T$, where ds/dt is the velocity in space at time t.

This agrees with $\Delta(\xi,\tau,X,T) = d(\xi,X)^2/(T-\tau)$ when the metric is fixed. For a varying metric it gives a reasonable notion of distance between points at different times. If we have two fixed metrics $\gamma_{ij}(x)$ and $G_{ij}(x)$ independent of t, with distances $\delta(\xi,X)$ and $D(\xi,X)$ along geodesics, then clearly

$$\delta(\xi,X)^2/(T-\tau) \leq \Delta(\xi,\tau,X,T) \leq D(\xi,X)^2/(T-\tau)$$

whenever $\gamma_{ij}(x) \leq g_{ij}(x,t) \leq G_{ij}(x)$.

We can now state our new Harnack inequality.

6.3 **Theorem.** *Suppose we have a solution of the Ricci flow equation on a compact surface with $R > 0$ for $0 < t \le T$. Then for any two points (ξ, τ) and (X, T) in space-time with $0 < \tau < T$ we have*

$$(e^{r\tau} - 1) R(\xi, \tau) \le e^{\Delta/4} (e^{rT} - 1) R(X, T)$$

where $\Delta = \Delta(\xi, \tau, X, T)$ is defined as before.

Proof. Let $L = \log R$ and let

$$Q = \frac{\partial L}{\partial t} - |DL|^2 = \Delta L + R - r.$$

Then using

$$2 |D_i D_j L - \frac{1}{2}(R - r) g_{ij}|^2 \ge Q^2$$

we compute

$$\frac{\partial Q}{\partial t} \ge \Delta Q + 2 \nabla L \cdot \nabla Q + Q^2 + rQ.$$

It follows from the maximum principle that

$$Q \ge -r\, e^{rt} / (e^{rt} - 1)$$

no matter how negative Q is to start, by comparing to the solution of the ordinary differential equation we would get if Q were constant in space. Now take any path γ from (ξ, τ) to (X, T) parameterized by time t for $\tau \le t \le T$, and compute as before

$$\frac{dL}{dt} = \frac{\partial L}{\partial t} + \frac{\partial L}{\partial s} \frac{ds}{dt}$$

$$L(X, T) - L(\xi, \tau) = \int_\tau^T \frac{dL}{dt}\, dt$$

$$\ge \int_\tau^T \left\{ |DL|^2 - \frac{re^{rt}}{e^{rt} - 1} + \frac{\partial L}{\partial s} \frac{ds}{dt} \right\} dt$$

$$\ge -\log (e^{rT} - 1) / (e^{r\tau} - 1) - \frac{1}{4} \int_\tau^T \frac{ds}{dt}^2 dt\,.$$

The infimum of the last integral over all such paths is the definition of Δ. The result follows by exponentiation.

7. The next step is rather unusual. We let

$$Z = \int Q\, R\, d\mu / \int R\, d\mu$$

and compute using the equation for Q that

$$\frac{dZ}{dt} \geq Z^2 + rZ\ .$$

Now if Z were ever to become positive, then it would blow up to infinity in a finite time. But we already know that the solution exists for all time. The only possible conclusion is that $Z \leq 0$! Using

$$Q = \frac{\Delta R}{R} - \frac{|DR|^2}{R^2} + (R - r)$$

this gives us the following result.

7.1 **Lemma.** *For any solution with $R > 0$ we have*

$$\int (R-r)^2 d\mu \leq \int \frac{|DR|^2}{R} d\mu.$$

Since

$$\frac{d}{dt}\int R \log R\, d\mu = \int (R-r)^2 d\mu - \int \frac{|DR|^2}{R} d\mu$$

we get the following suprising result.

7.2 **Theorem.** *For the Ricci flow on a compact surface with $R > 0$ the integral*

$$\int R \log R\, d\mu$$

is decreasing.

Note that $x \log x \geq -1/e$ is bounded below. Therefore R cannot be large except on a set where $\int R\, d\mu$ is small, which is a set whose Gauss image is small. Therefore this estimate by itself precludes the formation of a cone-like singularity. We think of this estimate as a statement about entropy. Since the integral of R is constant, it represents a probability measure. Then the integral of $R \log R$ is the negative of the entropy. The estimate says that entropy is increasing.

8. We now combine the Harnack inequality and the entropy estimate to conclude that R is bounded. Pick a point ξ at time τ where the curvature R is largest. Then wait for a time $T - \tau = 1/2 R_{\max}(\tau)$. During that time

$$\frac{d}{dt} R_{\max} \le R_{\max}^2$$

and so $R_{\max}(T) \le 2 R_{\max}(\tau)$. On the other hand

$$\frac{\partial}{\partial t} g_{ij} = (r - R) g_{ij}$$

so distances will grow at most by a constant factor (since $R_{\max} \ge r$ the time interval is bounded). Hence if $d(\xi, X)$ is the geodesic distance at time T we will have

$$\Delta(\xi, \tau, X, T) \le C\, d(\xi, X)^2 / (T - \tau).$$

Then (again using the bound on the time period, and assuming $\tau \ge 1$) the Harnack inequality gives

$$R(\xi, \tau) \le C\, R(X, T)$$

for all X in a ball around ξ of radius

$$\rho = \pi / \sqrt{R_{\max}(T)/2}\,.$$

On the other hand, if our surface is oriented then Theorem 5.9 in Cheeger and Ebin [2] tells us that the injectivity radius of M is at least ρ at time T. (If M is not oriented, pass to the double cover.) In the ball of radius ρ around ξ we have R comparable to $R(\xi, \tau) = R_{\max}(\tau)$, which is at least half of $R_{\max}(T)$ by the choice of T. Therefore if we integrate over the ball B of radius ρ around at time T

$$\int_B R \log R \, d\mu \ge c \log R_{\max}(T)$$

at time T for some $c > 0$. Then the entropy estimate shows $R_{\max}(T)$ is bounded, and hence $R_{\max}(\tau)$ is bounded. This is true for all $\tau \ge 1$, so R is bounded.

Once R is bounded, we get as before a lower bound on the injectivity radius, and since the volume is bounded this gives an upper bound on the diameter. Using the diameter bound, plus the fact that the growth of distances is bounded, we easily get for

$T-\tau \leq 1$ the estimate

$$\Delta(\xi,\tau,X,T) \leq C/(T-\tau)$$

and hence the Harnack inequality tells us that for $t \geq 1$ and any two points x and y we have

$$R(x,t) \leq C\, R(y,t+1).$$

Therefore we also get a lower bound on R.

8.1 **Theorem.** *If we have a solution to the Ricci flow equation with $R>0$ on a compact surface, then there exist constants $c>0$ and $C<\infty$ with $0<c \leq R \leq C$ for all time.*

8.2 **Corollary.** *All of the derivatives of the curvature remain bounded for all time also.*

Proof. Having control of the diameter, the volume, and the injectivity radius, we can control the Sobolev constant. From [3], Theorem 13.4, we have

$$\frac{d}{dt}\int |D^n R|^2 d\mu + 2\int |D^{n+1}R|^2 d\mu \leq C_n \int |D^n R|^2 d\mu$$

and using

$$\int |D^n R|^2 d\mu \leq \varepsilon \int |D^{n+1}R|^2 d\mu + C_n(\varepsilon)\int R^2 d\mu$$

we deduce

$$\frac{d}{dt}\int |D^n R|^2 d\mu + \int |D^n R|^2 d\mu \leq C_n$$

which shows that

$$\int |D^n R|^2 d\mu \leq C_n$$

for all n and all time. Then the Sobolev inequality gives supremum bounds for all derivatives.

9. We apply the lower bound on R to the evolution equation for M_{ij}. Recall from section 4 that we chose the potential function f to solve $\Delta f = R - r$, and let

$$M_{ij} = D_i D_j f - \frac{1}{2}\Delta f \cdot g_{ij}.$$

9.1 The equation for $|M_{ij}|^2$:

$$\frac{\partial}{\partial t}|M_{ij}|^2 = \Delta |M_{ij}|^2 - 2|D_k M_{ij}|^2 - 2R|M_{ij}|^2.$$

This follows from a straightforward calculation.

9.2 **Corollary.** *If* $R \geq c > 0$ *then*

$$|M_{ij}| \leq C\, e^{-ct}$$

for some constant C. *Hence* $M_{ij} \to 0$ *exponentially. This follows from the maximum principle.*

Next we consider a modification of the Ricci flow. Consider the equation

$$\frac{\partial}{\partial t} g_{ij} = 2M_{ij} = (r - R)g_{ij} - 2D_i D_j f.$$

This equation differs from the Ricci flow only by transport along a one parameter family of diffeomorphisms generated by the gradient vector field of the potential f. Since M_{ij} converges to zero exponentially, the modified metrics will converge as $t \to \infty$. We shall show that their derivatives also converge, and the limiting metric is smooth.

First note that the bound $0 < c \leq R \leq C$ on R still holds for the modified flow, since it differs only by a diffeomorphism. Next note that the metrics $g_{ij}(x,t)$ are all equivalent, since they converge. Then to prove convergence of the $g_{ij}(x,t)$ as $t \to \infty$, it suffices to show that all the covariant derivatives of M_{ij} go to zero exponentially.

To obtain the higher derivatives bounds on M_{ij} it is convenient to switch to complex notation. This happens by viewing the surface as a Kähler manifold of complex dimension one. On a Kähler manifold each real tensor, say T_{ij}, has complex components $T_{\alpha\beta}, T_{\alpha\bar\beta}, T_{\bar\alpha\beta}, T_{\alpha\bar\beta}$ chosen so that if $T = T_{ij}\, dx^i\, dx^j$ in real coordinates then

$$T = T_{\alpha\beta}\, dz^\alpha\, dz^\beta + T_{\alpha\bar\beta}\, dz^\alpha\, d\bar z^\beta + T_{\bar\alpha\beta}\, d\bar z^\alpha\, dz^\beta + T_{\alpha\bar\beta}\, d\bar z^\alpha\, d\bar z^\beta$$

in complex coordinates. Since we have only one dimension, all unbarred indices may be interchanged, as may all bared ones. Moreover any pair of a bared and unbarred index may be contracted. This allows us to represent any tensor as equivalent to several fully symmetric tensors with all indices unbarred or barred. These correspond to the irreducible representations of the Lie group S^1. We say that a tensor with k unbarred indices has weight k, whereas one with k barred indices has weight $-k$. We can then drop the indices, and regard the tensor of weight k as a complex function on the principal tangent bundle. The exterior derivatives of a tensor T are given by DT and $\bar{D}T$, where if T has weight k then DT has weight $k+1$ and $\bar{D}T$ has weight $k-1$. For example, if $T = T_{\alpha\beta}$ then $DT = D_\alpha T_{\beta\gamma}$ and $\bar{D}T = g^{\bar{\alpha}\beta} D_{\bar{\alpha}} T_{\beta\gamma}$.

It is now easy to derive the following formulas for a tensor T of weight k.

9.3 Formulas on a surface

$$D\bar{D}T - \bar{D}DT = -\frac{1}{2} k R T$$

$$\Delta T = D\bar{D}T + \bar{D}DT$$

$$D\Delta T = \Delta DT - \frac{1}{2} k\, DR \cdot T - \left(k + \frac{1}{2}\right) R\, DT$$

$$\frac{\partial}{\partial t} DT = D \frac{\partial}{\partial t} T + \frac{1}{2} k\, DR \cdot T$$

$$\frac{\partial}{\partial t} |T|^2 = \bar{T} \frac{\partial}{\partial t} T + T \frac{\partial}{\partial t} \bar{T} + k(R - r)\, |T|^2.$$

Using these, we compute

$$\frac{\partial f}{\partial t} = \Delta f + rf - b$$

$$\frac{\partial}{\partial t} Df = \Delta Df - \frac{1}{2} R\, Df + r\, Df$$

$$\frac{\partial}{\partial t} D^2 f = \Delta D^2 f - 2R\, D^2 f + r\, D^2 f$$

$$\frac{\partial}{\partial t} D^3 f = \Delta D^3 f - 2DR \cdot D^2 f - \frac{9}{2} R\, D^3 f + r\, D^3 f$$

$$\frac{\partial}{\partial t} D^4 f = \Delta D^4 f - 2D^2 R \cdot D^2 f - \frac{13}{2} DR \cdot D^3 f - 8R\, D^4 f + r\, D^4 f$$

and consequently

$$\frac{\partial}{\partial t} |D^2 f|^2 \le \Delta |D^2 f|^2 - 2R\ |D^2 f|^2$$

$$\frac{\partial}{\partial t} |D^3 f|^2 \le \Delta |D^3 f|^2 - (6R + r)\ |D^3 f|^2 + 4\ |DR|\ |D^2 f|\ |D^3 f|$$

$$\frac{\partial}{\partial t} |D^4 f|^2 \le \Delta |D^4 f|^2 - (12R + 2r)\ |D^4 f|^2$$

$$+ 4\ |D^2 R|\ |D^2 f|\ |D^4 f| + 13\ |DR|\ |D^3 f|\ |D^4 f|$$

and so on.

The importance here for us is just that $D^2 f$ is the complex form of

$$M_{ij} = D_i D_j f - \frac{1}{2} \Delta f \cdot g_{ij}$$

the trace-free part of the second covariant derivative. We already know that since $R \ge C > 0$ we have $|D^2 f|$ going to zero exponentially. It now follows easily since $|DR|$ is bounded that $|D^3 f|$ goes to zero exponentially; and since $|D^2 R|$ is bounded we also have $|D^4 f|$ going to zero exponentially. In fact all $|D^k f|$ go to zero exponentially.

However, if we integrate by parts we can bound the L^2 norms of $|D^k \bar{D}^l D^2 f|$ in terms of the L^2 norms of $|D^{m+2} f|$. Hence all the derivatives of $D^2 f$ go to zero in L^2. Since we have uniform control of the Sobolev constant independent of time, we therefore have all the derivatives of $M_{ij} = D^2 f$ going to zero exponentially in the supremum norm as $t \to \infty$. This proves that the solution of the modified equation converges exponentially to a limit metric with $M_{ij} = 0$.

10. A metric g_{ij} with $M_{ij} = 0$ is a soliton solution for the Ricci flow. It moves only by a diffeomorphism, so its shape remains unchanged. We shall show that on a compact surface there are no soliton solutions with $R > 0$, other than constant curvature where there is no motion. On a non-compact surface solitons with $R > 0$ do exist. For example, on the plane the metric

$$ds^2 = \frac{dx^2 + dy^2}{1 + x^2 + y^2}$$

is a soliton with $R > 0$ which flows by conformal dilation. It is asymptotic to a flat cylinder at infinity, with maximum curvature at the origin.

10.1 **Theorem.** *On a compact surface there are no soliton solutions other than constant curvature.*

Proof. A soliton solution for the Ricci flow consists of a metric g_{ij} and a vector field v_i such that g_{ij} flows along v_i under the Ricci deformation. This happens when

$$2\left\{R_{ij} - \frac{r}{n} g_{ij}\right\} = D_i v_j + D_j v_i$$

and on a surface this simplifies to

$$(R - r) g_{ij} = D_i v_j + D_j v_i .$$

A soliton of the Ricci flow has

$$\frac{\partial R}{\partial t} = \Delta R + R(R - r)$$

and for a soliton the minimum value of R is constant. Then at that point $\partial R / \partial t = 0$. This shows that $R \geq 0$, and by the strong maximum principle we even have $R > 0$. From the evolution of the tensor $|M_{ij}|^2$ in 9.1 we see from the maximum principle that if a soliton has $R > 0$ then it must have $M_{ij} = 0$. For at the point where $|M_{ij}|^2$ is largest the equation says it should decrease, while if it flows by a diffeomorphism the maximum is unchanged. Therefore the vector field v_i along which the soliton flows must be $D_i f$, the gradient of the potential function f.

The vector field v_i must be conformal, since flowing along v_i changes the metric g_{ij} conformally. Now very few conformal vector fields can be gradients of a function. In complex coordinates the conformal vector field is holomorphic, and hence is locally given by $v(z)\,\partial/\partial z$ for a holomorphic function $v(z)$. At a zero of v there will be a power series expansion

$$v(z) = a\, z^p + \ldots \qquad (a \neq 0)$$

and if $p > 1$ the vector field will have closed orbits in any neighborhood of zero. Now a gradient flow cannot have a closed orbit. Hence v has only simple zeros, and f is a Morse function on the surface with only maxima and minima, and no saddles. Then there is one maximum and one minimum, and the surface is a union of two discs, thus a sphere.

Consider a soliton solution on the sphere S^2. The gradient of f must be a holomorphic vector field. Then it has exactly two zeros, which we can take to be at 0 and ∞. If we take $z = u + iv$ as complex coordinate, the holomorphic vector field must be $c\, z\, \partial/\partial z$ for some complex number c.

10.2 **Lemma.** *If* $c\, z\, \partial/\partial z$ *is a gradient vector field then* c *is real.*

Proof. Write the metric as

$$ds^2 = g(u,v)\,(du^2 + dv^2).$$

Then $\nabla f = c\, z\, \partial/\partial z$ means that if $c = a + bi$ then

$$\frac{\partial f}{\partial u} = (au - bv)g \qquad \frac{\partial f}{\partial v} = (bv + au)g.$$

Taking the mixed partials $\partial^2 f/\partial u\,\partial v$ and equating them at the origin $u = v = 0$ gives $b = 0$, so c is real.

Consequently our soliton is defined on the cylinder and moves by translation down the cylinder. Let x and y be coordinates on the cylinder, with translation in the x direction being the flow and identifying $y \to y + 2\pi$. Since the gradient of f is just $a\,\partial/\partial x$ for a real constant a, if the metric is given by

$$ds^2 = g(x,y)\,(dx^2 + dy^2)$$

we get the equations

$$\frac{\partial f}{\partial x} = ag \qquad \frac{\partial f}{\partial y} = 0\,.$$

The second shows that $f = f(x)$ is a function of x only, and then the first shows that $g = g(x)$ is also a function of x only, if $a \neq 0$. (If $a = 0$ then f is constant and R is constant.)

Consider a metric on the cylinder regarded as a quotient of the xy plane by $y \to y + 2\pi$, and independent of y. Then we can write the metric as

$$ds^2 = g(x)(dx^2 + dy^2).$$

The condition that the metric extend as $x \to -\infty$ to a metric on the plane given by

$$u = e^x \cos y$$

$$v = e^x \sin y$$

is that $g(x)$ be a smooth function of e^{2x}, with no constant term. For

$$ds^2 = g(x)\, e^{-2x}(du^2 + dv^2)$$

and so $e^{-2x} g(x)$ must be a smooth function of $u^2 + v^2 = e^{2x}$. Likewise the condition that the metric extend as $x \to +\infty$ is that $g(x)$ be a smooth function of e^{-2x} with no constant term.

The curvature of the metric is given by

$$R = -\frac{1}{g}\left[\frac{g'}{g}\right]'$$

as can be easily computed, where prime denotes differentiation in x. If g is a soliton with velocity c moving by translation in x, then $g = g(x + ct)$ satisfies

$$\frac{\partial g}{\partial t} = (r - R)g$$

which becomes

(10.3) $$c\, g' = rg + \left[\frac{g'}{g}\right]'.$$

To solve this, we substitute $g = v'$, where v is determined up to a constant. By including the constant in v, this integrates to

(10.4) $$v'' - C\,v'^2 + r\,v\,v' = 0$$

which can be integrated again to

(10.5) $$v' = \frac{r}{c} v + \frac{r}{c^2}\left[1 - ke^{cv}\right]$$

with an arbitrary constant k. Now if we substitute $y = cv + 1$ and $u = rx/c$ we get the equation

$$\frac{dy}{du} = y - ke^{y-1}$$

whose solutions are given by

(10.6) $$u = \int \frac{dy}{y - ke^{y-1}} .$$

Now suppose we have a solution $g(x)$ where as $x \to -\infty$ we have an expansion

$$g(x) = b\,e^{\lambda x} + c\,e^{2\lambda x} + \ldots$$

in powers of $e^{\lambda x}$ with no constant term. Then $v(x)$ will have an expansion

$$v(x) = a + \lambda b\,e^{\lambda x} + 2\lambda c\,e^{\lambda x} + \ldots$$

and hence be bounded as $x \to -\infty$. Likewise if $g(x)$ has an expansion

$$g(x) = b\,e^{-\mu x} + c\,e^{-2\mu x} + \ldots$$

as $x \to +\infty$ then $v(x)$ will be

$$v(x) = a - b\mu e^{-\mu x} - 2c\mu e^{-2\mu x} - \ldots$$

and hence will also be bounded as $x \to +\infty$. If v is bounded so is y, while $u \to \pm\infty$ when x does. Therefore the denominator in the previous integral must have two zeros. This happens precisely when $0 < k < 1$.

Suppose then that $0 < k < 1$ and that the equation

$$y = k\,e^{y-1}$$

has two solutions $y = k - p < 1$ and $y = 1 + q > 1$.

10.8 **Lemma.** *We always have* $p < q$, *and* $p/q \to 1$ *as* $k \to 1$, *while* $p/q \to 0$ *as* $k \to 0$.

Proof. As $k \to 0$, $p \to 1$ and $q \to \infty$, so $p/q \to 0$. In general

$$e^{p+q} = \frac{1+q}{1-p}$$

and since

$$\frac{e^x - 1 - x}{x^2} \to \frac{1}{2} \quad \text{as} \quad x \to 0$$

letting $x = p + q$, using the previous expression, and using $p \to 0$ and $q \to 0$ as $k \to 1$, we get

$$\frac{e^{(p+q)} - 1 - (p+q)}{(p+q)^2} = \frac{p}{(1-p)(p+q)} \to \frac{1}{2} \quad \text{as} \quad k \to 1$$

which shows $p/q \to 1$. Finally since

$$e^{2x} = 1 + 2x + 2x^2 + \frac{4}{3}x^3 + \frac{2}{3}x^4 + \ldots$$

$$\frac{1+x}{1-x} = 1 + 2x + 2x^2 + 2x^3 + 2x^4 + \ldots$$

we can see that

$$e^{2x} < \frac{1+x}{1-x} \quad \text{for} \quad x > 0.$$

This means that $p \neq q$ for $0 < k < 1$. Since $p/q \to 0$ as $k \to 0$, we must always have $p/q < 1$ by continuity.

To return to our discussion, for each value of R between 0 and 1 we get a soliton solution on the cylinder with bounded diameter. We now examine the asymptotics as $x \to \pm\infty$. Near $y = 1 - p$ write $y = 1 - p + z$. Then we have a power series expansion

$$y - ke^{y-1} = pz - \frac{1}{2}(1-p)z^2 + \ldots$$

which integrates to an expansion of

$$u = \int \frac{dy}{y - ke^{y-1}} = \int \frac{dz}{pz - \frac{1}{2}(1-p)z^2 + \ldots}$$

starting out as

$$u = \frac{1}{p} \log z + \ldots$$

which gives in turn an expansion of z in powers of e^{pu}. This in turn gives an expansion of $g(x)$ in powers of $e^{\lambda x}$ with $\lambda = rp/c$ as $x \to -\infty$. Likewise we get an expansion of $g(x)$ in powers of $e^{-\mu x}$ where $\mu = rq/c$ as $x \to +\infty$. Then $\lambda/\mu = p/q$.

By an appropriate choice of k we can make the ratio $\lambda/\mu = p/q$ any number with $0 < \lambda/\mu < 1$. Then by an appropriate choice of the velocity c we can make λ and μ any numbers we want with $0 < \lambda < \mu$. Notice that we do not attain $\lambda = \mu$ except in the limiting case where $c = 0$. Thus to get a solution on S^2 we need $\lambda = \mu = 2$, which makes the velocity $c = 0$, so we just have the constant curvature solutions. The other solutions we have found exist on orbifolds.

10.9 **Corollary.** *On a compact surface with* $R > 0$ *the heat flow*

$$\frac{\partial}{\partial t} g_{ij} = (r - R) g_{ij}$$

converges exponentially to a constant curvature metric.

Proof. For the modified flow we have seen that the curvature R converges to its limiting value exponentially. But since there are no soliton solutions on S^2, we must have R converging to the constant r exponentially. This then implies that the unmodified flow will also converge exponentially.

References

[1] Cao, H.-D., *Deformation of Kähler metrics to Kähler-Einstein metrics on compact Kähler manifolds,* **Invent. Math.,** Vol. **81** (1985), pp. 359-372.

[2] Cheeger, J. and D. Ebin, *Comparison theorems in Riemannian geometry,* North-Holland, Amsterdam, 1975, p. 174.

[3] Hamilton, R. S., *Three-manifolds with positive Ricci curvature,* **J. Diff. Geom.,** Vol. **17** (1982), pp. 255-306.

[4] Hamilton, R. S., *Four-manifolds with positive curvature operator,* **J. Diff. Geom.,** Vol. **24** (1986), pp. 153-179.

[5] Li, P. and S.-T. Yau, *On the parabolic kernel of the Schrödinger operator,* **Acta Math.,** Vol. **156** (1968), pp. 153-201.

Department of Mathematics
University of California at San Diego
La Jolla, CA 92093

Contemporary Mathematics
Volume 71, 1988

CURVATURE AND COMPACT SPACELIKE SURFACES IN 4-DIMENSIONAL SPACETIMES

Demir N. Kupeli

ABSTRACT. The way a spacelike surface H sits in a 4-dimensional spacetime M may be measured by its mean null curvature and shape functions $\phi_H, \Omega_H \to \mathbb{R}$, respectively. The relation of the shape function of a spacelike surface to the curvature of a 4-dimensional spacetime is investigated. It is shown that the curvature of a 4-dimensional spacetime has a crucial influence on complete spacelike surfaces (in the induced metric) with average positive shape functions being diffeomorphic to S^2 or $\mathbb{R}P^2$ rather than a compact spacelike surface with Euler characteristic $\chi \leq 0$. Yet, no 4-dimensional flat spacetime admits a compact spacelike surface with average positive shape function and Euler characteristic $\chi \geq 0$. However, it is shown that the existence of a particular sequence of compact spacelike surfaces in a 4-dimensional spacetime yields curvature singularities regardless of the causality assumption and geodesic incompleteness. Furthermore, under some additional assumptions, such a curvature singularity in a perfect fluid spacetime yields existence of an incomplete null geodesic along which energy density goes to infinity.

1. PRELIMINARIES. An n-dimensional, connected, oriented, time oriented Lorentzian manifold M with metric g of signature $(-++\cdots+)$ is called a spacetime. A symmetric $(0,2)$-tensor field T on a spacetime is called a stress-energy tensor. A spacetime is said to obey the Einstein equation for the stress-energy tensor T if $\text{Ric} - \frac{1}{2}(\text{Sc})g = T$, where "Ric" is the Ricci tensor and "Sc" is the scalar curvature. A connected spacelike submanifold H of codimension 2 in M is called a spacelike surface. The second fundamental tensor L_u of a spacelike surface H in the null direction u orthogonal to H at $q \in H$ is defined by $L_u x = -(\nabla_x U)^T$, where $x \in T_qH$, U is an orthogonal

1980 Mathematics Subject Classification (1985 Revised). 83C75.

null extension of u to a neighborhood of q in H and $(\nabla_x U)^T$ is the component of $\nabla_x U$ tangent to H. The <u>second fundamental form</u> II_u of a spacelike surface H in the null direction u orthogonal to H at $q \in H$ is defined by $II_u(x,y) = g(L_u x, y)$, where $x, y \in T_q H$. The <u>mean null curvature function</u> $\phi_H : H \to \mathbb{R}$ of a spacelike surface H is defined by $\phi_H(q) = (\mathrm{tr} L_u)(\mathrm{tr} L_w)$, where u, w are future pointing null vectors orthogonal to H at q with $g(u,w) = -1$. A compact spacelike surface H with everywhere positive mean null curvature function ϕ_H is called a <u>closed trapped surface</u>. The <u>shape function</u> $\Omega_H : H \to \mathbb{R}$ of a spacelike surface H in a 4-dimensional spacetime is defined by $\Omega_H(q) = II_u(e_1,e_1) II_w(e_2,e_2) + II_u(e_2,e_2) II_w(e_1,e_1)$, where u, w are future pointing null vectors orthogonal to H at q with $g(u,w) = -1$ and $\{e_1, e_2\}$ are orthonormal eigenvectors of either L_u or L_w. The <u>surface gravity function</u> $K_H^{\perp} : H \to \mathbb{R}$ of a spacelike surface H is defined by $K_H^{\perp}(q) = K(H_q^{\perp})$, where $K(H_q^{\perp})$ is the curvature of the (timelike) fiber $H_q^{\perp}$ of the normal bundle $H^{\perp}$ of H at q. (See [1] and [2] for the geometric interpretations of the above definitions).

2. INTRODUCTION. In the literature, existence of a closed trapped surface H in a 4-dimensional spacetime M is considered to be related to the curvature of spacetime in that H is contained in such a strong gravitational field that the light rays (future directed null geodesics) emanating orthogonally from H are dragged back and in fact, are converging in average. (That is, H has an everywhere positive mean null curvature function). On the other hand, curvature of spacetime is manifested in gravitational tidal forces, and therefore, physically, it is expected that a gravitational field which is strong enough to give a rise of closed trapped surfaces also causes incomplete non-spacelike geodesics along which the gravitational tidal forces go to infinity as the incomplete ends of these geodesics are approached.

Mathematically, the Hawking-Penrose singularity theorem (cf. [3], p. 266) partially confirms the above expectation in showing that, existence of a closed trapped surface in a chronological spacetime yields existence of an incomplete non-spacelike geodesic under general assumptions. However, the theorem is very negative to provide any information about the curvature of spacetime along the predicted incomplete non-spacelike geodesic(s). (Also see section 2 of [2] for further discussion of the Hawking-Penrose theorem in flat spacetimes). Nevertheless, it seems that the appearence of closed trapped surfaces in the large curvature regions of spacetime (strong gravitational field), in particular, with small areas and large mean null curvature functions near the curvature singularities, is a common property of physically realistic spacetimes possessing curvature singularities (for example, Friedmann-Roberson-Walker, Kruskal, Reissner-Nordstrom and Kerr spacetimes). But, we also notice that most of these closed trapped surfaces have everywhere positive shape functions. In other words, there also appear compact spacelike surfaces with everywhere positive shape functions in the large curvature regions of physically realistic spacetimes, in particular, with small areas and large positive shape functions near the curvature singularities.

In this paper, we shall investigate the relation between the curvature of spacetime and existence of compact spacelike surfaces with average positive shape functions. We notice first that, in a 4-dimensional spacetime, average positivity of the shape function of a spacelike surface H may not be a consequence of large curvature only, but also the way H sits in the spacetime, and second, that every compact spacelike surface is a complete Riemannian surface in the induced metric. Therefore, it seems reasonable to investigate the relation of curvature to the compactness of a complete spacelike surface (in the induced metric) to understand the relation between the curvature and existence of compact spacelike surfaces with average positive shape functions. We show that curvature of a spacetime has a crucial influence on the existence of compact spacelike surfaces with average positive shape functions being diffeomorphic to S^2 or $\mathbb{RP}^2$ rather than a compact spacelike surface with Euler characteristic $\chi \leq 0$. Yet, no flat spacetime contains a compact spacelike surface with average positive shape function and Euler characteristic $\chi \geq 0$. (That is, the curvature of a 4-dimensional spacetime seems to have an influence on the topology of compact spacelike surfaces with average positive shape functions rather than their existence). Then we show that, in the case of existence of a particular sequence of compact spacelike surfaces with small areas and arbitrarily large average shape functions (which exists in physically realistic spacetimes), some components of the curvature necessarily go to infinity along a sequence of points in the spacetime. Furthermore, under some additional assumptions, such a curvature singularity in cosmological circumstances yields existence of an incomplete null geodesic along which energy density goes to infinity as its incomplete end is approached.

3. CURVATURE AND THE SHAPE FUNCTION. In achieving the results of this section, we make use of the results of Cohn-Vossen on complete Riemannian surfaces by making estimates on the induced curvature K_H of a spacelike surface H in a 4-dimensional spacetime using Gauss-Codazzi equation. Since the proofs of the results below are available in [1] and [2], we shall only state those which we need for our discussion. Recall that, if we consider existence of compact spacelike surfaces diffeomorphic to S^2 or $\mathbb{RP}^2$ with average positive shape functions is a consequence of non-trivial gravitational field then it is natural to expect that such surfaces cannot appear in 4-dimensional flat spacetimes. Indeed, the following theorem confirms this expectation. We shall always assume that the integrals in the resuts below exist.

THEOREM 3.1: Let H be a spacelike surface in a 4-dimensional flat spacetime. If $\int_H \Omega_H > 0$ then H cannot be diffeomorphic to either of S^2, $\mathbb{RP}^2$, $S^1 \times S^1$ and $\mathbb{RP}^2 \# \mathbb{RP}^2$.

Proof: See theorem 3.1 in [2]. ∎

We now show the influence of non-trivial curvature to the existence of compact spacelike surfaces diffeomorphic to S^2 or $\mathbb{RP}^2$.

THEOREM 3.2: Let H be a complete spacelike spacelike surface (in the induced metric) in a 4-dimensional spacetime M obeying the Einstein equation for the stress-energy tensor T. Assume that $K_H^{\perp}(q) + \mathrm{Ric}(u,w) + T(u,w) > \Omega_H(q)$ at each $q \in H$, where $\{u,w\}$ is a pair of future pointing null vectors orthogonal to H at q with $g(u,w) = -1$. Then,

a) H is diffeomorphic to S^2 iff $\int_H [K_H^{\perp} + \mathrm{Ric}(u,w) + T(u,w)] > 2\pi + \int_H \Omega_H$.

b) H is diffeomorphic to $\mathbb{RP}^2$ iff H is not orientable.

Proof: See theorem 3.2 in [2]. ∎

REMARK 3.3: It is shown in [3] that, a compact, acausal spacelike surface in the domain of dependence of an achronal set has trivial normal bundle, and therefore, is orientable. That is why, case (b) of the above theorem seems to be a rare situation which usually does not occur in physically realistic spacetimes unless there exists causality violations.

We now state two special cases of the above theorem for black hole and cosmological circumstances.

COROLLARY 3.4: Let H be a complete spacelike surface (in the induced metric) in a 4-dimensional spacetime M which obeys the Einstein equation for the stress-energy tensor T satisfying $\mathrm{tr}T = 0$. Assume that

$K_H^{\perp}(q) + 2T(u,w) > \Omega_H(q)$ at each $q \in H$, where $\{u,w\}$ is a pair of future pointing null vectors orthogonal to H at q. Then,

a) H is diffeomorphic to S^2 iff $\int_H [K_H^{\perp} + 2T(u,w)] > 2\pi + \int_H \Omega_H$.

b) H is diffeomorphic to $\mathbb{RP}^2$ iff H is not orientable.

Proof: See corollary 3.5 in [2]. ∎

The assumptions of the above corollary are expected to be satisfied by the spacetimes obeying the Einstein equation for an electromagnetic stress-energy tensor. In particular, Kruskal, Reissner-Nordstrom and (charged) Kerr black holes satisfy the assumptions of the above corollary. Also see [5] for a discussion of the surface gravity in connection with the black hole thermodynamics.

DEFINITION 3.5: A spacetime M is called weakly spatially isotropic with respect to a future pointing unit timelike vector $z \in T_qM$ if $R(x,z)z = kx$ for every $x \in z^{\perp}$, where $k \in \mathbb{R}$, $z^{\perp}$ is the orthogonal complement of $\operatorname{span}\{z\}$ and R is the curvature tensor.

COROLLARY 3.6: Let H be complete spacelike surface (in the induced metric) in a 4-dimensional spacetime M which obeys the Einstein equation for the perfect fluid stress-energy tensor $\hat{T} = (\rho + p)Z \otimes Z + p\hat{g}$, where Z is a future directed unit timelike vector field on M, ρ (energy density) and p (pressure) are continuous functions on M and $\hat{g}$ is the $(2,0)$-tensor field physically equivalent to metric tensor g (see [6], p. 65). Assume that

1) H is orthogonal to the vector field Z

2) M is weakly spatially isotropic with respect to Z_q at each $q \in H$

3) $\rho(q) > 3\Omega_H(q)$ at each $q \in H$.

Then,

a) H is diffeomorphic to S^2 iff $\int_H \rho > 6\pi + \int_H \Omega_H$.

b) H is diffeomorphic to $\mathbb{R}P^2$ iff H is not orientable.

Proof: See corollary 3.4 in [2]. ∎

The assumptions of the above corollary are expected to be satisfied by infinitesimally spatially isotropic spacetimes (see [7], [8] and [9]). In particular, the Friedmann-Robertson-Walker spacetimes satisfy the assumptions of the above corollary when the time arrow is reversed.

REMARK 3.7: The above theorem and its corollaries show existence of an inequality which involves the area, average shape function and some curvature components for a complete spacelike surface being diffeomorphic to S^2. Note that however, even though the existence of spacelike surfaces diffeomorphic to S^2 with everywhere average positive shape functions does not seem to indicate existence of "large curvature regions" of spacetime which can be considered strong gravitational field. In fact, if H is a compact spacelike surface in a 4-dimensional spacetime then, from Gauss-Codazzi equation and Gauss-Bonnet theorem, $\int_H [K_H^{\perp} + Ric(u,w) + T(u,w)] = 2\pi\chi(H) + \int_H \Omega_H$ (see [2]). Therefore, the existence of a compact spacelike surface H does not seem indicate large curvature components unless $\int_H \Omega_H$ is a large positive number or vise versa.

In the next theorem and its special cases, we study the relation between the curvature of spacetime and the shape functions of compact spacelike surfaces whose areas are bounded from above.

THEOREM 3.8: Let $\{H_i\}$ be an infinite sequence of compact spacelike surfaces with $\chi(H_i) \geq N$ and $area(H_i) \leq C$ in a 4-dimensional spacetime M obeying the Einstein equation for the stress-energy tensor T, where $N \leq 2$ is some integer and $C > 0$. Then,

$\lim_{i\to\infty}\int_{H_i}\Omega_{H_i} = \infty$ iff $\lim_{i\to\infty}\int_{H_i}[K^{\perp}_{H_i} + Ric(u_i,w_i) + T(u_i,w_i)] = \infty$, where $\{u_i w_i\}$ is a pair of future pointing null vectors orthogonal to H_i at each point $r_i \in H_i$ with $g(u_i,w_i) = -1$. In particular, if either of the above conditions holds then there exists a sequence of points $\{q_i\}$ with $q_i \in H_i$ such that $\lim_{i\to\infty}[K^{\perp}_{H_i}(q_i) + Ric(u_{q_i},w_{q_i}) + T(u_{q_i},w_{q_i})] = \infty$, where $\{u_{q_i},w_{q_i}\}$ is a pair of future pointing null vectors orthogonal to H_i at $q_i \in H_i$ with $g(u_{q_i},w_{q_i}) = -1$.

Proof: See theorem 3.6 in [2]. ▪

As before, the above theorem has two special cases for black hole and cosmological circumstances.

COROLLARY 3.9: Let $\{H_i\}$ be an infinite sequence of compact spacelike surfaces with $\chi(H_i) \geq N$ and $area(H_i) \leq C$ in a 4-dimensional spacetime M which obeys the Einstein equation for a stress-energy tensor T satisfying $trT = 0$, where $N \leq 2$ is some integer and $C > 0$. Then, $\lim_{i\to\infty}\int_{H_i}\Omega_{H_i} = \infty$ iff $\lim_{i\to\infty}\int_{H_i}[K^{\perp}_{H_i} + 2T(u_i,w_i)] = \infty$, where $\{u_i,w_i\}$ as in theorem 3.8. In particular, if either of the above conditions holds then there exists a sequence of points $\{q_i\}$ with $q_i \in H_i$ such that $\lim_{i\to\infty}[K^{\perp}_{H_i}(q_i) + 2T(u_{q_i},w_{q_i})] = \infty$, where $\{u_{q_i},w_{q_i}\}$ as in theorem 3.8.

Proof: See corollary 3.8 in [2]. ▪

COROLLARY 3.10: Let $\{H_i\}$ be an infinite sequence of compact spacelike surfaces with $\chi(H_i) \geq N$ and $area(H_i) > C$ in a 4-dimensional spacetime M obeying the Einstein Equation for the perfect fluid strees-energy tensor $\hat{T} = (\rho + p)Z \otimes Z + p\hat{g}$, where $N \leq 2$ is some integer and $C > 0$. Assume that, for each i, H_i is orthogonal to Z and M is weakly spatially isotropic with respect to Z_{r_i} for every $r_i \in H_i$. Then,

$\lim_{i\to\infty} \int_{H_i} \Omega_{H_i} = \infty$ iff $\lim_{i\to\infty} \int_{H_i} \rho = \infty$. In particular, if either of the above conditions holds then there exists a sequence of points $\{q_i\}$ with $q_i \in H_i$ such that $\lim_{i\to\infty} \rho(q_i) = \infty$.

Proof: See corollary 3.7 in [2]. ∎

The assumptions of the above corollaries are satisfied by Kruskal, Reissner-Nordstrom black holes and Friedmann-Robertson-Walker cosmological spacetimes, respectively. At this point, it is natural to expect existence of incomplete non-spacelike geodesics terminating at this curvature singularity as in the above spacetimes. The proposition below partly confirms this expectation in cosmological circumstances with some additional asssumptions. Since this proposition is not avaliable in [2], we shall provide its proof.

PROPOSITION 3.11: In addition to the assumptions of corollary 3.10, assume also that,

1) M is strongly causal.

2) Every "TIP" in M has a generator which lies on its boundary.

3) Each point $\bar{q} \in \bar{M}$ ($= M \cup \partial_c M$) has a neighborhood $\bar{V}$ in $\bar{M}$ such that ρ can be extendend to a continuous function $\bar{\rho} : \bar{V} \to \mathbb{R} \cup \{\infty\}$, where $\bar{M}$ is the causal complition of M with its Hausdorff topology (see [10]) and $\mathbb{R} \cup \{\infty\}$ has order topology.

4) The sequence $\{H_i\}$ is contained in a compact subset $\bar{V}$ of $\bar{M}$.

5) $\lim_{t\to\varepsilon} |g(\dot{\gamma}, Z_\gamma)| \geq K > 0$ for every future inextendable null geodesic $\gamma : [0,\varepsilon) \to M$, where $\varepsilon \leq \infty$.

6) The pressure p is bounded from blow.

Then, M contains an incomplete null geodesic $\beta : [0,\varepsilon) \to M$ such that $\lim_{t\to\varepsilon} \rho(\beta(t)) = \infty$.

Proof: From assumptions (2) and (4), the sequence $\{q_i\}$ in corollary 3.10 converges to a point $\bar{q} \in \partial_c M$, where $\partial_c M$ is the causal boundary of M. Let $\beta : [0,\varepsilon) \to M$ be a null geodesic generator of $I^-(\bar{q})$ lying on $\partial I^-(\bar{q})$, where $\varepsilon \leq \infty$. Then, from assumption (3), $\lim_{t\to\varepsilon} \rho(\beta(t)) = \infty$. To show β is incomplete (that is, $\varepsilon < \infty$), it suffices to show that, if $\beta : [0,\varepsilon) \to M$ is complete then it contains a conjugate point to "0" in contradiction with that β lies on the boundary of a "TIP". Assume β is complete. Then, since $Ric(\dot\beta,\dot\beta) = T(\dot\beta,\dot\beta) = (\rho(\beta) + p(\beta))(g(\dot\beta,Z))^2$, from assumptions (5) and (6), $\lim_{t\to\infty} Ric(\dot\beta,\dot\beta) = \infty$ and therefore, $\int_0^\infty Ric(\dot\beta.\dot\beta)dt = \infty$. Thus, β contains a conjugate point to "0" from theorem 3.12 in [11]. ∎

The assumptions of the above theorem are satisfied by the Friedmann-Robertson-Walker cosmological spacetimes when the time arrow is reversed.

REMARK 3.12: In proposition 3.11, notice that the strong causality of M is necessary for the definition of causal boundary of M. In particular, if M is globally hyperbolic (strong cosmic censorship) then M satisfies the condition (2) (see [10]). Also see [12] for a discussion of condition (3) in connection with the Big-Bang singularity. Recall also that a photon is considered to be a future directed null geodesic in General Relativity. The energy of a photon β with respect to an instantaneous observer $z \in T_qM$ (that is, z is a future pointing unit timelike vector) is defined by $E_z = -g(\dot\beta,z)$ (see [6], p. 130). Therefore, in condition (5), it is assumed that the photons are not fading away with respect to the instantaneous observers in Z.

Presently, I don't have a similiar result as in proposition 3.11 for black hole circumstances. The main difficulty in proving such a result in Ricci flat spacetimes appears to be the crucial involvement of conjugate points to the shear tensor along null geodesics because of the complete break down of null isotropy in the spacetime. (See [9] and [11] for further discussion of conjugate points in such spacetimes).

ACKNOWLGEMENTS

I would like to expess my deepest gratitute to my thesis advisor, Professor John A. Thorpe, for extremely useful discussions on the subject and, for his comments and encourgement. I also owed a considerable debt to Professor Detlef Gromoll, for calling my attention to the results of Cohn-Vossen and, for a number of useful discussions on complete Riemannian surfaces. I would also like to thank Professor Roger Penrose, for an inspiring discussion about the assumptions of the last proposition.

BIBLIOGRAPHY

1. Kupeli, D.N., "On Null Hypersurfaces and Spacelike Surfaces in Spacetimes", Ph.D. Thesis, State University of New York at Stony Brook, Stony Brook, (1985).

2. Kupeli, D.N., "Curvature and Closed Trapped Surfaces in 4-Dimensional Spacetimes", To Appear, Gen. Rel. Grav.

3. Hawking, S.W. and Ellis, G.F.R., The Large Scale Structure of Spacetime, Cambridge University Press, New York, 1973.

4. Kupeli, D.N., "Null Cut Loci of Spacelike Surfaces", To Appear, Gen. Rel. Grav.

5. Hawking, S.W. "The Analogy Between Black Hole Mechanics and Thermodynamics", Ann. NY Acad. Sci., 224, (1973)

6. Sachs, R.K. and Wu, H., General Relativity for Mathematicians, Springer-Verlag, New York, 1977.

7. Karcher, H., "Infinitesimale Charakterisierung von Friedmann Universen", Arch. Math., 38, (1982).

8. Harris, S.G., "A characterization of Robertson-Walker Spaces by Null Sectional Curvature", Gen. Rel. Grav., 17, (1985).

9. Kupeli, D.N., "On Null Isotropy in Spacetimes", Submitted to Gen. Rel. Grav.

10. Geroch, R., Kronheimer, E.H. and Penrose, R., "Ideal Points in Spacetime", Proc. R. Soc. Lond., A.327, (1972).

11. Kupeli, D.N., "On Existence and Comparison of Conjugate Points in Riemannian and Lorentzian Geometry", To Appear, Math. Ann.

12. Penrose, R., "Singularities and Time-Asymmetry" in General Relativity, ed. Hawking and Israel, Cambridge University Press, New York, 1979.

DEPARTMENT OF MATHEMATICS
UNIVERSITY OF ALABAMA AT BIRMINGHAM
BIRMINGHAM, AL 35294

Contemporary Mathematics
Volume **71**, 1988

SPACETIME SINGULARITIES FROM HIGH MATTER DENSITIES

Richard Schoen

ABSTRACT

It is generally believed that if a sufficient amount of matter is compressed into a small region, then the resulting gravitational effects will be strong enough to cause the system to collapse to a black hole. In recent work (with S.-T. Yau) we have proven a theorem which provides mathematical support for this belief. More specifically, the theorem states that if the matter density within a given bounded spacelike region satisfies a certain inequality, and if the radius (suitably defined) of this region satisfies a closely related inequality, then somewhere in or near this region, there must be a closed trapped surface. As shown by Hawking and Penrose, if a spacetime contains a closed trapped surface, then the formation of a black hole is almost inevitable. The details of our theorem and its proof appear in the following reference:

1. R. Schoen and S.-T. Yau, "The Existence of a Black Hole Due to Condensation of Matter", Common Math Phys <u>90</u>, 575 (1983).

Richard Schoen
Department of Mathematics
University of California (San Diego)
La Jolla, CA 92109

0271-4132/88 $1.00 + $.25 per page

Contemporary Mathematics
Volume **71**, 1988

METRIC SINGULARITY PHENOMENA IN PSEUDO-RIEMANNIAN GEOMETRY

Marek Kossowski

ABSTRACT. Singularities in metrics play a significant role in pseudo-Riemannian Geometry. Here we describe a simple metric singularity phenomenon and observe that the fundamental constructions of differential geometry can still be carried out (i.e. there exists a weak form of the metric connection). This yields a natural tensor defined at points of degeneracy. We then briefly describe several interpretations of this tensor.

The objective of this work is to identify and study the differences between Pseudo-Riemannian and Riemannian Geometry. An essential distinction is the existence of submanifolds which inherit degenerate metrics. For example let $(V, \langle\ \rangle^{(r,s)})$ denote a real vector space with non-degenerate bilinear form of type (r, s) [i.e., r plus signs and s minus signs]. To simplify exposition we assume $r \geq s$, and let $i : N \longrightarrow V$ be a smooth immersion of a compact manifold.

0271-4132/88 $1.00 + $.25 per page

Theorem: If $\dim(N) \geq r$, then there exists a point $p \in N$ where the induced metric degenerates, (i.e. $\dim (T_pN \cap T_pN^{\perp}) \geq 1$).

Proof: Choose a null vector W in V so that $W^{\perp}$ is a degenerate hyperplane with induced metric of type $(r-1, 0)$. Now $\langle W, \rangle^{(r,s)}$ restricts to a smooth real-valued function on N and hence has a critical point p. At this point T_pN lies in $W^{\perp}$ and thus must inherit a degenerate metric. □

Of course there is a similar phenomena in general Pseudo-Riemannian manifolds. The author has observed that there is a connection-like object associated to such a degenerate metric. This "dual connection" leads to a natural intrinsic tensor at points of metric degeneracy. We will now describe this construction.

Keeping $(N, i^* \langle\,\rangle)$ in mind we define a manifold with metric to be a pair $(N, \langle\,\rangle)$ where $\langle\,\rangle$ is any smooth section of the symmetric 2-tensor bundle over N. Given such a $(N, \langle\,\rangle)$ with $\langle\,\rangle_p$ a degenerate bilinear form, we get a subspace Rad_p of T_pN which is $\langle\,\rangle_p$ - orthogonal to all of T_pN. We will denote the sets of smooth vector fields and 1-forms on N by $V^{\infty}(N)$ and $\Lambda^1(N)$, respectively.

Definition Given $(N, \langle\,\rangle)$, a C^{∞} dual connection on N is a map

$$\square : V^{\infty}(N) \times V^{\infty}(N) \longrightarrow \Lambda^1(N)$$

$$X, Y \longmapsto \square_X Y$$

which satisfies

a) $\square$ is bilinear over $\mathbf{R}$;

b) for all $f \in C^{\infty}(N)$ and $X, Y \in V^{\infty}(N)$

1) $\square_{fx}Y = f\,\square_{x}Y$

2) $\square_{x}fY = x(f)\langle Y,\_\rangle + f\,\square_{x}Y$.

The <u>torsion</u> of $\square$ is the tensor

$$\mathrm{Tor}(X,Y,Z,) = \square_{X}Y(Z) - \square_{T}X(Z) - \langle [X,Y],Z\rangle.$$

Here $\square_X Y(Z)$ denotes the pairing of the 1-form $\square_X Y$ with the vector field Z. $\square$ is <u>compatible</u> with $\langle\,\rangle$ if $X\langle Y,Z\rangle = \square_X Y(Z) + \square_X Z(Y)$.

<u>Lemma</u>. (Fundamental Lemma of Pseudo Riemannian geometry). Given $(N, \langle\,\rangle)$ with $\langle\,\rangle$ a smooth section of the symmetric two-tensor bundle, there exists a unique torsion-free, compatible, dual connection.

<u>Proof:</u> As in the classical case we use the relations of compatibility and Tor = 0 to write

$$2\square_X Y(Z) = X\langle Y,Z\rangle + Y\langle Z, X\rangle - Z\langle X,Y\rangle$$
$$+ \langle [X,Y],Z\rangle - \langle [X,Z],Y\rangle - \langle [Y,Z],X\rangle.$$

One then shows that this object satisfies a) and b).$\square$

Observe that if $\langle\,\rangle$ is non-degenerate then $\square_X Y(Z) = \langle \nabla_X Y, Z\rangle$, where ∇ is the classical Levi-Civita connection.

Now let $R \in \mathrm{Rad}_p$ and X_p, $Y_p \in T_pN$. We define

$$II_p(X,Y,R) = \square_{X_p} Y(R).$$

<u>Proposition</u>

a) II_p is a tensor on $T_pN \times T_pN \times \mathrm{Rad}_p$

b) $II_p(X,Y,R) = II_p(Y,X,R)$.

Proof:

a) Observe that $\square_{X_p} fY(R) = X(f) \langle Y,R\rangle_p + f\square_{X_p} Y(R)$, and that $\langle Y,R\rangle_p = 0$

b) Observe that $0 = \langle R,[X,Y]\rangle_p = \square_{X_p} Y(R) - \square Y_p X(R)$. □

It is ironic that the nineteenth century differential geometers were confounded by the fact that the metric connection is not a tensor. Here we see that if the metric degenerates, then part of the connection does in fact become a tensor. (This may be taken as a short proof that God does not favor Mathematicians.)

It is natural to ask how the tensor II is related to the local differential geometry of the submanifold. The most primitive information contained in II is transversality conditions on the metric singularity. (See paper 1 for details.) The more subtle interpretations of II are the content of the following papers:

(1) Fold Singularities in Pseudo-Riemannian Geodesic Tubes. (To appear in The Proceedings of the AMS)

Let $i: N \longrightarrow (M, \langle\,\rangle)$ denote a C^∞ immersion of smooth manifolds. M will come equipped with a smooth non-degenerate Pseudo-Riemannian metric. If $N^\perp$ denotes the orthogonal bundle over N, and $N^\perp_p := \{v_q \in T_qM \mid i(p) = q, \text{ and } v_q \perp i_*(T_pN)\}$, then the geodesic tube map $i^\perp: N^\perp \longrightarrow M$ is defined by $v_p \longmapsto \exp_{i(p)} v$. If

(M, ⟨ ⟩) is Riemannian, (i.e. ⟨ ⟩ is positive definite) then $i^{\perp}$ is a local diffeomorphism on a neighborhood of any point on the zero section of $N^{\perp}$. If (M, ⟨ ⟩) is not definite, this is not always the case. In particular if $N_p^{\perp} \cap T_pN$ is non-trivial vector space then it is clear that $(i^{\perp})_*$ cannot have full rank at $o_p \in N^{\perp}$. Here we show that in the case where $N_p^{\perp} \cap T_pN$ is one-dimensional, the tensor II determines when the geodesic tube map has a simple fold singularity at o_p, (i.e. a $S_{1,0}$ singularity).

(2) Pseudo-Riemennian Metric Singularities and the Extendibility of Parallel Transport. (To appear in <u>The Proceedings of the AMS.</u>)

This paper is a sequel to paper 1 above. We are given a C^∞ immersion $i:N \to (M, \langle\ \rangle)$, and $p \in N$ is a point where $N_p^{\perp} \cap T_pN$ is one-dimensional. In paper 1 we showed that there is a tensor II_p: $T_pN \times T_pN \times Rad_p \to R$ **<u>intrinsic</u>** to (N, I* ⟨ ⟩) which determines an **<u>extrinsic</u>** feature of the immersion. Here we show that II controls the following two intrinsic properties. First, II determines which pairs of vector fields X,Y on N have the property that the intrinsic covariant derivative $\nabla_X Y$ extends smoothly to all of N. Second, given a curve in N containing p, II_p determines which parallel vector fields along the curve extend smoothly through p. As an application we locally characterize product and flat metric singularities.

(3) The Intrinsic Conformal Structure of a Light-like Hypersurface in M^4. (In preparation.)

A hypersurface in M^4 is light-like if it is everywhere tangent to an ambient light cone. In this setting the tensor II provides a simple intuitive link with surface theory in Euclidean space E^3, and is used to give global characterizations of the simplest light-like hypersufaces. This also enables several pathologies in light-like hypersurfaces to be identified.

It is also natural to ask if this metric degeneracy phenomena can be linked to global properties of submanifolds in real Minkowski space $M^m = V^{(m-1,1)}$. This is the content of papers 4, 5, and 6 below. The tensor II plays a more subtle role here.

(4) The S^2-valued Gauss Maps of a Space-like Surface in M^4. (In preparation)

An immersed 2-manifold M in M^4 is space-like if the induced metric is everywhere positive definite. The Grassmann manifold, $G^2(M^4)$, of positive definite oriented 2-planes in M^4 may be viewed as a subset of $S^2 \times S^2$. Hence the Gauss map $g: M \longrightarrow G_2(M^4)$ projects to give two S^2-valued Gauss maps, $\pi_i \cdot g = g_i: M \longrightarrow S^2$ $(i = 1,2)$. (The degree of these g_i is related to the pathology mentioned in 3 above.) Now choosing a splitting s of $M^4 = E^3 + E^-$ endows the two spheres with area forms and M with two curvature functions K_i^s, (which are directly related to the II tensors of the two light-like hypersurface generated by M.) In this paper we produce several geometric inequalities involving the K_i^s which are valid for any splitting of M^4.

(5) A Gauss-Bonnett-Hopf-Poincare Type Formula for Compact Hypersurfaces in M^m. (In preparation)

Consider a compact oriented hypersurface H in real Minkowski space M^4. By imposing an appropriate transversality condition on this degeneracy, we get a compact codimension-2 (in M^m) submanifold S(H) on which $TH \cap TH^{\perp}$ is one dimensional. Let

$$H^{\perp} = \{v_p \in T_pM^m | p \in H, v_p \in T_pH^{\perp}\}$$

denote the orthogonal line bundle over H. Since $H^{\perp} \cap TH$ has nontrivial intersection on S(H) we see that $H^{\perp}$ has some unusual twisting when compared with the normal bundle. Let $\mathrm{Orth}^c(H) = TM/H^{\perp}$ denote the co-orthogonal, bundle over H (rank = dim(H)). Choosing an appropriate sphere S^{m-1} in M^m we may view a nonzero section N of $H^{\perp}$ as a classifying map for $\mathrm{Orth}^c(H)$; that is as a type of "Gauss map for H". With the natural transversality conditions imposed one expects a relationship between the degree of N, and invariants of the bundle $TH/H^{\perp}$ over S(H). One also expects that these invariants can be realized by intrinsic differential geometric data. A complication lies in the observation that there may be secondary degeneracies in S(H) itself, which in turn produce a stratification of S(H). So there should be a more refined relationship for each of the different stratifications types.

(Fortunately there are only a finite number of C^∞ open types.) Here we establish such a relationship for the simplest stratification.

(6) Fary-Milnor and Gauss-Bonnett type Theorems for Surfaces in M^3. (In preparation)

Here we use the observations of 4 and 5 above to yield results in the spirit of classical theorems concerning curves and surfaces in Euclidean space E^3.

Marek Kossowski
Rice University
Department of Mathematical Sciences
Houston, Texas 77251

Department of Mathematics
University of South Carolina
Columbia, South Carolina 29208

Contemporary Mathematics
Volume **71**, 1988

COVARIANT QUANTIZATION OF DYNAMICAL SYSTEMS WITH CONSTRAINTS

Karel V. Kuchař

ABSTRACT. Generally relativistic field theories lead to constraints which close in a characteristic way under Poisson brackets. To remain consistent at the quantum level, the constraint operators must be factor ordered so that the Poisson brackets go over into commutators without acquiring terms which would spoil the closure. The problem of finding a consistent factor ordering is resolved for finite dimensional gauge systems; these mimic many features of relativistic field theories. The proposed quantization process is covariant under all relevant transformations which keep the physical content of the classical theory intact. As a result, it does not matter whether one reduces the gauge theory to the physical theory at the classical level and then quantizes the classical theory, or whether one quantizes the classical theory first and then reduces it to the quantum physical theory.

As people say in my native Bohemia, one speaks the language and one drinks beer. This led a prominent Czech linguist to comment, in a sonnet he wrote in praise of the Czech language:[1] "No wonder then that what one speaks is beer." I admit that the "Litera scripta manet" of the Romans is more noble than the Czech beer talk, but it has shortcomings of its own. The petrified written word loses the freshness of a spoken one and a talk polished for conference proceedings becomes just another article, i's dotted and t's crossed, tiresome to follow. This is why I do not see much sense in repeating pedantically the details of what I have written elsewhere. Rather, I want this account to serve the same purpose as my beer talk at Santa Cruz: to be a brief exposition of what I feel should be preserved when quantizing a classical theory with constraints. I thus leave the text broken into short sections reflecting the content of individual transparencies, to keep the staccato rhythm of the points which I want to emphasize. To fill the gaps, the readers (as well as the listeners on that late June morning) should go to the <u>real</u> papers which I quote at the end of this report.

53B20, 53B50, 58A17, 58G30, 70G05, 70G05, 70G35, 70H05, 81D05, 81C10, 83C45.

GEOMETRODYNAMICS

The changes of the spatial metric $g_{ab}(x)$, $x \in \Sigma$, in the general theory of relativity are generated by the Hamiltonian which is a linear combination of an infinite system of super-Hamiltonian $H(x)$ and supermomentum $\Pi_a(x)$ constraints. The super-Hamiltonians are quadratic functions of the momenta $p^{ab}(x)$ conjugate to $g_{ab}(x)$ and they generate the dynamical evolution of the geometry under a normal displacement of the hypersurface Σ in the spacetime manifold. The supermomenta are linear homogeneous functions of the momenta $p^{ab}(x)$ and they generate a gauge change of the metric $g_{ab}(x)$ under spatial diffeomorphisms, DiffΣ, of Σ.
Schematically,

(1)

CONSTRAINTS	DEPENDENCE ON p's	GENERATE
SUPER-HAMILTONIAN $H(x)[g,p]$	QUADRATIC INHOMOGENEOUS	DYNAMICS
SUPERMOMENTUM $\Pi_a(x)[g,p]$	LINEAR HOMOGENEOUS	GAUGE (DiffΣ)

The consistency of the evolution is ensured by the Poisson bracket relations

(2a) $$\{\Pi_a(x), \Pi_b(x')\} = \Pi_b(x)\ \delta_{,a}(x,x') - (ax \leftrightarrow bx') \quad ,$$

(2b) $$\{\Pi_a(x), H(x')\} = H(x)\ \delta_{,a}(x,x') \quad ,$$

(2c) $$\{H(x), H(x')\} = g^{ab}(x)\ \Pi_a(x)\ \delta_{,b}(x,x') - (x \leftrightarrow x') \quad ,$$

which imply that the data remain on the constraint surface

(3) $$H(x) = 0 = \Pi_a(x) \quad .$$

Two features of the closure relations (2) are important. First, the explicit occurrence of the canonical variable $g_{ab}(x)$ in Eq.(2c) means that Eqs.(2) do not represent a Lie algebra. (Note, however, that Eqs.(2a) alone represent a Lie algebra, namely LDiffΣ.) Second, the constraints linear and quadratic in the momenta close in a characteristic fashion: two linear constraints close again into a linear constraint, a linear constraint closes with a quadratic constraint into a quadratic constraint, while two quadratic constraints close into a linear one.

All these features of the general theory of relativity as a dynamical theory (called geometrodynamics) are shared by other theories covariant under spacetime diffeomorphisms: in particular, by the string theory and by parametrized field theories on a given Riemannian background.

When quantizing such theories, one attempts to replace the classical constraints by operators and the constraint surface (3) by the set of physical states Ψ which are annihilated by the constraint operators. The quantization is consistent only if the Poisson bracket relations (2) can be carried over into commutator relations without acquiring extra terms which would spoil the closure. It is not a priori clear whether there exists a factor ordering of the constraints which achieves this goal. The factor ordering problem became notorious in the general theory of relativity;[2] in string theories, it is believed that it has a solution only in spaces with a certain number of dimensions.[3] It motivates the following query into what it means to quantize a classical system subject to constraints.

FINITE DIMENSIONAL MODELS

In field theories, the factor ordering problem is coupled to renormalization. To face it in a pure form, one should turn to finite dimensional models. The simplest of these are gauge theories,[4] which lack the super-Hamiltonian constraints. Instead, the dynamical evolution is provided by a true Hamiltonian

$$H = \frac{1}{2} G + U + Y = \frac{1}{2} G^{AB}(Q)P_A P_B + U^A(Q)P_A + Y(Q) \tag{4}$$

which is assumed to be a quadratic function of the momenta P_A conjugate to the coordinates Q^A in a 'big' configuration manifold $\mathcal{M}$. A number of linear constraints associated with the gauge invariance,

$$\pi_\alpha = 0 \quad , \quad \text{with } \pi_\alpha := \phi^A_\alpha(Q)P_A \quad , \tag{5}$$

play the same role as the supermomenta do in geometrodynamics. They close in a similar way,

$$\{\pi_\alpha, \pi_\beta\} = C^\gamma{}_{\alpha\beta}(Q)\pi_\gamma \quad , \tag{6}$$

except that, to mimic a prominent feature of the total geometrodynamical system, they are not required to represent a Lie algebra. The coefficients $C^\gamma{}_{\alpha\beta}(Q)$ in Eq.(6), unlike their counterparts in Eq.(2a), are thus allowed to depend on the coordinates Q^A. Equation (6) then resembles Eq.(2c), which also contains a coefficient depending on a canonical coordinate, namely, on the metric $g_{ab}(x)$. Moreover, the Hamiltonian must be an observable, i.e., its Poisson brackets with linear constraints must vanish on the constraint surface,

$$\{H, \pi_\alpha\} = H^\beta_\alpha(Q,P)\pi_\beta \quad . \tag{7}$$

This is a counterpart of Eq.(2b). Note, however, that in geometrodynamics the Poisson bracket between a quadratic super-Hamiltonian and a linear supermomentum closes into a quadratic super-Hamiltonian, whereas in a gauge theory the Poisson bracket between a quadratic Hamiltonian and a linear gauge constraint closes into a linear gauge constraint. Indeed, there is no quadratic constraint into which such a bracket could close.

One may think that when the gauge theory is parametrized, i.e., when the time t and its conjugate momentum P_t are adjoined to the rest of the canonical variables Q^A, P_A and the Hamiltonian H is replaced by the super-Hamiltonian constraint $P_t + H$, the closure scheme should switch to the geometrodynamical pattern (2b). However, it does not. With H denoting now the super-Hamiltonian constraint, one still has an equation of the form (7).

The geometrodynamical pattern (2b) appears in finite dimensional theories only when one studies parametrized relativistic systems (like a relativistic particle moving in a given spacetime background) endowed with additional gauge degrees of freedom.[5] The super-Hamiltonian constraint H of such systems (which restricts the spacetime momentum of the system onto the mass shell) has a Poisson bracket with the linear gauge constraints (5) which in general closes into both the quadratic and the linear constraints,

$$\{H, \pi_\alpha\} = \Lambda(Q)H + H^\beta_\alpha(Q,P)\pi_\beta \quad ; \tag{8}$$

for particular models, $H^\beta_\alpha = 0$ and Eq.(8) reduces to the geometrodynamical form.

Quantization of such relativistic models, however, presents an additional problem to that of a consistent factor ordering. The construction of a Hilbert space requires a selection of positive energy states out of all the states which are annihilated by the constraint operators. One thus needs to identify an energy observable $E = W^A(Q)P_A$ which has weakly vanishing Poisson brackets with all the constraints and hence remains conserved in the dynamical evolution. I shall discuss the factor ordering of relativistic systems in a forthcoming paper[6] but I do not have enough time to say anything else about them here. Moreover, geometrodynamics lacks what one may call the energy observable[7] and it is thus not clear whether the construction of an inner product has any relevance for geometrodynamics.

None of the finite dimensional models above is rich enough to possess several super-Hamiltonians whose Poisson brackets would close into linear gauge constraints, as in the geometrodynamical equation (2c). In fact, it is quite difficult to construct a non-trivial finite dimensional system with this property. The only models which I know are reducible to a collection of completely independent (non-interacting) relativistic particles.

The preceding discussion indicates that finite dimensional systems can model some, though not all, characteristic properties of canonical geometrodynamics or other generally relativistic field theories. It is easy to find systems whose constraints close on the constraint surface without necessarily representing a Lie algebra, but it is quite difficult to replicate the finer details of the geometrodynamical scheme (2). I shall thus take only the simplest of all such models, namely, a finite-dimensional gauge theory (4)-(7), and discuss the factor ordering problem within its framework.

WHAT DO I WANT TO ACHIEVE?

The Hamiltonian (4) of a gauge theory drives the system along a trajectory in the big phase space Q^A, P_A. Much of that space, however, is physically irrelevant. Only those trajectories which lie on the constraint surface (5) matter, and even there the physical points correspond to the whole orbits generated by the constraints (5). The space of orbits inherits the symplectic structure of the big phase space and the orbits can thus be identified using the physical canonically conjugate variables q^a, p_a. Because the Hamiltonian (4) is a classical observable, Eq.(7), it reduces in the space of the orbits to a physical Hamiltonian

$$h = \frac{1}{2} g^{ab}(q)p_a p_b + u^a(q)p_a + y(q) \tag{9}$$

which depends only on the physical variables.

This reduction of a gauge theory to the physical theory does not depend at the classical level on any one of the following transformations:

1) Point transformations in the big phase space,

(10) $$Q^A \to Q^{A'} = Q^{A'}(Q^B), \quad P_A \to P_{A'} = Q^B_{A'}(Q^{C'})P_B \quad ,$$

where $Q^B_{A'} := \partial Q^B / \partial Q^{A'}$;

2) Point transformations in the physical phase space,

(11) $$q^a \to q^{a'} = q^{a'}(q^b), \quad p_a \to p_{a'} = q^b_{a'}(q^{c'})\, p_b \quad ,$$

with $q^b_{a'} := \partial q^b / \partial q^{a'}$;

3) Mixing of constraints,

(12) $$\pi_\alpha \to \pi_{\alpha'} = \Lambda^\beta_{\alpha'}(Q^C)\pi_\beta \quad ;$$

4) Change of the (quadratic) Hamiltonian within the same equivalence class modulo the constraints,

(13) $$H \to \bar{H} = H + M^{\alpha A}(Q^B)P_A\pi_\alpha + M^\alpha(Q^B)\pi_\alpha \quad .$$

I want to propose a canonical quantization of finite dimensional gauge systems which would be covariant under all of the enumerated transformations. Moreover, I require that the proposed scheme solves the factor ordering problem in such a way that it reproduces the quantized physical theory. In other words, it should not matter whether one reduces the gauge theory to the physical theory at the classical level and then quantizes the physical theory, or whether one quantizes the gauge theory first and then reduces the quantum gauge theory to the quantum physical theory.[8] Finally, the solution to the factor ordering problem should be expressible in a language which does not require an explicit construction of the physical variables q^a, p_a and h from the big phase space variables Q^A, P_A and H. Otherwise, the proposed algorithm would have little practical use and one could hardly hope to apply it to infinitely dimensional systems.

In brief, my aim is to develop a canonical quantization of finite dimensional gauge systems which is

- covariant under transformations (1)-(4),
- solves the factor ordering problem,

and

- reproduces the quantized physical theory
- without explicitly finding physical variables.

The words which I have used to describe my objectives reveal that, in comparison with other approaches to the same problem, I want to emphasize some ideas and to play down others. An impressive structure of generally relativistic field theories is the quadratic nature of the 'dynamical' constraints and the linear nature of the 'gauge' constraints in the generalized momenta. To talk about the polynomial structure of constraints and/or observables in the generalized momenta requires a separation of the momenta from the coordinates: i.e., one must emphasize the cotangent bundle structure of the phase space. This also explains why I limit attention to the point transformations (1) and (2) and do not consider all canonical transformations. Further, when I allow an arbitrary mixing of constraints (3), I cannot simultaneously require that the constraints represent the generators of a Lie group. This is reflected in their closure relations (6) which contain, in general, structure functions $C^{\gamma}{}_{\alpha\beta}(Q)$ instead of structure constants. I am thus trying to play down a possible group theoretical origin of the constraints and I am replacing the concept of a gauge group by the concept of a gauge algebra, considered as some subalgebra (6) of the Poisson algebra of dynamical variables which are linear and homogeneous in the canonical momenta.

The most important switch of emphasis, however, occurs in the attitude towards the role of the Hilbert space in the quantization process. The standard attitude is that the constraints should be represented by self-adjoint operators acting on some Hilbert space, albeit maybe only one with an indefinite metric. In my mind, the introduction of a Hilbert space so early in the quantization process is neither a mathematical necessity nor a physically motivated requirement. Mathematically, to give a meaning to constraints and observables as differential operators, the only thing which one needs is that the states Ψ form a suitable function space, say, a $\mathcal{F} = C^{\infty}(\mathcal{M}, \mathbb{C})$ space. Physically, the Hilbert space structure is forced on us only by the statistical interpretation of the theory, i.e., it is only the space of <u>physical</u> states which should be endowed with an inner product. Borrowing terminology from an old scholastic dispute between realists and nominalists, I can say that the operators representing constraints and observables do not need to be self-adjoint on the big function space <u>ante rem</u>, i.e., <u>before</u> the imposition of constraints, but only on the small Hilbert space $\mathcal{H}_0$ of physical states constructed <u>post rem</u>, i.e., <u>after</u> the imposition of constraints. It is mainly this relaxation of attitude toward the Hilbert space structure which enables me to present a resolution of the factor ordering problem which satisfies all my requirements. Mine is thus a 'nominalistic' rather than a 'realistic' solution of the problem.

Let me again briefly summarize my position in a table:

EMPHASIZED	PLAYED DOWN
Cotangent bundle structure	
Point transformations	Canonical transformations
Gauge algebra	Gauge group
Hilbert space structure <u>post rem</u>: after the imposition of constraints	Hilbert space structure <u>ante rem</u>: before the imposition of constraints

SPLITTING

I shall now sketch how to implement the objectives of the last section.[9] There are two structures which are immediately given to me in a gauge theory:

I) The gauge algebra $\mathcal{V}$ which is a closed subalgebra of the Poisson algebra of dynamical variables linear in the momenta. In that algebra, I arbitrarily select a basis $\pi_\alpha = \phi^A_\alpha(Q)P_A$, whose coefficients yield $C \leq N$ (by virtue of Eq.(6) surface-forming) vector fields

$$\phi^A_\alpha(Q) \quad . \tag{14}$$

II) An equivalence class of observables (13) generating the dynamics. Out of that class, I arbitrarily select a representative Hamiltonian (4) whose kinetic piece yields me a contravariant (possibly degenerate) metric

$$G^{AB}(Q) \quad . \tag{15}$$

The vector fields (14) and the metric (15) are the building blocks of the whole edifice of the canonical quantization of gauge systems.

The elements of the gauge algebra generate one-parameter groups of canonical transformations which, on the constraint surface (5), form the C-dimensional orbits. Any set of n independent functions $q^a(Q)$, $a = 1, \cdots, n$, on $\mathcal{M}$, which are constant along the orbits,

$$\{q^a(Q), \pi_\alpha\} = 0, \quad \text{i.e.,} \quad q^a{}_{,A}\,\phi^A_\alpha = 0 \quad , \tag{16}$$

can serve as the physical coordinates. Their gradients

$$Q^a_A := q^a{}_{,A} \tag{17}$$

enable me to project contravariant vectors (and tensors) into the physical space. In particular, from the 'big' metric (15) I can extract the physical metric

$$g^{ab} := G^{AB}Q^a_A Q^b_B \tag{18}$$

which I assume to be non-degenerate and, indeed, positive definite. The two metrics, G^{AB} and g_{ab}, help me to convert the n covectors Q^a_A into n vectors

(19) $$Q^A_a := G^{AB} Q^b_B g_{ba}$$

which complement the C vectors ϕ^A_α into a basis in $T_Q\mathcal{M}$:

(20) $$\phi^A_\alpha(Q) \ , \quad Q^A_a(Q) \quad .$$

Dual to the basis (20) I have the cobasis

(21) $$\phi^\alpha_A(Q) \ , \quad Q^a_A(Q)$$

defined by the standard orthonormality relations.

The bases (20) and (21) enable me to split any vector or covector field into a physical piece and a gauge piece. In particular, I can identify the momentum

(22) $$p_a = Q^A_a(Q) P_A$$

as a standard representative of the equivalence class of observables which are canonically conjugate to the physical position q^a.

By splitting the vector potential part $U := U^A P_A$ and the kinetic part $G := G^{AB} P_A P_B$ of the Hamiltonian (4), I arrive at the decomposition

(23) $$U = u^a p_a + \upsilon^\alpha \pi_\alpha$$

and

(24) $$G = g^{ab} p_a p_b + \gamma^{\alpha\beta} \pi_\alpha \pi_\beta$$

whose coefficients are given by the projections

(25) $$u^a := U^A Q^a_A \quad , \qquad \upsilon^\alpha := U^A \phi^\alpha_A \quad ,$$
$$g^{ab} := G^{AB} Q^a_A Q^b_B \quad , \qquad \gamma^{\alpha\beta} := G^{AB} \phi^\alpha_A \phi^\beta_B \quad .$$

By virtue of Eq. (7), u^a and g^{ab} can depend on the coordinates Q^A only through the physical coordinates q^a. A similar statement holds for the scalar potential $Y(Q)$ in the Hamiltonian (4): namely, $Y(Q) = y(q(Q))$. This spells out how to get the physical Hamiltonian (9) from the 'big' Hamiltonian (4).

QUANTIZATION

I can explain now how to obtain a consistent factor ordering of the constraints and of the Hamiltonian which has all the desired features of the last section.[10] The trick is to order the gauge parts and the physical parts separately. The constraints are replaced by the directional derivatives

(26) $$\hat{\pi}_\alpha = -i\, \phi^A_\alpha(Q)\partial_A \quad .$$

This carries the Poisson algebra (6) homomorphically into the commutator algebra, with the constraint operators on the right of the structure functions:

(27) $$\frac{1}{i}\,[\hat{\pi}_\alpha, \hat{\pi}_\beta] = C^\gamma{}_{\alpha\beta}(Q)\hat{\pi}_\gamma \quad .$$

The gauge piece υ of the linear observable U is ordered in the same way,

(28) $$\hat{\upsilon} = \upsilon^\alpha(Q)\hat{\pi}_\alpha \quad ,$$

but the physical piece u is complemented by a physical divergence term,

(29) $$\hat{u} = -\,i(u^a\partial_a + \frac{1}{2}\,\mathrm{div}_g\mathbf{u}) \quad ,$$

which will ultimately ensure that $\hat{u}$ is self-adjoint on the physical space. Here,

(30) $$\mathrm{div}_g\mathbf{u} := |g|^{-\frac{1}{2}}\,(|g|^{\frac{1}{2}}\,u^a)_{,a} \quad ,$$

while $\partial_a = Q^A_a\partial_A$ is the directional derivative along the vector field Q^A_a. Under such a mixed ordering, the Poisson algebra of any two linear observables U and V is still homomorphically mapped into the quantum commutator algebra:

(31) $$\{U,V\} = W \;\Rightarrow\; \frac{1}{i}\,[\hat{U},\hat{V}] = \hat{W} \quad .$$

Next comes the ordering of the kinetic energy observable G. I order the physical piece as one would order the Laplace-Beltrami operator:

(32) $$\hat{g} = -\,\text{"}\Delta_g\text{"} := -\,|g|^{-\frac{1}{2}}\,\partial_a\,|g|^{\frac{1}{2}}\,g^{ab}\,\partial_b \quad .$$

The quotation marks serve as a reminder that the directional derivatives $\partial_a := Q^A_a\partial_A$ are not, strictly speaking, the partial derivatives with respect to q^a. They reduce to them only if ∂_a acts on a function of physical coordinates.

Similarly, I can order $\hat{\gamma}$ as one would order the Laplace-Beltrami operator "Δ_γ", taking due account of the fact that the differentiations $\partial_\alpha := \phi^A_\alpha\partial_A$ are taken in a non-holonomic basis. Alternatively, I can forget the gauge part γ altogether because it does not influence the physical results.

All the differential operators which I have introduced are assumed to act on the space $\mathcal{F} = C^\infty(\mathcal{M}, \mathbb{C})$. I define the subspace $\mathcal{F}_0 \subset \mathcal{F}$ of the physical states by the requirement that the physical states $\Psi \in \mathcal{F}_0$ be annihilated by all the operators $\hat{\nu} = \nu^{\alpha}(Q)\hat{\pi}_{\alpha}$ from the quantized gauge algebra or, alternatively, by the basis operators (26). As a consequence, the physical states can depend on Q^A only via the physical coordinates q^a:

$$\Psi(Q) = \psi(q(Q)) \quad . \tag{33}$$

This enables me to define an inner product of two physical states Ψ_1 and Ψ_2 by

$$< \Psi_1 , \Psi_2 > := \int d^n q \; |g(q)|^{\frac{1}{2}} \; \psi_1^*(q) \; \psi_2(q) \quad . \tag{34}$$

The physical Hilbert space $\mathcal{H}_0$ is then obtained from the physical space $\mathcal{F}_0$ by selecting elements of $\mathcal{F}_0$ of finite norm (34) and completing them with respect to this norm.

I can now gather all the quantized pieces of the 'big' Hamiltonian together into the Hamilton operator $\hat{H}$. The gauge pieces begin always with one $\hat{\pi}_{\alpha}$ operator on the right and the ∂_a operator in the physical pieces, when acting on a physical state $\Psi \in \mathcal{F}_0$, reduces to a differentiation with respect to the physical coordinates,

$$\partial_a \Psi = \partial\psi(q)/\partial q^a \qquad \forall\Psi \in \mathcal{F}_0 \quad . \tag{35}$$

As a result, the action of the 'big' Hamilton operator on a physical state reduces to that of the physical Hamiltonian

$$\hat{h} := -\frac{1}{2}\Delta_g - i(u^a\partial_a + \frac{1}{2}\,\mathrm{div}_g \mathbf{u}) + y(q) \tag{36}$$

on the physical state:

$$\hat{H}\,\Psi = \hat{h}\,\psi \quad . \tag{37}$$

This implies that $\hat{H}$ is self-adjoint in the inner product (34) and that

$$\frac{1}{i}\,[\hat{H}, \hat{\pi}_{\alpha}]\,\Psi = 0 \quad \forall\Psi \in \mathcal{F}_0 \quad ; \tag{38}$$

This is a quantum version of the classical condition (7) that H be an observable. Equations (27) and (38) ensure that the factor ordering problem is consistently resolved. One can easily generalize the quantization prescription $H \to \hat{H}$ so that it applies, with similar consequences, to any classical observable which is at most quadratic in the momenta P_A.

FACTOR ORDERING WITHOUT SPLITTING

I have introduced a factor ordering of the linear and quadratic variables by splitting their physical parts from the gauge parts. Now I must show how to proceed without identifying the physical variables q^a, p_a and without any splitting.[10]

This is done with the help of the Levi-Civita pseudotensor of the physical space considered as a tensor in the big space. Construct an object

(39) $$\phi_{A_1\cdots A_n} := \delta_{A_1\cdots A_n B_1\cdots B_C}\,\phi_1^{B_1}\cdots\phi_C^{B_C}$$

orthogonal to the orbits ($\delta_{A_1\cdots A_N}$ is the alternating symbol) and normalize it by its magnitude $\|\phi\|$,

(40) $$\|\phi\|^2 := \frac{1}{n!}\,\phi_{A_1\cdots A_n}\,G^{A_1B_1}\cdots G^{A_nB_n}\,\phi_{B_1\cdots B_n}\ ,$$

into

(41) $$\epsilon_{A_1\cdots A_n} := \|\phi\|^{-1}\,\phi_{A_1\cdots A_n}\ .$$

One can prove that $\epsilon_{A_1\cdots A_n}$ is the Levi-Civita pseudotensor $\epsilon_{a_1\cdots a_n}$ of the physical space lifted up to the big space:

(42) $$\epsilon_{A_1\cdots A_n} = \epsilon_{a_1\cdots a_n}\,Q^{a_1}_{A_1}\cdots Q^{a_n}_{A_n}\ .$$

However, in the construction (39)-(41) of $\epsilon_{A_1\cdots A_n}$, there is no need to find the projectors Q^a_A .

The first service which the Levi-Civita pseudotensor provides is an invariant characterization of the inner product (34). I choose an arbitrary n-dimensional surface σ transverse to the orbits, and the Levi-Civita pseudotensor endows it with the measure

(43) $$d\sigma := \epsilon_{A_1\cdots A_n}\,dQ^{A_1}\wedge\cdots\wedge dQ^{A_n}\ .$$

I define then the inner product of two arbitrary physical states $\Psi_1\ ,\ \Psi_2 \in \mathcal{F}_o$ as

(44) $$<\Psi_1\ ,\ \Psi_2> := \int_\sigma d\sigma\ \Psi_1^*(Q)\ \Psi_2(Q)\ .$$

One can show that this inner product does not depend on σ and that it reduces to the old expression (34).

The second service which the Levi-Civita pseudotensor provides is the factor ordering of the linear and quadratic observables which does not require any splitting. One can express the physical divergence of the physical part u^a of U^A in the form

(45) $$\mathrm{div}_g\mathbf{u} = \frac{1}{(n-1)!}\,\epsilon^{A_1\cdots A_n}\,\partial_{A_1}(U^A\epsilon_{AA_2\cdots A_n})$$

(the indices are raised by the big metric G^{AB}) and use it in the expression (29) which, together with (28), yields the operator

$$\hat{U} = - i(U^A \partial_A + \frac{1}{2} \mathrm{div}_g \mathbf{u}) \quad . \tag{46}$$

Similarly, one can rewrite the expression (32) for the Laplace-Beltrami ordering of the physical part of the kinetic energy,

$$\hat{g} = - \frac{1}{(n-1)!} \epsilon^{AA_2 \cdots A_n} \partial_A \epsilon^B{}_{A_2 \cdots A_n} \partial_B \quad . \tag{47}$$

Forgetting about its gauge part γ, I can collect the kinetic, linear and potential pieces of $\hat{H}$ into an expression which is factor ordered without a previous identification of the physical variables q^a and p_a.

THE QUANTUM WELL OF ORVIETO

I shall illustrate how the formalism works in a simple gauge system.[11,4] Take a non-relativistic particle moving in a three-dimensional Euclidean space E^3 with Cartesian coordinates $Q^A = (X,Y,Z)$ and let the translation group T(1) act on E^3 as a gauge group by helical motions

$$\begin{aligned} X(\tau) &= X(0) \cos\tau + Y(0) \sin\tau \quad , \\ Y(\tau) &= - X(0) \sin\tau + Y(0) \cos\tau \quad , \\ Z(\tau) &= Z(0) + \tau \quad . \end{aligned} \tag{48}$$

The orbits (48) form an infinite system of spiral staircases which nowhere meet, surrounding the central Z = 0 column. Drawing but two of them for the sake of visualization, one gets the design of St. Patrick's well built in Orvieto by Pope Clement VII de' Medici to supply the town with water in case of siege. I call the model the well of Orvieto; people interested in canonical quantization are always besieged by armies of particle physicists.

The orbits (48) can be considered as points of the physical configuration space and labeled by the physical coordinates $q^a = (r,\theta)$,

$$r = R, \quad \text{with } R := (X^2 + Y^2)^{1/2} \quad , \tag{49}$$

$$\theta = (\Theta - Z) \operatorname{Mod} 2\pi, \quad \text{with } \Theta := \operatorname{arctg} \frac{X}{Y} \quad . \tag{50}$$

The generator

$$\phi^A = dQ^A(\tau)/d\tau = (Y, -X, 1) \tag{51}$$

of the group action is a Killing vector of the Euclidean metric $G^{AB} = \delta^{AB}$ and hence the constraint $\pi = \phi^A P_A$ has a vanishing Poisson bracket with the

Hamiltonian $H = \frac{1}{2}\delta^{AB}P_AP_B$:

$$\{H, \pi\} = 0 \quad . \tag{52}$$

The metric

$$g^{ab} = \text{diag}\,(1,\ 1 + r^{-2}) \tag{53}$$

induced in the two-dimensional physical configuration space is curved, but regular and complete. It becomes flat for $r \ll 1$ and again for $r \gg 1$, where it corresponds to a cylindrical surface with embedding radius 1. The complete embedding diagram has the shape of a cylindrical vessel curving rather abruptly into a flattened bottom at $r = 0$. The quantum states of the physical system can be described by scalar functions $\psi(q)$ with the inner product (34) and the physical Hamiltonian is given by the Laplace-Beltrami operator Δ_g with respect to the metric (53):

$$\hat{h} = -\frac{1}{2}\Delta_g = -\frac{1}{2}r^{-1}(1+r^2)^{\frac{1}{2}}\,\partial_r\, r(1+r^2)^{-\frac{1}{2}}\,\partial_r - \frac{1}{2}(1+r^{-2})\,\partial_\theta^{\,2} \quad . \tag{54}$$

I shall now quantize the well of Orvieto in the big space E^3.[12] In cylindrical coordinates $Q^A = (R,\Theta,Z)$ I have

$$\phi^A = (0,\ 1,\ 1) \quad , \qquad G^{AB} = \text{diag}(1,\ R^{-2},\ 1) \quad . \tag{55}$$

This yields the Levi-Civita pseudotensor

$$\epsilon_{AB} = R(1 + R^2)^{-\frac{1}{2}} \times \begin{Vmatrix} 0, & 1, & -1 \\ -1, & 0, & 0 \\ 1, & 0, & 0 \end{Vmatrix} \tag{56}$$

and leads to the Hamilton operator

$$\begin{aligned} \hat{H} &= -\frac{1}{2}\epsilon^{AC}\,\partial_A\,\epsilon^{B}{}_{C}\,\partial_B \\ &= -\frac{1}{2}R^{-1}\,(1 + R^2)^{\frac{1}{2}}\,\partial_R\,R(1 + R^2)^{-\frac{1}{2}}\,\partial_R \\ &\quad -\frac{1}{2}R^2\,(1 + R^2)^{-1}\,(\partial_Z - R^{-2}\,\partial_\Theta)^2 \quad . \end{aligned} \tag{57}$$

It is obvious that $\hat{H}$ commutes with the gauge constraint

$$\hat{\pi} = -\,i(\partial_\Theta + \partial_Z) \quad . \tag{58}$$

This amounts to the quantum version of the classical equation (52),

$$\frac{1}{i}\,[\hat{H},\ \hat{\pi}] = 0 \quad , \tag{59}$$

and guarantees that the factor ordering of $\hat{H}$ and $\hat{\pi}$ is consistent.

When acting on a physical state

$$\Psi(R,\ \Theta,\ Z) = \psi(r = R,\ \theta = \Theta - Z) \quad , \tag{60}$$

the second term in (57) yields $-\frac{1}{2}(1+r^2)\,r^{-2}\,\partial_\theta{}^2$ and $\hat{H}$ thus reduces back to the physical Hamiltonian (54). This shows that the results obtained from the quantum gauge theory comply with our physical expectations.

This cannot be said about the naive way in which most people would be tempted to quantize the Orvieto well model. It seems only natural to order the classical Hamiltonian $H = \frac{1}{2}\,\delta^{AB}P_A P_B$ as the Laplace-Beltrami operator Δ_G with respect to the flat metric $G^{AB} = \delta^{AB}$. Such a naive ordering is still consistent, Eq.(59), but it yields physical predictions which are different from those of the quantized physical theory.[12] In particular, in the presence of a potential consistent with the action of the gauge group, the Hamiltonian with the physical ordering (57) of its kinetic part has a different spectrum than the naively ordered Hamiltonian. Here one sees that consistency alone does not uniquely determine the factor ordering.

Simple as the Orvieto well model may be, it has an interesting field-theoretical application.[13] In scalar electrodynamics, the action of the gauge group on the real and imaginary parts of the scalar field can be considered as a rotation in the field plane while its action on the electromagnetic potential has a character of a translation. One can thus express the joint action of the gauge transformations of the 'first' and 'second' kind on the total field space as an infinite number of replicas of the action of T(1) on E^3 by helical motions. Standard quantization of scalar electrodynamics corresponds to the 'naive' factor ordering. One may wonder whether the 'physical' ordering leads to higher order corrections which could be in principle detectable. Unfortunately, the difference between the two orderings is obscured by renormalization problems. One would like to see how to implement the 'physical' factor ordering advocated in this talk, with its strong geometric flavor, in field theoretic systems, and whether such an ordering casts light on the problem which has motivated all these investigations, namely, the problem of the canonical quantization of gravity.

ACKNOWLEDGMENTS

I want to thank Drs. Charles Torre and Christopher Stephens for their comments on the draft of this report. Support from the NSF Grant PHY 85-03653 to the University of Utah and from the Institute for Theoretical Physics, University of California in Santa Barbara, under NSF Grant PHY 82-17853, supplemented by funds from NASA, are gratefully acknowledged.

BIBLIOGRAPHY

1. P. Eisner, Sonety Kněžně, J. Podroužek, Praha, 1945.

2. J. Anderson, in: Eastern Theoretical Conference, edited by M.E. Rose, Gordon and Breach, New York, 1963.

   J. Schwinger, Phys. Rev., 130 (1963), 1253; 132 (1963), 1317.

   P.A.M. Dirac, in: Contemporary Physics: Trieste Symposium 1968, edited by A. Salam, International Atomic Energy Agency, Vienna, 1969, p. 539.

   B.S. DeWitt, Phys. Rev., 160 (1967), 1113.

   A. Ashtekar, Phys. Rev. Lett., 57, (1986), 2244.

3. R.C. Brower, Phys. Rev., D6 (1972), 1655;

   S. Hwang, Phys. Rev., D28 (1983), 2614;

   R. Marnelius, Nuc. Phys., B211 (1983), 14.

4. K. Kuchař, Phys. Rev., D34, 3031 (1986).

5. A. Ashtekar and M. Stillerman, J. Math. Phys., 27 (1986), 1319.

6. K. Kuchař, to be submitted to the Phys. Rev. D.

7. K. Kuchař, J. Math. Phys., 22 (1981), 2640.

8. A. Ashtekar and G. Horowitz, Phys. Rev., D26 (1982), 3342, discuss a quantization scheme which does not satisfy this requirement.

9. For the details, see Ref. 4.

10. K. Kuchař, Phys. Rev., D34, 3044 (1986).

11. B.S. DeWitt, in: General Relativity: An Einstein Centenary Survey, edited by S.W. Hawking and W. Israel, Cambridge University Press, Cambridge, 1979, footnote on p. 729.

12. K. Kuchař, Phys. Rev. D35, scheduled for publication in the 15 January 1987 issue.

13. I. Bialynicki-Birula, work in progress, private communication.

DEPARTMENT OF PHYSICS
UNIVERSITY OF UTAH
SALT LAKE CITY, UTAH 84112

Contemporary Mathematics
Volume **71**, 1988

PROBLEMS ON DIFFEOMORPHISMS ARISING FROM QUANTUM GRAVITY

John L. Friedman[1] and Donald M. Witt[2]

In classical general relativity, two metrics or two sets of tensor fields are considered physically equivalent if they differ by the action of a diffeomorphism. The analogue in the canonical approach[3] to quantum gravity of this "general covariance" is an invariance of state vectors under diffeomorphisms [3]; but the invariance extends only to diffeos that can be continuously deformed to the identity and which, in the asymptotically flat case, approach the identity at spatial infinity [4-11]. Two diffeos are *isotopic* if one can be deformed to the other via a continuous family of diffeos -- if, in the space Diff of diffeos endowed with the C^∞ topology, they are connected. Then for a compact manifold N, the "homeotopy" group π_0(Diff) of isotopically inequivalent diffeomorphisms of N acts on the quantum space of state vectors associated with N. For asymptotically flat 3-geometries, one may restrict consideration to metrics with a fixed asymptotic behavior and to diffeos that respect that asymptotic form. Two classes of such diffeos can act nontrivially on the state space: those that asymptotically correspond to elements of the symmetry group at spatial infinity (e.g., rotations and translations) and those that, while trivial at infinity, are not isotopic to the identity. The change in a state vector under diffeos in the first class determines the momentum and angular momentum of the state. Elements in the second class include the diffeo corresponding to a rotation by 2π at spatial infinity and diffeos that permute identical prime factors of the topology. If, in the spirit of geometrodynamics, one regards a prime factor of the topology as a particle -- a topological geon -- then the action of these

---

1980 Mathematics Subject Classification (1985 Revision). 57R50, 83C45, 83E05.

[1]Research supported in part by the National Science Foundation under Grant No. PHY 8303597.

[2]Research supported in part by a grant from the Graduate School, The University of Wisconsin-Milwaukee.

[3]A review of the canonical approach to quantum gravity with a bibliography is given in [1]. The superspace of Riemannian 3-geometries is discussed in [2].

0271-4132/88 $1.00 + $.25 per page

latter diffeos on a state vector determines whether the geons have integral or half-integral values of angular momentum and even or odd (or other) statistics.

Diffeos that approach the identity at infinity can be regarded as fixing (at least) a point and a frame of the compact manifold obtained by adding a point at infinity. The amount of structure that is to be fixed depends on the precise definition of asymptotic flatness at spatial infinity, but because the space Diff_e of diffeos that fix a frame e at x has the same homotopy type as the space of diffeos that fix an entire disk about the point x, the group of inequivalent diffeos is isomorphic to $\pi_0(\mathrm{Diff}_e)$, independent of the details of asymptotic flatness.

The diffeo $\mathcal{R}$ corresponding to a 2π-rotation plays an extraordinary role in both the physical and mathematical aspects of the theory. In quantum gravity, it is responsible for the appearance of half-integral angular momentum (double-valued representations of the rotation group) in a theory where the only dynamical variable is the metric. On the mathematical side, it turns out to provide the only example of a diffeo that is homotopic but not isotopic to the identity. (This is discussed below.) By topologists, $\mathcal{R}$ is termed a rotation parallel to a sphere.

Given a submanifold with topology $S^2\times I$ of a 3-manifold M, one can define $\mathcal{R}$ as a diffeo that is the identity outside the submanifold and which rotates successive spheres of the copy of $S^2\times I$ by an angle that varies smoothly between 0 and 2π. Explicitly, if $\{x^i\}$ is a chart on M for which the submanifold is the region with $r < |x| < 2r$, one could set

$$\mathcal{R}(x) = R^i_j(\phi)x^j ,$$

on the submanifold, where $R^i_j(\phi)$ is the rotation matrix associated with a rotation by ϕ about some axis and ϕ is a smooth function on M that vanishes inside the sphere $|x| = r$ and has the value 2π outside $|x| = 2r$.

One might wonder how in the asymptotically flat context the relevant group of diffeos is $\pi_0(\mathrm{Diff}_e)$, while only the apparently smaller group $\pi_0(\mathrm{Diff})$ can have nontrivial action on the space of quantum states of compact manifold. The answer is not difficult to understand, and it provides a key to the structure of Diff for manifolds that are not prime. In representing an isolated system by an asymptotically flat geometry, one idealizes a situation in which a 3-manifold with topology $M - D$ (where D is a 3-disk and M is compact) is separated by a nearly flat region from a larger space which we will take to be compact with topology $M \# N$. If a rotation parallel to a sphere about a disk D in N is not isotopic rel. D to the identity in N, then it appears that $\pi_0(\mathrm{Diff}_e(M))$ is a subgroup of $\pi_0(\mathrm{Diff}(M\# N))$ [12]. This is known to be true if M and N are prime [13] and recent work [14,15] may allow a proof for arbitrary M and N.

Then for M part of a larger, generic universe, $\pi_0(\mathrm{Diff}_e(M))$ is a subgroup of $\pi_0(\mathrm{Diff})$ for the larger space. Any 3-manifold N can be written as a connected sum of its prime factors M_i, $N \simeq M_1 \# \ldots \# M_n$. The homeotopy group $\pi_0(N)$ is then generated by

(i) rotations parallel to the separating sphere of an irreducible factor or to the belt sphere of a handle,

(ii) the other elements of $\pi_0(\mathrm{Diff}_e)$ for each prime factor M_1,

(iii) permutations of diffeomorphic prime factors, and

(iv) slides.

A slide can be visualized as the result of moving a prime factor M_i along a loop ℓ of $N - M_i$ with ℓ not homotopic to zero. In analogy with R, a slide might be called a rotation parallel to a torus: one rotates concentric tori about ℓ through an angle that varies from 0 to 2π as one moves outward. (One may also slide one end of a handle.)

In the remainder of this article, we will discuss the structure of the group $\pi_0(\mathrm{Diff}_e)$ for some spherical spaces. Because detailed formal computations are available [10,11], we will present a more intuitive picture. In particular, we hope in Section 1 to show the reader that for many of the spherical spaces (and for other 3-manifolds whose geometrical structure is known explicitly) the group of inequivalent diffeos can be understood with surprisingly little knowledge of topology. In Section 2 we seek similarly to illustrate the proof of existence of a diffeo that is homotopic but not isotopic to the identity [10]. Because of recent results of Boileau and Otal, the present theorem is stronger than that given in Ref. 10. Although the discussion is somewhat more technical than that of Section 1, the lack of an isotopy can be seen in a fairly simple-minded way. We do not know of an intuitive description of the homotopy and would like to encourage topologists to find one.

1. HOMEOTOPY GROUPS OF SPHERICAL SPACES.

The group $SO(4) \simeq \dfrac{SU(2)\times SU(2)}{Z_2}$ acts on $S^3 \simeq SU(2)$ by $SU(2)$-multiplication: (g_1, g_2): $g \to g_1 g\, g_2^{-1}$. A spherical space is a manifold of the form S^3/H, where H is a finite subgroup of $SO(4)$ acting freely on S^3 [16,17]. Since $SU(2) \subset SO(4)$ (set $g_2 = \mathrm{Id}$), all finite subgroups of $SU(2)$ give rise to spherical spaces. Each spherical space with $H \subset SU(2)$ can be constructed from opposite faces of a solid polyhedron, and $\pi_0(\mathrm{Diff}_e)$ may then be identified with the isometry group of the polyhedron. Apart from the cyclic groups Z_p (which provide the lens spaces) the finite subgroups of $SU(2)$ are the double coverings in $SU(2)$ of the finite subgroups of $SO(3)$, namely, T^*, the 24 element covering of the tetrahedral group; O^*, the 48 element covering of the octahedral group; I^*, the 120 element covering of the icosahedral group; and the family D^*_{4m}, the 4m-element coverings of the dihedral groups.

The octahedral space, S^3/T^*, is constructed from a solid octahedron as in Fig. 1. The identification of a pair of (shaded) faces is shown, and identification of the other pairs of faces are implied by the octahedral symmetry. The spaces S^3/O^* and S^3/I^* are similarly obtained from a truncated cube and a solid dodecahedron. Each space S^3/D^*_{4m} is constructed from a 2m-sided prism: The top and bottom are identified after a relative rotation by π/m, and opposite rectangular faces are identified after a rotation by $\pi/2$ as in Fig. 2.

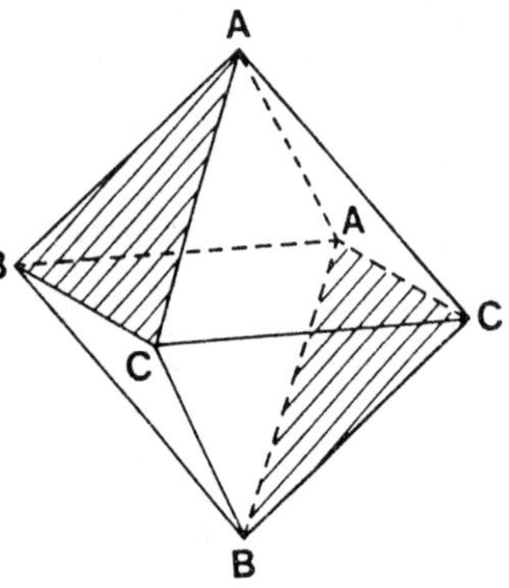

Fig. 1. Octahedron space, S^3/T^*.

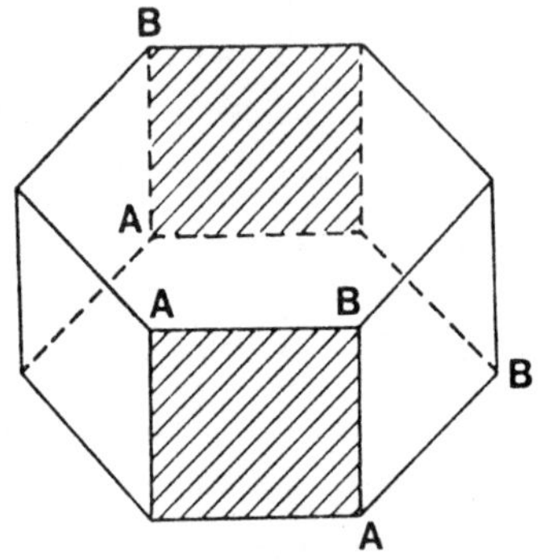

Fig. 2. A prism manifold, S^3/D^*_{12}.

The lens spaces $L(p,q)$ are constructed from a disk by identifying the top and bottom hemispherical surfaces after a relative rotation by $2\pi(q/p)$.

Two examples will illustrate how these constructions are obtained. In each case one looks for a unit cell associated with the action of the n-element group H. That is, one finds a cell whose n-images cover the 3-sphere and intersect only at their faces. Then S^3/H is constructed from the unit cell by suitable identification of boundary points.

A rotation by θ about the $\hat{n}$-axis will be denoted by $R(\theta\hat{n})$, and corresponding elements of SU(2) will be written $\pm u(\frac{\theta}{2}\hat{n})$. Given a basis $\{i\sigma_i\}$ for the Lie algebra of SU(2), one could set $u(\frac{\theta}{2}\hat{n}) = \exp(i\frac{\theta}{2}\hat{n}\cdot\vec{\sigma})$.

Example 1: The lens space $S^3/Z_2 \simeq SO(3)$. The orbit of a point g in $S^3 \simeq SU(2)$ is $\pm g$. The unit cell in SU(2) is the ball surrounding the identity, 1, bounded by the sphere of points halfway along geodesics joining 1 to -1. Because points of the boundary sphere are mapped to diametrically opposite points, S^3/Z_2 is isomorphic to a ball with diametrically opposite boundary points identified.

Example 2: The prism manifold S^3/D^*_{12}. One denotes by D_6 the 6-element (orientation preserving) symmetry group of the equilateral triangle (equivalently, the symmetry group of the prism with 3 vertical faces). Its double covering in SU(2) is D^*_{12}. The group D_6 consists, apart from the identity, of rotations by π about each of the three bisectors, together with rotations by $2\pi/3$ and $4\pi/3$ about the $\hat{k}$-axis perpendicular to the triangle. If we denote by $\hat{n}_i$ the unit vectors along the three bisectors, the elements of D^*_{12} are ± 1, $u(\pm \frac{\pi}{2} \hat{n}_i)$, $u(\pm \frac{\pi}{3} \hat{k})$, and $u(\pm \frac{2\pi}{3} \hat{k})$. Along each of the six directions $\pm\hat{n}_i$ (call these directions horizontal), and along the two vertical directions $\pm\hat{k}$, geodesics from the origin meet points of its orbit (the other elements of D^*_{12}). To build a unit cell about the identity, construct six vertical faces bisecting the six horizontal geodesics joining 1 to the points $u(\pm \frac{\pi}{2} \hat{n}_j)$; and two horizontal faces bisecting the two vertical geodesics joining 1 to $u(\pm \frac{\pi}{3} \hat{k})$. In this way one obtains the boundary of a 6-sided prism, whose opposite faces are mapped into one another by the action of D^*_{12}. The prism, with its boundary points thus identified, is the spherical space of Fig. 2. An essentially identical discussion shows that $S^3/D^*_{2(2n+1)}$ is constructed from a prism of $2n+1$ sides by identifying opposite faces.

A diffeo can be isotopic in Diff_x to the identity only if it acts as the identity on the fundamental group $\pi_1(M,x)$, since a homotopy of the diffeo to the identity would homotope the image of each element of π_1 back to the element itself. For a spherical space the fundamental group $\pi_1(S^3/H,1)$ is H itself, and if one regards the space as a polyhedron with opposite faces identified, generators of π_1 are curves joining identified points of the polyhedron's boundary. Each symmetry of the polyhedron permutes these generators and thus cannot be isotopic (or homotopic) to the identity. Consequently, from the construction of the prism manifold in Example 2, one already knows that the symmetry group D_{12} of a prism with 6 vertical faces is a subgroup of $\pi_0(\mathrm{Diff}_x)$. In fact, there are no additional elements:[4] $\pi_0(\mathrm{Diff}_x) \simeq D_{12}$, and because each element can be realized as an isometry $\pi_0(\mathrm{Diff}_x) \simeq \pi_0(\mathrm{Isom}_x)$.

When one fixes a frame as well as a point, a new feature arises. The diffeo R associated with a 2π rotation will in general not be isotopic to

---

[4] The proof that there are no other elements of $\pi_0(\mathrm{Diff}_x)$ is straightforward [8], relying on the fiber bundle exact sequence for the bundle $\mathrm{Diff} \to M$, with fiber Diff_x, where the projection maps $\psi \in \mathrm{Diff}$ to $\psi(x) \in M$; and one must know [18] that when no points are fixed, $\pi_0(\mathrm{Diff})$ is no larger than $\pi_0(\mathrm{Isom}) \approx Z_2$.

the identity, although it has no effect on the fundamental group of the 3-manifold. For orientable 3-manifolds, only R appears to have this property: Any other diffeo shows itself homotopically inequivalent to the identity by its nontrivial action on $\pi_1(M)$ [19]. We are not, however, aware of a proof of this statement with "homotopically" replaced by "isotopically."

The groups $\pi_0(\mathrm{Diff})$ have been found for all spherical spaces except S^3/H with $H = T^*$, I^*, and extensions of T^* and I^* by cyclic groups [18,20-24]. These results are consistent with Hatcher's proposed generalization of the Smale conjecture, that for all spherical spaces Diff and Isom have the same homotopy type. Whenever this is true, in particular when $\pi_0(\mathrm{Diff}) = \pi_0(\mathrm{Isom})$ and $\pi_1(\mathrm{Diff}) = \pi_1(\mathrm{Isom})$, one can explicitly calculate $\pi_0(\mathrm{Diff}_x)$ and $\pi_0(\mathrm{Diff}_e)$ [11]. For all spherical spaces of the form S^3/H with $H \subset SU(2)$, one finds $\pi_0(\mathrm{Diff}_x) = \pi_0(\mathrm{Isom}_x)$, the symmetry group of the polyhedron from which one builds the space. For all but the lens spaces, $\pi_0(\mathrm{Diff}_e)$ is an extension of $\pi_0(\mathrm{Diff}_x)$ by a Z_2 generated by R.

2. HOMOTOPY IS NOT ISOTOPY FOR DIFFEOMORPHISMS OF 3-MANIFOLDS

For the open 3-manifold, R^3, reflection through the origin provides a simple example of a diffeo that is homotopic but not isotopic to the identity. The homotopy is given by $\psi_t(\vec{x}) = t\vec{x}$, $-1 \le t \le 1$, clearly a continuous family of continuous maps, but not an isotopy since at $t = 0$, R^3 is mapped to a point. It is curious that the diffeo associated in canonically quantized gravity with a 2π-rotation provides the only known example for *closed* 3-manifolds of a diffeo that is isotopic but not homotopic to the identity. A theorem due to Hendricks [25] (see also [26]) characterizes the small set of 3-manifolds on which R is homotopic to the identity and an explicit list is given in [10]. As we will see, a straightforward argument shows that for some of these (the spaces listed below) R is not isotopic to the identity.

THEOREM: R is homotopic but not isotopic to the identity rel. D for S^3/H, where $H = D^*_{4m}$, $D^*_{4m} \times Z_p$, $D'_{2^k m}$, or $D'_{2^k m} \times Z_p$, with m odd.

COROLLARY: Let M be the connected sum of any two of the factors listed in the Theorem, and let R be a rotation parallel to a sphere enclosing one of the factors. Then R is homotopic but not isotopic to the identity.

REMARK: If the Poincaré conjecture is true, the spaces listed in the Theorem are the only prime 3-manifolds for which R is homotopic but not isotopic to the identity.

These results are stronger than the corresponding versions in Ref. [10]. (Theorem 2.2 and Corollary 2.2) because recent work by Boileau and Otal [27] shows that Diff and Isom have the same homotopy type for D^*_{4m}. Our Theorem relies on this result - in particular, on the fact that $\pi_1(\mathrm{Diff}) \simeq \pi_1(\mathrm{Isom}) \simeq Z_2$

for D^*_{4m}. The proof of the Theorem is most easily visualized for these spaces, and they are the ones not included in Ref. 10.

The key is to ask what an isotopy of R to the identity does to a frame fixed by R. Let e be a point in the bundle E of oriented frames on $M = S^3/H$. A diffeo ψ drags the frame e at $x \in M$ to ψe at $\psi(x)$. One defines a projection p: Diff $\to$ E by $p(\psi) = \psi e$. The image in E of a closed curve ψ_t of diffeos with $\psi_0 = \psi_1 = 1$ is then the closed curve $p(\psi_t) = \psi_t e$ in E.

Now the fundamental group of the frame bundle is

$$\pi_1(E) \simeq \pi_1(M) \times Z_2 \simeq H \times Z_2 ,$$

where the Z_2 is generated by a frame rotation path, $\theta \to R(\theta\hat{n})e$, $0 \leq 0 \leq 2\pi$.

LEMMA: If the diffeo R is isotopic to 1 rel e, then $(1,-1) \in \mathrm{Im}(p_*)$, where $p_*: \pi_1(\mathrm{Diff}) \to \pi_1(E)$ is the map induced by p.

PROOF: Let R_θ be a diffeo that rotates a family of concentric spheres about $x \in M$ by an angle that varies from θ to zero as one moves outward. (If $\{y^i\}$ is a chart about x, one could choose the diffeo $y^i \to R[f(y)\theta\hat{n}]^i{}_j y^j$, where f is a function that decreases smoothly from 1 to zero as one moves outward and R^i_j the rotation matrix associated with the rotation R). Under the path $\theta \to R_\theta$ of diffeos, the orbit of e is the path of frame rotations $\theta \to R(\theta\hat{n})$. If $R \sim 1$ rel e, there is a path ψ_t of diffeos that untwist without rotating e. Then the product path R followed by ψ is a closed curve of diffeos for which the orbit of e is again the path of frame rotations. Thus $p_*(c) = (1,-1) \in \pi_1(E) \simeq H \times Z_2$. □

We can now see why $\pi_1(\mathrm{Diff}) \simeq \pi_1(\mathrm{Isom}) \simeq Z_2$ implies that R cannot be isotopic to the identity:

PROOF OF THEOREM: If $R \sim 1$ rel. e, then p_*, regarded as a map from Z_2 to $H \times Z_2$ has image $\{(1,1), (1,-1)\}$.

But $\mathrm{Im}(p_*)$ contains another element: Consider the path of diffeos $\hat{\psi}_\phi$, $0 \leq \phi \leq \pi$, mapping $[g] \in SU(2)/H$ to $[g\, u(\hat{n}\phi)]$, for some fixed direction $\hat{n}$. Because D^*_{4m} acts on SU(2) on the left, ψ_ϕ is well defined and the orbit $\hat{c}$ of $[g]$ is a closed curve whose lift to SU(2) runs from g to -g. Thus $p_*(\hat{c}) = (-1,1)$ or $(-1,-1)$, and if R were isotopic to the identity, $\mathrm{Im}(p_*)$ would be larger than Z_2, a contradiction. For the remaining spaces, the proof of the Theorem is given in Ref. [10] and again relies on the fact that Diff and Isom have the same homotopy type [24]. □

We expect that at least for orientable manifolds, the only examples of diffeos that are homotopic but not isotopic to the identity are rotations parallel to the separating sphere of a prime factor in the list given in the Corollary above, but the question remains open [28, 29].

BIBLIOGRAPHY

1. K. Kuchar, "Canonical methods in quantization," in Quantum Gravity 2, an Oxford Symposium, eds. C. J. Isham, R. Penrose and D. W. Sciama, Clarendon, Oxford, 1981, 329-376.

2. A. E. Fischer, "The theory of superspace," in Relativity, eds. M. Carmeli, S. Fickler, and L. Witten, Plenum, New York, 1970.

3. P. W. Higgs, Phys. Rev. Lett. 1 (1958), 373-5.

4. J. L. Friedman and R. D. Sorkin, "Spin 1/2 from gravity," Phys. Rev. Lett., 44 (1980), 1100-1103.

5. C. J. Isham, "Topological θ-sectors in canonically quantized gravity," Phys. Lett., 106B (1981), 188-192.

6. C. J. Isham, "θ-states induced by the diffeomorphism group in canonically quantized gravity," in Quantum Structure of Space and Time, eds. M. J. Duff and C. J. Isham, Cambridge Univ. Press, Cambridge (1982).

7. J. L. Friedman and R. D. Sorkin, "Half-integral spin from quantum gravity," Gen. Rel. and Grav., 14 (1982), 615-620.

8. R. D. Sorkin, "Remarks on diffeomorphisms as particle symmetries," in General Relativity and Gravitation, eds. B. Bertotti, F. de Felice, and A. Pascolini, Consiglio Nazionale Della Ricerche, Rome, 1983.

9. R. D. Sorkin, "Introduction to topological geons," in Topological Properties and Global Structure of Space-Time, Plenum, 1986.

10. J. L. Friedman and D. M. Witt, "Homotopy is not isotopy for homeomorphisms of 3-manifolds," Topology, 25 (1986), 35-44.

11. D. M. Witt, "Symmetry groups of state vectors in canonical quantum gravity," J. Math. Phys., 27 (1986), 573-592.

12. E. César de Sa and C. Rourke, "The homotopy type of homeomorphisms of 3-manifolds," Bull. Am. Math. Soc. (new series), 1 (1979), 251-254.

13. A. Hatcher, "On the diffeomorphism group of $S^1 \times S^2$," Proc. Am. Math. Soc., 83 (1981), 427-430.

14. H. Hendriks and F. Laudenbach, "Diffeomorphismes des sommes connexes en dimension trois," Topology, 23 (1984), 423-443.

15. D. McCullough, "Mappings of reducible 3-manifolds," in Geometric and Algebraic Topology, Banach Center Publication v. 18, Polish Scientific, Warsaw, 1986.

16. H. Seifert and W. Threlfall, A Textbook of Topology, Academic Press, New York, 1980.

17. J. A. Wolf, Spaces of Constant Curvature, Publish or Perish, Berkeley, 1977.

18. J. H. Rubinstein, "On 3-manifolds that have finite fundamental group and contain Klein bottles," Trans. Am. Math. Soc., 251 (1979), 129-137.

19. H. Hendriks, "Obstruction theory in 3-dimensional topology," Bull. Am. Math. Soc., 83 (1983), 737-8.

20. C. Hodgson and J. H. Rubinstein, "Involutions and diffeomorphisms of lens spaces," Proc. Canadian Math. Cong., 1982, to appear.

21. J. H. Rubinstein and J. S. Birman, "Homeotopy groups of some non-Haken manifolds," Proc. Lond. Math. Soc. (3), 49 (1984), 517-536.

22. K. Asano, "Homeomorphisms of prism manifolds," Yokohama Math. J., 26 (1978), 19-25.

23. F. Bonahon, "Difféotopies des espaces lenticulaires," Topology, 22 (1983), 305-314.

24. N. V. Ivanov, "Homotopy of spaces of automorphisms of some three-dimensional manifolds," Sov. Math. Dokl., 20 (1979), 47; 20 (1979), No. 6, vii.

25. H. Hendriks, "Applications de la théorie d'obstruction en dimension 3," Bull. Soc. Math. Fr. Memo., 53 (1977), 81-196.

26. S. Plotnick, "Equivariant intersection forms, knots in S^4, and rotations in 2-spheres," Trans. Am. Math. Soc., 296 (1986), 543-55.

27. M. Boileau and J.-P. Otal, "Groupe des diffeotopies des certaines varietes de Siefert," C.R. Acad. Sci. Paris, Serie 1, 303 (1986), 19-22.

28. F. Laudenbach, "Topology in dimension trois: homotopie et isotopie," Astérisque, 12 (1974), 1-152.

29. P. Scott, "Homotopy implies isotopy for some Seifert fibre spaces," Topology, 22 (1985), 341-351.

DEPARTMENT OF PHYSICS
UNIVERSITY OF WISCONSIN
MILWAUKEE, WI 53201

and

DEPARTMENT OF PHYSICS
UNIVERSITY OF CALIFORNIA
SANTA BARBARA, CA 93106

Contemporary Mathematics
Volume **71**, 1988

INITIAL VALUE DECOMPOSITION OF THE SPACETIME DIFFEOMORPHISM GROUP

Philip B. Yasskin[1]

ABSTRACT. The spacetime diffeomorphism group $\mathcal{D}M$ of a product spacetime $M = S \times T$ is written as the partial (and possibly local) doubly semi-direct product $\mathcal{D}M = \mathcal{D}_T S \times \mathcal{D}_S T$ of the time dependent spatial diffeomorphism group $\mathcal{D}_T S$ and the space dependent temporal diffeomorphism group $\mathcal{D}_S T$. Further, $\mathcal{D}_T S$ is the semi-direct product $\mathcal{D}_T S = \mathcal{D}S \times \mathcal{D}_T S^\circ$ of the spatial diffeomorphism group $\mathcal{D}S$ and the group $\mathcal{D}_T S^\circ$ of time dependent spatial diffeomorphisms which leave the initial surface fixed. Similarly, $\mathcal{D}_S T$ is the doubly semi-direct product $\mathcal{D}_S T = F(S, \mathbf{R}) \times \mathcal{D}^\circ_S T$ of the space dependent time translation group $F(S, \mathbf{R})$ and the group $\mathcal{D}^\circ_S T$ of space dependent temporal diffeomorphisms which leave the initial surface fixed. Then, in the initial value formulation of a classical relativistic field theory, $\mathcal{D}S$ and $F(S, \mathbf{R})$ are the initial freedom, while $\mathcal{D}_T S^\circ$ and $\mathcal{D}^\circ_S T$ are the evolutionary gauge freedom. The doubly semi-direct product is a generalization of the semi-direct product using two actions and appears to be a new construction in elementary group theory.

The motivation for this research comes from the initial value formulation of classical relativistic field theories with gauge freedoms. In principal, one would like to understand the space of solutions of the spacetime (4-dimensional) field equations modulo the spacetime gauge freedoms. In the initial value formulation of these theories[1-5], one divides the field equations into constraints on the initial data and evolution equations for this data. Two sets of initial data satisfying the constraints are regarded as gauge equivalent if they evolve into gauge equivalent

---

1980 Mathematical Subject Classification (1985) Revision):
58D05, 70G5 83C40, 70H05.
[1]Supported in part by NSF Grant #DMS-8504338

spacetime solutions. These are the gauge freedoms in the initial data. In addition, there may be some arbitrariness in the evolution equations which is regarded as the gauge freedom in the evolution.

One is naturally led to ask whether and how the spacetime gauge group splits into an initial gauge group and an evolutionary gauge group. The evolutionary gauge freedom is straightforward to identify by performing a Bergmann-Dirac[1-4] analysis of the theory. However, if the spacetime gauge group includes the spacetime diffeomorphism group (or even just certain subgroups) then the initial gauge freedom appears to be much more difficult to identify. This is because initial data at one instant of time may be evolved into initial data at a second instant of time which may then be gauge transformed (via a spacetime diffeomorphism) into another set of initial data at the first instant of time which is then gauge equivalent to the original initial data. Thus the initial gauge freedom appears to be intimately intermixed with the evolution equations. The purpose of this paper is to isolate the initial and evolutionary parts of the spacetime diffeomorphism group using geometrical and group theoretical techniques without having to explicitly use any field equations.

Throughout, we assume that **spacetime** M is the product $M = S \times T$ of a 3-manifold S, called **space,** and a 1-manifold T (usually $\mathbf{R}$), called **time,** possibly with boundaries. (Note: This is always true for a globally hyperbolic spacetime[6,7].) We let $\mathcal{D}M$ denote the **spacetime diffeomorphism group;** i.e. the set of C^∞-diffeomorphisms $\eta: M \to M$ under composition. (If necessary, we might impose further restrictions on the diffeomorphisms in $\mathcal{D}M$ such as spatial or temporal orientability or asymptotic conditions at spatial or temporal infinity. All of the following holds regardless of such restrictions unless specifically mentioned.) Similarly, we let $\mathcal{D}S$ and $\mathcal{D}T$ denote the **spatial** and **temporal diffeomorphism groups,** respectively (with the corresponding restrictions, if necessary).

We will also need to study the **time dependent spatial diffeomorphism group,** $\mathcal{D}_T S$, which consists of those C^∞-maps, $\sigma: S \times T \to S$, whose restriction to each **"instant of time"** $t \in T$, given by

$$\sigma_t: S \to S: \sigma_t(x) = \sigma(x,t)$$

belongs to $\mathcal{D}S$. Notice that $\mathcal{D}_T S$ is a group under **pointwise composition,** defined as follows: For σ_1 and $\sigma_2 \in \mathcal{D}_T S$, let $\sigma_1 \bullet \sigma_2$ be the C^∞-map

$$\sigma_1 \bullet \sigma_2: S \times T \to S: (\sigma_1 \bullet \sigma_2)(x,t) = \sigma_1(\sigma_2(x,t), t)$$

whose restriction $(\sigma_1 \bullet \sigma_2)_t$ to each instant t is just the composition

$$(\sigma_1 \bullet \sigma_2)_t = \sigma_{1t} \circ \sigma_{2t}$$

of the restrictions σ_{1t} and σ_{2t} (which justifies the name pointwise composition).

Similarly, we will study the **space dependent temporal diffeomorphism group**, $\mathcal{D}_S T$, consisting of those C^∞-maps, $\tau: S \times T \to T$, whose restriction to each **"point of space"** $x \in S$, given by

$$\tau_x: T \to T: \quad \tau_x(t) = \tau(x,t)$$

belongs to $\mathcal{D}T$. Again notice that $\mathcal{D}_S T$ is a group under **pointwise composition**, where for τ_1 and $\tau_2 \in \mathcal{D}_S T$,

$$\tau_1 \cdot \tau_2: S \times T \to T: \quad (\tau_1 \cdot \tau_2)(x,t) = \tau_1(x, \tau_2(x,t))$$

so that its restriction to each point x is

$$(\tau_1 \cdot \tau_2)_x = \tau_{1x} \circ \tau_{2x}.$$

The principal result of this research is that $\mathcal{D}M$ is partially (and sometimes locally) isomorphic to $\mathcal{D}_T S \times \mathcal{D}_S T$. (The notions of partial and local isomorphism will be clarified later.) This is obvious at the infinitesimal level of Lie algebras since an element of the Lie algebra of $\mathcal{D}M$ is a vector field on $M = S \times T$ which may be uniquely decomposed as the sum of a vector field tangent to the S-foliation (in the Lie algebra of $\mathcal{D}_T S$) and a vector field tangent to the T-foliation (in the Lie algebra of $\mathcal{D}_S T$).

At the macroscopic level of Lie groups, the partial (or local) isomorphism is

$$i: \mathcal{D}M \to \mathcal{D}_T S \times \mathcal{D}_S T: \quad i(\eta) = (\pi_S \circ \eta, \pi_T \circ \eta)$$

where $\pi_S: S \times T \to S$ and $\pi_T: S \times T \to T$ are the natural projections. This is clarified by the commuting diagram:

$$\begin{array}{ccc}
& & S \\
& \nearrow \pi_S \circ \eta & \uparrow \pi_S \\
S \times T & \xrightarrow{\eta} & S \times T \\
& \searrow \pi_T \circ \eta & \downarrow \pi_T \\
& & T
\end{array}$$

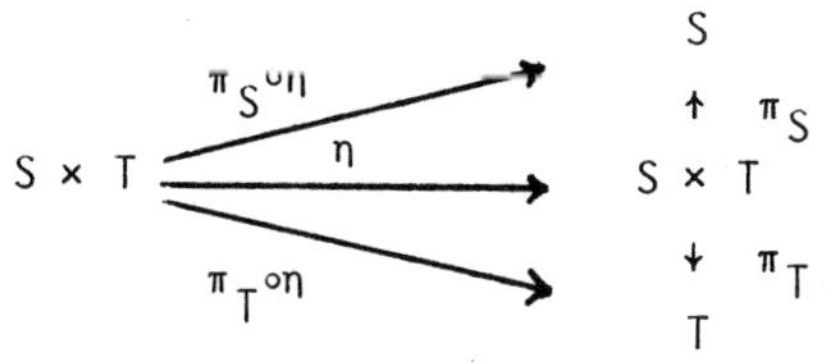

Similarly, the inverse of i is

$$i^{-1}: \mathcal{D}_T S \times \mathcal{D}_S T \to \mathcal{D}M: \quad i^{-1}(\sigma,\tau) = \sigma \times \tau$$

where $\sigma \times \tau$ is the Cartesian product of maps. This is clarified by the commuting diagram:

$$\begin{array}{ccc}
& & S \\
& \nearrow \sigma & \uparrow \pi_S \\
S \times T & \xrightarrow{\sigma \times \tau} & S \times T \\
& \searrow \tau & \downarrow \pi_T \\
& & T
\end{array}$$

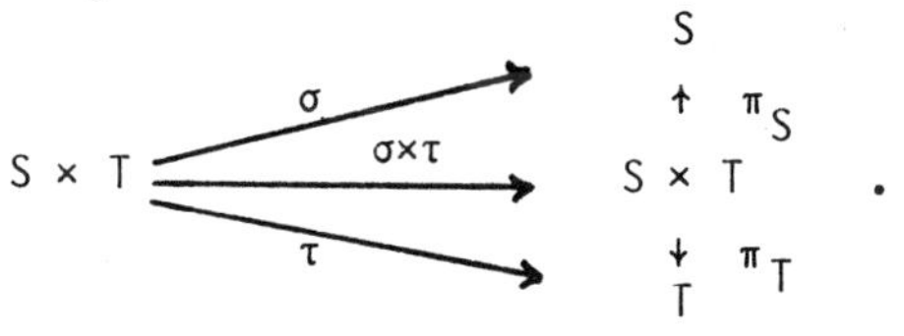

There are two problems with these definitions. First, for $\eta \in \mathcal{D}M$, although $\pi_S \circ \eta$: $S \times T \to S$ and $\pi_T \circ \eta$: $S \times T \to T$ are C^∞-maps, their restrictions $(\pi_S \circ \eta)_t$: $S \to S$ and $(\pi_T \circ \eta)_x$: $T \to T$ may not be invertible for all $t \in T$ and $x \in S$. Hence $i(\eta)$ may not belong to $\mathcal{D}_T S \times \mathcal{D}_S T$. Second, for $\sigma \in \mathcal{D}_T S$ and $\tau \in \mathcal{D}_S T$, although $\sigma \times \tau$: $S \times T \to S \times T$ is a C^∞-map from M to M, it may not be invertible. Hence $i^{-1}(\sigma,\tau)$ may not belong to $\mathcal{D}M$. These two problems are simply defined away, as follows: We say that $\eta \in \mathcal{D}M$ is **decomposable** if $\pi_S \circ \eta \in \mathcal{D}_T S$ and $\pi_T \circ \eta \in \mathcal{D}_S T$, and let $(\mathcal{D}M)_{decomp}$ denote the set of decomposable elements of $\mathcal{D}M$. Similarly, we say that $(\sigma,\tau) \in \mathcal{D}_T S \times \mathcal{D}_S T$ is **composable** if $\sigma \times \tau \in \mathcal{D}M$, and let $(\mathcal{D}_T S \times \mathcal{D}_S T)_{comp}$ denote the set of composable elements of $\mathcal{D}_T S \times \mathcal{D}_S T$. Then Then i and i^{-1} define an isomorphism of sets:

$$(\mathcal{D}M)_{decomp} \cong (\mathcal{D}_T S \times \mathcal{D}_S T)_{comp}.$$

We can now clarify the notions of partial and local isomorphisms. A **partial bijection** f: $X \to Y$ from a set X to a set Y is a bijection f: $U \to V$ from a subset $U \subseteq X$ to a subset $V \subseteq Y$. A **partial homeomorphism** f: $X \to Y$ from a topological space X to a topological space Y is a homeomorphism f: $U \to V$ from a subspace $U \subseteq X$ to a subspace $V \subseteq Y$ with the relative topologies. Such a **partial homeomorphism** is called **local** if U and V are open. A **partial isomorphism** f: $G \to H$ from a group G to a group H is a bijection f: $U \to V$ from a subset $U \subseteq G$ to a subset $V \subseteq H$ such that the identities $e_G \in G$ and $e_H \in H$ satisfy $e_G \in U$, $e_H \in V$ and $f(e_G) = e_H$, and further, if g_1, g_2 and $g_1 \cdot g_2 \in U$ then $f(g_1 \cdot g_2) = f(g_1) \cdot f(g_2)$. A **partial isomorphism** f: $G \to H$ between topological groups G and H is a partial isomorphism of groups whose underlying bijection f: $U \to V$ is also a homeomorphism from the subspace $U \subseteq G$ to the subspace $V \subseteq H$ with the relative topologies. Such a **partial isomorphism** of topological groups is called **local** if U and V are open.

So far, we have shown that i: $\mathcal{D}M \to \mathcal{D}_T S \times \mathcal{D}_S T$ is a partial bijection. If we impose the C^k-topology (or the H^s-Sobolev topology) upon $\mathcal{D}M$, $\mathcal{D}_T S$ and $\mathcal{D}_S T$ then it can be shown that i is a partial homeomorphism. Further, to determine when i is a local homeomorphism we need to determine when $(\mathcal{D}M)_{decomp}$ and $(\mathcal{D}_T S \times \mathcal{D}_S T)_{comp}$ are open. Since $(\mathcal{D}M)_{decomp}$ consists of those diffeomorphisms η for which $(\pi_S \circ \eta)_t$ is invertible for all $t \in T$ and $(\pi_T \circ \eta)_x$ is invertible for all $x \in S$, and since $(\mathcal{D}_T S \times \mathcal{D}_S T)_{comp}$ consists of those pairs (σ,τ) for which $\sigma \times \tau$ is invertible, an inverse function theorem argument shows that i is a local homeomorphism if S and T are compact (or alternatively, if appropriate asymptotic conditions are imposed upon $\mathcal{D}M$, $\mathcal{D}S$ and $\mathcal{D}T$).

It remains to discuss the group structure and to see in what sense i

is a partial isomorphism of topological groups. So far, we have a group structure for $\mathcal{D}M$, $\mathcal{D}_T S$ and $\mathcal{D}_S T$, but we do not have a group structure on $\mathcal{D}_T S \times \mathcal{D}_S T$. If i were a bijection then we could use i and the group structure on $\mathcal{D}M$ to induce a group structure on $\mathcal{D}_T S \times \mathcal{D}_S T$. However, since i is only a partial bijection, we can only use it to induce a partial group structure on $\mathcal{D}_T S \times \mathcal{D}_S T$. (A **partial group** is a set with a product rule for some pairs of elements which is associative whenever defined, has an identity element and has an inverse of each element. A **partial topological group** G is **local** if the set of pairs of elements for which the product is defined, is an open subset of $G \times G$.) Hence the induced partial product rule on $\mathcal{D}_T S \times \mathcal{D}_S T$ is given as follows: If (σ_1, τ_1) and $(\sigma_2, \tau_2) \in (\mathcal{D}_T S \times \mathcal{D}_S T)_{comp}$ and if $i^{-1}(\sigma_1, \tau_1) \circ i^{-1}(\sigma_2, \tau_2) \in (\mathcal{D}M)_{decomp}$ then

$$\begin{aligned}(\sigma_1, \tau_1) \cdot (\sigma_2, \tau_2) &= i[i^{-1}(\sigma_1, \tau_1) \circ i^{-1}(\sigma_2, \tau_2)] \\ &= (\pi_S \circ (\sigma_1 \times \tau_1) \circ (\sigma_2 \times \tau_2), \pi_T \circ (\sigma_1 \times \tau_1) \circ (\sigma_2 \times \tau_2)) \\ &= (\sigma_1 \circ (\sigma_2 \times \tau_2), \tau_1 \circ (\sigma_2 \times \tau_2)).\end{aligned}$$

Consequently, i is a partial isomorphism of partial topological groups. Further, if S and T are compact (or alternatively, if appropriate asymptotic conditions are imposed upon $\mathcal{D}M$, $\mathcal{D}S$ and $\mathcal{D}T$) then i is a local isomorphism of local topological groups.

Although the above formula for the partial product rule on $\mathcal{D}_T S \times \mathcal{D}_S T$ is sufficient to define a partial group structure, it is interesting to ask whether the product rule on $\mathcal{D}_T S \times \mathcal{D}_S T$ can be expressed in terms of the product rules on $\mathcal{D}_T S$ and $\mathcal{D}_S T$, so that $\mathcal{D}_T S \times \mathcal{D}_S T$ could be regarded as some type of product of the subgroups $\mathcal{D}_T S$ and $\mathcal{D}_S T$. This can, in fact, be done. What occurs is a generalization of the semi-direct product which I call the second-order, matched, doubly semi-direct product:

Recall that if G and H are groups then their **direct product,** $G \times H$, has the product rule

$$(g_1, h_1) \cdot (g_2, h_2) = (g_1 \cdot g_2, h_1 \cdot h_2).$$

If in addition r is a **right action** of G on H; i.e.

$$r_{g_1} \circ r_{g_2} = r_{g_2 \cdot g_1} \quad \text{and} \quad r_{e_G} = id_H,$$

which **preserves the identity** of H; i.e.

$$r_g\, e_H = e_H,$$

and **acts by group homomorphisms**; i.e.

$$r_g(h_1 \cdot h_2) = (r_g h_1) \cdot (r_g h_2),$$

then the **semi-direct product** $G \times^1_r H$ with r **acting on the first factor** has the product rule

$$(g_1, h_1) \cdot (g_2, h_2) = (g_1 \cdot g_2, [r_{g_2} h_1] \cdot h_2).$$

There are other versions of the semi-direct product using a left action and/or letting the action act on the second factor and/or using an action of H on G.

The doubly semi-direct product has two actions instead of one. Since there are many different doubly semi-direct products, I will give the definitions of one version of each of the three simplest types. For all three types, let ρ be a right action of H on G which preserves the identity of G and let r be a right action of G on H which preserves the identity of H. Then the **first-order, matched, doubly semi-direct product** $G \, {}^{1}{}^{1}_{\rho}{\times}^{1}_{r} \, H$ with ρ and r acting on the first factors, has the product rule

$$(g_1, h_1) \cdot (g_2, h_2) = ([\rho_{h_2} g_1] \cdot g_2, [r_{g_2} h_1] \cdot h_2).$$

This product rule is associative iff ρ and r act by group homomorphisms, r is **constant on the orbits** of ρ, and ρ is constant on the orbits of r; i.e.

$$r_{\rho_h g} = r_g \quad \text{and} \quad \rho_{r_g h} = \rho_h.$$

Similarly, the **first-order, mixed, doubly semi-direct product** $G \, {}^{1}{}^{1}_{\rho}{\times}^{2}_{r} \, H$ with ρ acting on the first factor and r acting on the second factor, has the product rule

$$(g_1, h_1) \cdot (g_2, h_2) = ([\rho_{h_2} g_1] \cdot g_2, h_1 \cdot [r^{-1}_{g_1} h_2]).$$

This product rule is associative iff ρ and r are **intertwined** according to the following formulas:

$$r_g(h_1 \cdot h_2) = (r_g h_1) \cdot (r^{-1}_{\rho_{h_1}(g^{-1})} h_2),$$

and

$$\rho_h(g_1 \cdot g_2) = (\rho_{r^{-1}_{g_2} h}\, g_1) \cdot (\rho_h g_2).$$

Notice that ρ and r do not act by group homomorphisms but rather that the corresponding formulas are modified by (intertwined with) the opposite action. Finally, the **second-order, matched, doubly semi-direct product** $G \, {}^{1}{}^{2}_{\rho}{\times}^{1}_{r} \, H$ with ρ and r acting on the first factor, has the product rule

$$(g_1, h_1) \cdot (g_2, h_2) = ([\rho_{r^{-1}_{g_2} h_2}\, g_1] \cdot g_2, [r_{\rho^{-1}_{h_2} g_2}\, h_1] \cdot h_2)$$

which is associative iff ρ and r are **intertwined** according to:

$$r_g(h_1 \cdot h_2) = (r^{-1}_{\rho^{-1}_{h_2}(g^{-1})} h_1) \cdot (r_g h_2),$$

and

$$\rho_h(g_1 \cdot g_2) = (\rho^{-1}_{r^{-1}_{g_2}(h^{-1})} g_1) \cdot (\rho_h g_2).$$

It is obvious that there are many other versions of these types of doubly semi-direct products and it would appear that there are many other types of doubly semi-direct products with more complicated product rules and intertwining conditions. It is an interesting mathematical question of whether there is a complete list. I don't know the answer.

Returning to the problem at hand, it turns out that the product rule on $\mathcal{D}_T S \times \mathcal{D}_S T$ is that of the partial, second-order, matched, doubly semi-direct product $\mathcal{D}_T S \;{}^{1^2 1}_{\rho}\!\!\times_r\; \mathcal{D}_S T$ where (i) ρ is the partial right action of $\mathcal{D}_S T$ on $\mathcal{D}_T S$ given by

$$(\rho_\tau \sigma)(x,t) = \sigma(x,\tau(x,t))$$

for $\tau \in \mathcal{D}_S T$ and $\sigma \in \mathcal{D}_T S$ and (ii) r is the partial right action of $\mathcal{D}_T S$ on $\mathcal{D}_S T$ given by

$$(r_\sigma \tau)(x,t) = \tau(\sigma(x,t),\, t)$$

for $\sigma \in \mathcal{D}_T S$ and $\tau \in \mathcal{D}_S T$. The actions are partial in that the restricted maps $(\rho_\tau \sigma)_t$ and $(r_\sigma \tau)_x$, given by

$$(\rho_\tau \sigma)_t(x) = \sigma_{\tau_x t} x \quad \text{and} \quad (r_\sigma \tau)_x(t) = \tau_{\sigma_t x} t,$$

may not be invertible for all $t \in T$ and $x \in S$. The doubly semi-direct product is partial because the product rule is only defined for some pairs of elements since it involves the inverses of $\rho_\tau \sigma$ and $r_\sigma \tau$. Once again, everything becomes local if S and T are compact (or if appropriate asymptotic conditions are imposed upon $\mathcal{D}M$, $\mathcal{D}S$ and $\mathcal{D}T$).

Note: The notions of partial and local group actions generalize the well-known concept of a local one-parameter group of diffeomorphisms of a manifold.

We finally return to the original problem of decomposing $\mathcal{D}M$ into initial and evolutionary parts. So far we have that $\mathcal{D}M$ is the partial (or local) doubly semi-direct product $\mathcal{D}_T S \;{}^{1^2 1}_{\rho}\!\!\times_r\; \mathcal{D}_S T$. We now decompose $\mathcal{D}_T S$ and $\mathcal{D}_S T$ into initial and evolutionary parts. Please notice that so far the decomposition has been completely symmetric between S and T, and has been independent of the dimensions of S and T. From now on we will need to use the fact that T is one-dimensional and that $t = t_0$ (usually $= 0$) is a choice of a preferred point in the interior of T (the initial time).

We first decompose $\mathcal{D}_T S$. First notice that $\mathcal{D}S$ may be identified with the subgroup of $\mathcal{D}_T S$ consisting of the time independent spatial diffeomorphisms via the embedding map

$$(\hat{\ }): \mathcal{D}S \to \mathcal{D}_T S: \quad s \longmapsto \hat{s} \quad \text{where} \quad \hat{s}_t(x) = s(x).$$

In addition, let $\mathcal{D}_T S^\circ$ denote the subgroup of $\mathcal{D}_T S$ consisting of those time dependent spatial diffeomorphism which restrict to the identity on S at time t_0; i.e.

$$\mathcal{D}_T S^\circ = \{\sigma \in \mathcal{D}_T S \mid \sigma_{t_0} = id_S\}.$$

Next, let $ad_{\hat{\ }}$ denote the left action of $\mathcal{D}S$ on $\mathcal{D}_T S^\circ$ given by

$$ad_{\hat{s}}\, \sigma = \hat{s} \bullet \sigma \bullet \hat{s}^{-1}$$

where $s \in \mathcal{D}S$, $\sigma \in \mathcal{D}_T S^\circ$ and the product is the pointwise composition of $\mathcal{D}_T S$. Notice that $ad_{\hat{s}}\, \sigma \in \mathcal{D}_T S^\circ$ since

$$(ad_{\hat{s}}\, \sigma)_{t_0} = (\hat{s} \bullet \sigma \bullet \hat{s}^{-1})_{t_0} = \hat{s}_{t_0} \circ \sigma_{t_0} \circ \hat{s}_{t_0}^{-1}$$

$$= s \circ id_S \circ s^{-1} = id_S.$$

Then it may be checked that $\mathcal{D}_T S$ can be decomposed as the semi-direct product

$$\mathcal{D}S = \mathcal{D}S \times^1_{ad_{\hat{\ }}} \mathcal{D}_T S^\circ$$

where the projections are

$$P_0: \mathcal{D}_T S \to \mathcal{D}S: \quad P_0\sigma = \sigma_{t_0}$$

and

$$P_1: \mathcal{D}_T S \to \mathcal{D}_T S^\circ: \quad P_1\sigma = \sigma_{t_0}^{-1} \circ \sigma.$$

Notice that $P_1\sigma \in \mathcal{D}_T S^\circ$ since

$$(P_1\sigma)_{t_0}(x) = (\sigma_{t_0}^{-1} \circ \sigma)(x, t_0) = \sigma_{t_0}^{-1}(\sigma_{t_0} x) = x.$$

Thus, each $\sigma \in \mathcal{D}_T S$ can be uniquely written as

$$\sigma = (P_0\sigma) \circ (P_1\sigma) = (\widehat{P_0\sigma}) \bullet (P_1\sigma).$$

Next we decompose $\mathcal{D}_S T$. We need to assume that there is an action α of the one-dimensional translation group ($\mathbb{R}$ under addition) on T such that (i) α acts simply transitively on the orbit of t_0, and (ii) the orbit of t_0 under α coincides with the orbit of t_0 under $\mathcal{D}T$. Since T is a one-dimensional manifold, possibly with boundary, it is diffeomorphic to either $\mathbf{R}$, $[-1,1]$, $(-1,1]$ or S^1. For $\mathbf{R}$, t_0 may be taken as 0, and α may be taken as the standard action of the translation group (addition). Similarly, we parametrize $[-1,1]$ or $(-1,1]$ as $\mathbf{R} \cup \{-\infty, +\infty\}$ or $\mathbf{R} \cup \{+\infty\}$ respectively (say via the diffeomorphism "tanh"). Then t_0 may be taken as 0, and α may be taken as the translation group on $\mathbf{R}$ with $\{-\infty, +\infty\}$ or $\{+\infty\}$ as fixed points respectively. Similarly, we parametrize S^1 as $\mathbf{R} \cup \{\infty\}$ (say by regarding S^1 as $[-1,1]$ with the endpoints identified and then using the diffeomorphism "tanh"), and may again take t_0 to be 0. To proceed further for S^1, we need to restrict $\mathcal{D}T$ to mean those diffeomorphisms which leave ∞ fixed. Then the orbit of $t_0 = 0$ is $\mathbf{R}$ and α may be taken as the translation group on $\mathbf{R}$ with ∞ as a fixed point. In all four cases, $t_0 = 0$, the orbit of t_0 is $\mathbf{R}$, and α is addition on $\mathbf{R}$ with the compliment of $\mathbf{R}$ (namely ϕ, $\{-\infty, +\infty\}$, $\{+\infty\}$, or $\{\infty\}$) consisting of fixed points.

Next, notice that the set of C^∞-functions from S into R, denoted $F(S, \mathbf{R})$, may be regarded as the group of space dependent time translations. As such, $F(S, \mathbf{R})$ may be embedded into $\mathcal{D}_S T$ via the embedding map

$$(\hat{\ }): F(S, \mathbf{R}) \to \mathcal{D}_S T: f \longmapsto \hat{f} \quad \text{where} \quad \hat{f}(x,t) = \alpha_{f(x)} t = t + f(x).$$

Also, let $\mathcal{D}_S^\circ T$ denote the subgroup of $\mathcal{D}_S T$ consisting of those space dependent temporal diffeomorphisms which leave the initial surface $t = t_0 = 0$ fixed; i.e.

$$D_S^\circ T = \{\tau \in \mathcal{D}_S T \mid \tau_x(0) = 0\}$$

Next, let λ denote the left action of $\mathcal{D}_S^\circ T$ on $F(S, \mathbf{R})$ given by

$$(\lambda_\tau f)(x) = \tau_x[f(x)]$$

where $\tau \in \mathcal{D}_S^\circ T$, and $f \in F(S, \mathbf{R})$. Also let r denote the right action of $F(S, \mathbf{R})$ on $\mathcal{D}_S^\circ T$ given by

$$(r_f \tau)_x(t) = \tau_x[t + f(x)] - \tau_x[f(x)]$$

where $f \in F(S, \mathbf{R})$ and $\tau \in \mathcal{D}_S^\circ T$. Notice that $r_f \tau \in \mathcal{D}_S^\circ T$ since $(r_f \tau)_x(0) = 0$. Then it may be checked that $\mathcal{D}_S T$ can be decomposed as the first-order, mixed, doubly semi-direct product

$$\mathcal{D}_S T = F(S, \mathbf{R}) \; {}_{\lambda}^{2}\!\times_{r}^{1\,1} \; \mathcal{D}_S^\circ T$$

where the projections are

$$P_0: \mathcal{D}_S T \to F(S, \mathbf{R}): (P_0 \tau)(x) = \tau_x(0)$$

and

$$P_1: \mathcal{D}_S T \to \mathcal{D}_S^\circ T: (P_1 \tau)_x(t) = \tau_x(t) - \tau_x(0).$$

Thus, each $\tau \in \mathcal{D}_S T$ can be uniquely written as

$$\tau = (\hat{P_0}\tau) \cdot (P_1\tau).$$

To summarize, we have shown that $\mathcal{D}M$ is partially (or locally) isomorphic to the partial (or local), second-order, matched, doubly semi-direct product $\mathcal{D}_T S \; {}^{1}_{\rho}\overset{2}{\times}{}^{1}_{r} \; \mathcal{D}_S T$. Further, $\mathcal{D}_T S$ is the semi-direct product $\mathcal{D}S \times^{1}_{\mathrm{ad}_\wedge} \mathcal{D}_T S^\circ$ and $\mathcal{D}_S T$ is the first-order, mixed, doubly semi-direct product $F(S, \mathbf{R}) \; {}^{2}_{\lambda}\overset{1}{\times}{}^{1}_{r} \; \mathcal{D}^\circ_S T$. Then $\mathcal{D}S$ and $F(S, \mathbf{R})$ constitute the *initial gauge group*, while $\mathcal{D}_T S^\circ$ and $\mathcal{D}^\circ_S T$ constitute the *evolutionary gauge group*. $\mathcal{D}S$ is the usual *spatial diffeomorphism freedom* in the choice of initial data. $F(S, \mathbf{R})$ is the *space dependent time translation group* which specifies the *slicing freedom to choose the initial spacelike surface* for the specification of initial data. $\mathcal{D}_T S^\circ$ specifies the *freedom in the evolution equations to make a time dependent spatial diffeomorphism*. An element σ in $\mathcal{D}_T S^\circ$ satisfies the *initial condition* $\sigma_{t_0} = \mathrm{id}_S$ and has its *time derivative* specified by the *shift* vector field. Finally, $\mathcal{D}^\circ_S T$ specifies the *slicing freedom in the evolution equations to choose the spacelike surfaces of simultaneity* regarded as *instants of time*. An element τ in $\mathcal{D}^\circ_S T$ satisfies the *initial condition* $\tau_x(0) = 0$ and has its *time derivative* specified by the *lapse* function.

I hope that this decomposition of the spacetime diffeomorphism group will prove useful in the study of the initial value formulation of classical relativistic field theories.

BIBLIOGRAPHY

1. P.A.M. Dirac, "Generalized Hamiltonian Dynamics," Can. J. Math., 2 (1950) 129.

2. P. G. Bergmann and J. Goldberg, "Dirac Bracket Transformations in Phasespace," Phys. Rev., 98 (1955) 531.

3. P.A.M. Dirac, Lectures on Quantum Mechanics, Belfer Graduate School of Science Monograph Series #2 (1964).

4. E.C.G. Sudarshan and N. Mukunda, Classical Dynamics: A Modern Perspective, (Wiley, New York, 1974).

5. M. Gotay, J. Isenberg, J. Marsden, R. Montgomery, J. Sniatycki, and P.B. Yasskin, "Momentum Maps and the Hamiltonian Treatment of Classical Field Theories with Constraints," in preparation.

6. R. Geroch, "Domain of Dependence," J. Math. Phys., 11 (1970) 437-449.

7. S.W. Hawking and G.F.R. Ellis, The Large Scale Structure of Space-Time, (Cambridge Univ. Press, 1973).

DEPARTMENT OF MATHEMATICS
TEXAS A&M UNIVERSITY
COLLEGE STATION, TX 77843-3368.

Contemporary Mathematics
Volume **71**, 1988

INCLUSION OF FERMIONS IN THE WAVE FUNCTION OF THE UNIVERSE

P.D. D'Eath[1] and J.J. Halliwell

ABSTRACT: The proposal of Hartle and Hawking for the wave function of the universe is reviewed in the context of Hawking's massive scalar field model. By considering the example of a Dirac field in this background, it is then shown how the proposal may be extended to allow for the inclusion of fermions, described by Grassmann variables using the holomorphic representation. The fermionic state is most naturally specified using non-local boundary conditions on a spacelike hypersurface. The main properties of the resulting quantum state are described.

1. INTRODUCTION

In quantum cosmology one is interested not only in the laws governing the evolution of the universe, but also in understanding the initial or boundary conditions which might uniquely determine the state of the universe. One particularly attractive proposal for the quantum state of the universe, based on a Feynman path integral, has been made by Hartle and Hawking [11]. This talk will review in section 2 the Hartle-Hawking proposal in the context of a simple background model for the large-scale structure of the universe–a closed Friedmann model driven by a massive scalar field [12],[13]. A considerable amount of work has been carried out in exploring the consequences of this proposal for a variety of more general cosmological models containing bosonic fields. Here we outline in section 3 the way in which this can be extended to include fermions, based on the more detailed presentation [4].

In practice one needs a description of fermions, since matter fields are fermionic. But the inclusion of fermions is also interesting as a question of principle, since one would like to treat fermions on the same footing as bosonic fields, even though they

1980 Mathematics Subject Classification (1985 Revision) 83C45.
[1] Supported in part by NSF Grant PHY-85-06686.

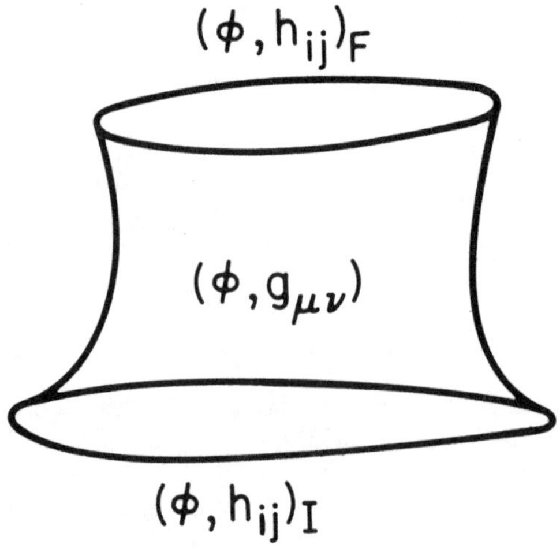

Fig. 1. An infilling field configuration $(\phi, g_{\mu\nu})$ between data on the initial and final surfaces.

are described most naturally by anti-commuting (Grassmann) quantities. A unified treatment can be given with the help of the holomorphic representation for fermions [6]. This requires a careful study of the boundary conditions natural for fermions, which turn out to be somewhat different from the more familiar conditions natural in the bosonic case. In section 3 we discuss the quantum state determined by the Hartle-Hawking proposal for a Dirac field in the cosmological background model of section 2, and comment on the issues raised.

2. QUANTUM COSMOLOGY: THE HARTLE-HAWKING PROPOSAL

Quantum field theory may be described in terms of the amplitude to go from data prescribed on an initial spacelike hypersurface to data prescribed on a final surface. Suppose for example that our fields consist only of gravity coupled to a massive scalar field ϕ. Then the amplitude K_{FI} to go from initial data, here given by the field ϕ and intrinsic spatial metric h_{ij} on an initial spacelike 3-surface I, to corresponding data on a final surface F is given formally by a Feynman path integral

$$K_{FI} = \int_{Lorentzian} \exp(iI) \mathcal{D}\phi \mathcal{D}g_{\mu\nu} \tag{2.1}$$

over all scalar fields ϕ and Lorentzian 4-metrics $g_{\mu\nu}$ which fill in between the data

(Fig. 1) and agree with them on the boundaries. Here I is the Lorentzian action of the infilling fields. Since the integrand is an analytic function of the fields, it is often assumed that the contour of integration in (2.1) may be rotated, giving also formally

$$K_{FI} = \int_{Euclidean} \exp(-\hat{I}) \mathcal{D}\phi \mathcal{D}g_{\mu\nu} \tag{2.2}$$

where the integral is now taken over all infilling scalar fields ϕ and Euclidean-signatured 4-metrices $g_{\mu\nu}$, and $\hat{I}$ is the Euclidean action.

Regarding the amplitude K_{FI} as a functional of the final data $(\phi, h_{ij})_F$, one can as in ordinary quantum mechanics build up the most general physical wave-functional $\psi(\phi, h_{ij})$ by superposition over initial data. Such a physical state will automatically obey the quantum constraint equations

$$\mathcal{H}_i \psi = 0 \quad , \tag{2.3}$$

$$\mathcal{H}_\perp \psi = 0 \quad . \tag{2.4}$$

Here $\mathcal{H}_i$ and $\mathcal{H}_\perp$ are quantum operator versions of the classical constraints of the theory, described in more detail in [10], [13]. The first-order equation (2.3) simply describes the invariance of ψ under spatial coordinate transformations applied to its arguments $\phi(x)$ and $h_{ij}(x)$. The second-order equation (2.4), known as the Wheeler-DeWitt equation, contains the dynamics of the theory.

Which of these many possible quantum states should one consider? Hartle and Hawking [11] have proposed, in the case that h_{ij} describes a positive-definite metric on a compact 3-surface, a choice singled out naturally by the path integral (2.2), in which the initial surface is shrunk to nothing and one defines

$$\psi(\phi, h_{ij}) = \int_{\substack{compact \\ Euclidean}} \exp(-\hat{I}) \mathcal{D}\phi \mathcal{D}g_{\mu\nu} \quad . \tag{2.5}$$

Here the integral is taken over all scalar fields ϕ and positive-definite 4-metrics $g_{\mu\nu}$ defined on compact 4-manifolds bounded by a single 3-surface, and agreeing with the data (ϕ, h_{ij}) specified on that surface (Fig. 2). Physically, this choice may be motivated by a property of the analogous path integral in the non-relativistic

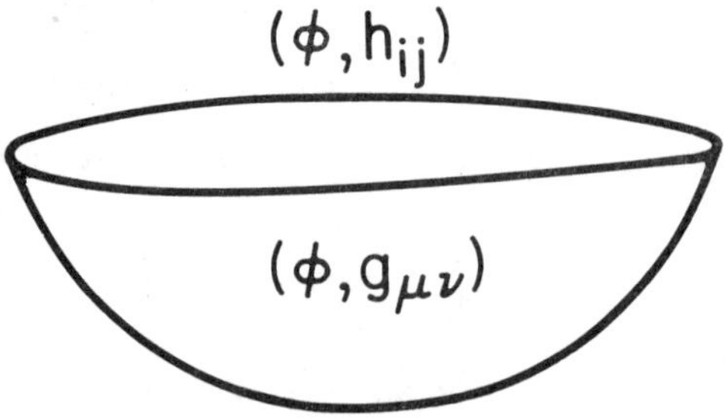

Fig. 2. An infilling field configuration $(\phi, g_{\mu\nu})$ contributing to the Hartle-Hawking wave function $\psi(\phi, h_{ij})$.

quantum mechanics of a single particle [10]. If one takes for $\hat{I}$ the Euclidean action of a particle path which starts at position x_o at time 0 and proceeds backwards indefinitely in imaginary time to a configuration of least action, and then integrates over such paths, one recovers the ground-state wave function $\psi_o(x_o)$. The expression (2.5) gives a close analogue of this construction, and may be regarded as defining a "ground state wave-function of the universe." Whether our universe corresponds to (2.5) is of course a matter for investigation.

One simple model in which this proposal has been examined is the massive scalar field model of Hawking [12][13]. Here the Lorentzian geometry and scalar field are restricted to have the form

$$ds^2 = \sigma^2(-N^2(t)dt^2 + e^{2\alpha(t)}d\Omega_3^2) \quad , \tag{2.6}$$

$$\phi = \phi(t) \quad . \tag{2.7}$$

where $\sigma^2 = 2/3\pi m_p^2$ in terms of the Planck mass m_p, and $d\Omega_3^2$ is the metric on a unit 3-sphere. The basic dynamical variables are $\alpha(t)$, $\phi(t)$ and their conjugate momenta.

The lapse function $N(t)$ plays the role of a Lagrange multiplier in the Hamiltonian and does not appear as an argument of the wave function $\psi(\alpha, \phi)$. The complete quantum mechanics of this system is contained in the Wheeler-DeWitt equation (2.4), which (with a particular choice of factor-ordering) gives

$$\left[\frac{\partial^2}{\partial\alpha^2} - \frac{\partial^2}{\partial\phi^2} + V(\alpha,\phi)\right]\psi(\alpha,\phi) = 0 \quad , \tag{2.8}$$

(2.9) $$V(\alpha,\phi) = e^{6\alpha}M^2\phi^2 - e^{4\alpha} \quad ,$$

where M is the mass of the scalar field.

To describe the properties of the hyperbolic equation (2.8), it is convenient to make the change of variables

(2.10) $$x = e^{\alpha}\cosh\phi \quad , \quad y = e^{\alpha}\sinh\phi \quad .$$

The characteristics are again lines at 45° in the xy plane (Fig. 3). The domain

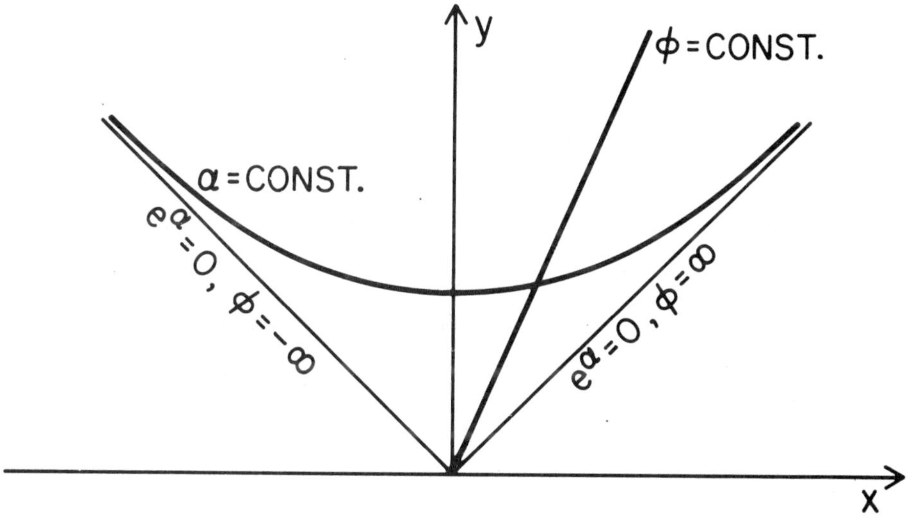

Fig. 3. Coordinates for the Wheeler-DeWitt equation (2.8).

$e^{\alpha} > 0$ of interest is bounded by the two null lines $e^{\alpha} = 0, \phi = \pm\infty$ emerging from the origin. Surfaces of constant α are spacelike hyperbolae within this light cone, while surfaces of constant ϕ are straight lines through the origin. The Hartle-Hawking proposal gives boundary conditions on $\psi(\alpha,\phi)$ on the light cone of the origin; the wavefunction may then be found by integrating forward given these data.

The Hartle-Hawking path-integral solution of (2.8) has a semi-classical interpretation. In the region $V < 0$ ("small 3-geometries") sketched in Fig. 4, one finds that through a typical point (α,ϕ) there is a solution to the Euclidean coupled Einstein-scalar classical field equations, describing a smooth Riemannian 4-geometry

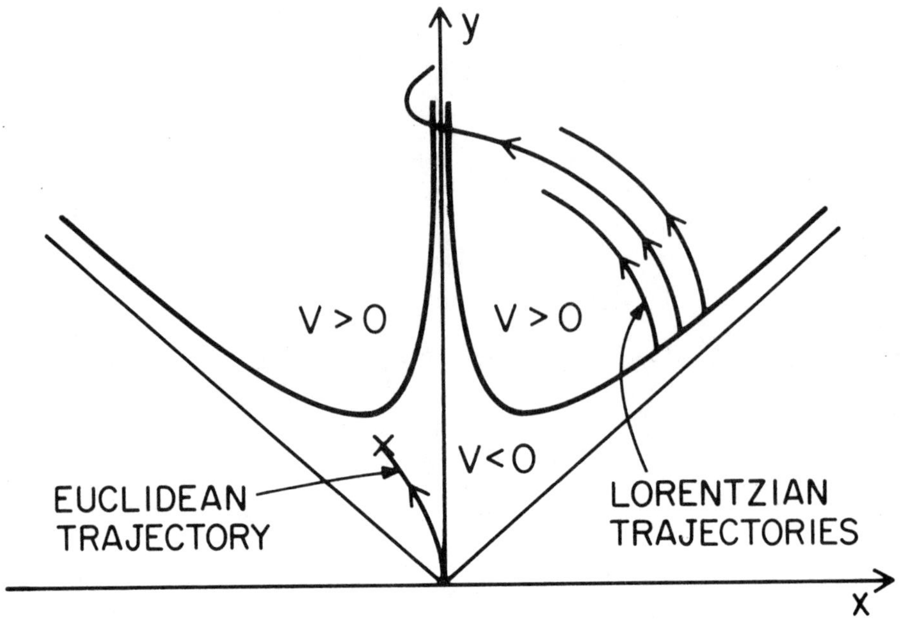

Fig. 4. Classical trajectories in the Euclidean and Lorentzian regions $V < 0$, $V > 0$.

bounded by a 3-sphere of radius e^{α}, carrying a scalar field ϕ (as in Fig. 2). In Fig. 4 such a classical solution is represented by a line proceeding to the origin. One expects such Euclidean classical solutions to provide the dominant contribution to the path integral (2.5) in this region. The wave function will behave exponentially, corresponding to a region forbidden classically to Lorentzian geometries.

For $V > 0$ ("large 3-geometries") ψ is oscillatory, and the dominant contribution to the path integral arises from a Lorentzian classical solution passing through the point (α, ϕ). The region $V > 0$ is thus filled with the trajectories of a family of Lorentzian classical solutions, some of which are sketched in Fig. 4. Each such classical geometry expands from a minimum radius at $V = 0$, and for large $|\phi|$ is well approximated by an inflationary (de Sitter) model. At late times in its evolution, such a classical solution turns into a radiation- or matter-dominated universe.

This quantum cosmological model is very restricted, due to the high degree of symmetry of the metric (2.6). A more realistic model is provided by allowing in all the degrees of freedom of the gravitational and scalar fields, regarded as perturbations on the Robertson-Walker background (2.6), (2.7). This has been treated by Halliwell and Hawking [9], using a decomposition of the perturbations into eigenmodes on the background. Their principal conclusions, based on the Hartle-Hawking proposal, are:

1. The wave functions of the linearized perturbations start in their ground state at early times in a Lorentzian background, *i.e.*, in the Lorentzian region near the surface $V = 0$ of Fig. 4. This provides further motivation for regarding the Hartle-Hawking proposal as defining the ground state of the universe.

2. The evolution of a wavefunction $\psi^{(n)}$, corresponding to a particular mode, along a Lorentzian classical trajectory of the background model (Fig. 4), is given by a Schrödinger equation

$$i\frac{\partial \psi^{(n)}}{\partial t} = H_2^{(n)} \psi^{(n)} \quad , \tag{2.11}$$

where t measures proper time in the background, and $H_2^{(n)}$ is a second-variation Hamiltonian. As it evolves, the wave function stays approximately in its ground state until it "freezes" after the mode goes outside the horizon during the de Sitter inflationary period. Later, in the matter- or radiation-dominated era, the mode re-enters the horizon in a highly excited state. Thus, even though the universe starts in a "ground state," a great deal of structure is generated at later times.

3. This leads to an approximately scale-free spectrum of density perturbations, of a kind suitable to account for the formation of galaxies.

3. INCLUSION OF FERMIONS

Suppose that one now wishes to treat fermion fields within this framework, say by including a Dirac field. It is convenient to work with 2-component spinors, in which case the Dirac field is described by two spinors ϕ_A, χ^A and their hermitian conjugates $\bar{\phi}_{A'}, \bar{\chi}^{A'}$ $(A = 0, 1; A' = 0', 1')$. Because spinors are being used, the metric $g_{\mu\nu}$ must be replaced by the tetrad $e^a_\mu (a = 0,1,2,3)$ or equivalently its spinor version $e^{AA'}_\mu$. The fermion fields are Grassmann quantities anti-commuting among themselves, but commuting with boson fields.

The complete Einstein-scalar-Dirac action can be written down, and takes the form

$$I = I_V + I_B \quad , \tag{3.1}$$

the sum of a volume contribution I_V and a surface contribution I_B (see [4] for details). The form of the term I_B depends on the boundary conditions being imposed at any bounding hypersurfaces. Just as in pure gravity [8], I_B is needed in order that the variational condition $\delta I = 0$, subject to the specified boundary data, should correctly lead to the classical field equations. This action can then be decomposed canonically, following the Hamiltonian treatment of Nelson and Teitelboim[14]. Note that, since the Dirac action is of first order in derivatives of fermion fields, half of the fields $\phi_A, \chi^A, \bar{\phi}_{A'}, \bar{\chi}^{A'}$ must be regarded as coordinates and half as momenta. Hence, in a classical or quantum boundary-value problem analogous to that depicted in Fig. 1, one might for example specify $(e_i^{AA'}, \phi, \phi^A, \chi^A)$ at the final surface, and $(e_i^{AA'}, \phi, \tilde{\phi}^{A'}, \tilde{\chi}^{A'})$ at the initial surface. Here specifying $e_i^{AA'}$ corresponds to specifying the spatial metric $h_{ij} = -e_{AA'i}e_j^{AA'}$ in the previous example, and only half of the fermionic variables are specified at each surface. At this point it is necessary to free the primed fermion fields from the restriction of being the hermitian conjugates of the unprimed fields, regarding them instead as independent quantities, and they are henceforth written with a tilde instead of a bar. Otherwise one would (for example) expect a badly-posed classical boundary-value problem for the fermion fields, obeying first-order equations, if they were completely specified at both the initial and final surfaces.

The amplitude K_{FI} to go from specified initial to final data is again given formally by a path integral

$$K_{FI} = \int_{Lorentzian} \exp(iI) De_\mu^{AA'} \; D\phi \; D\phi_A \; D\tilde{\phi}_{A'} \; D\chi^A \; D\tilde{\chi}^{A'} \quad , \tag{3.2}$$

where the infilling fields in the integration should agree with the specified data on the bounding hypersurfaces, and where Berezin integration[2] is being used for the fermionic integrations. The boundary term I_B of (3.1) plays a futher role in (3.2), ensuring that when the amplitude K_{JI} to go to data on on intermediate hypersurface J is composed with the amplitude K_{FJ} by summing over the data on J, one recovers the amplitude K_{FI}; thus it ensures that the quantum field theory is described by a Markov process. As in the bosonic case of section 2, one can regard K_{FI} as a functional of the final data (say), and hence build up the general physical Grassmann-valued wave functional $f(e_i^{AA'}, \phi, \phi^A, \chi^A)$ by superposition.

At least two questions are immediately raised by this description. How does one interpret such a Grassmann-valued wave function in terms, say, of familiar particle states, and how can the Hartle-Hawking proposal be extended to include fermions? We can consider these questions in the context of the background model of section 2, by treating the Dirac field as a perturbation in the presence of the background geometry[4]. The fermion fields can be given a decomposition with respect to spinor harmonics on the 3-sphere, as:

$$\phi_A = (2\pi)^{-1} e^{-3\alpha/2} \sum_{np} \left[M_{np}(t) \rho_A^{np}(\underline{x}) + \widetilde{R}_{np}(t) \bar{\sigma}_A^{np}(\underline{x}) \right] \quad , \tag{3.3}$$

$$\tilde{\phi}_{A'} = (2\pi)^{-1} e^{-3\alpha/2} \sum_{np} \left[\widetilde{M}_{np}(t) \bar{\rho}_{A'}^{np}(\underline{x}) + R_{np}(t) \sigma_{A'}^{np}(\underline{x}) \right] \quad , \tag{3.4}$$

$$\chi_A = (2\pi)^{-1} e^{-3\alpha/2} \sum_{np} \left[S_{np}(t) \rho_A^{np}(\underline{x}) + \widetilde{T}_{np}(t) \bar{\sigma}_A^{np}(\underline{x}) \right] \quad , \tag{3.5}$$

$$\tilde{\chi}_{A'} = (2\pi)^{-1} e^{-3\alpha/2} \sum_{np} \left[\widetilde{S}_{np}(t) \bar{\rho}_{A'}^{np}(\underline{x}) + T_{np}(t) \sigma_{A'}^{np}(\underline{x}) \right] \quad , \tag{3.6}$$

Here the pre-factors $(2\pi)^{-1} e^{-3\alpha/2}$ have been chosen to simplify the Dirac or anti-commutator brackets among pairs of Grassmann variables such as $M_{np}(t), \widetilde{M}_{np}(t)$, which are conjugate. The spinor fields $\rho_A^{np}(\underline{x})$ and $\sigma_{A'}^{np}(\underline{x})$ on the 3-sphere, labelled by indices n and p, are harmonics of the 3-dimensional Dirac operator with eigenvalues of a particular sign. Together with the hermitian conjugates $\bar{\sigma}_A^{np}(\underline{x})$ and $\bar{\rho}_{A'}^{np}(\underline{x})$, which have eigenvalues of the opposite sign, they provide a basis for the space of spinors such as ϕ_A and $\tilde{\phi}_{A'}$ on S^3. The harmonic decomposition (3.3) - (3.6) is of course very similar to that familiar for a Dirac field in flat space-time.

Instead of describing a quantum state by a wave functional such as $f(e_i^{AA'}, \phi, \phi^A, \chi^A)$, one could alternatively take as fermionic arguments any one out of each of the conjugate pairs M_{np} and $\widetilde{M}_{np}, R_{np}$ and $\tilde{R}_{np}$, etc. It turns out, as discussed below, that it is most natural in formulating the Hartle-Hawking proposal to use the non-local fermionic variables corresponding to one particular sign of the Dirac eigenvalues, and thus to consider a wave functional of the form $f(\alpha, \phi; M_{np}, R_{np}, S_{np}, T_{np})$. Strictly, one should include the perturbations of the gravitational and scalar fields also, but these decouple from the fermionic perturbations at lowest order, and can be treated separately following [9].

Conjugate variables such as $\widetilde{M}_{np}$ are represented by

$$\widetilde{M}_{np} \rightarrow \frac{\partial}{\partial M_{np}} \tag{3.7}$$

i.e. by using the holomorphic representation [6]. The total fermionic Hamiltonian can then be decomposed into a sum of terms of the form

$$H_n = N\left[-\nu + \nu(x\frac{\partial}{\partial x} + y\frac{\partial}{\partial y}) + m(yx + \frac{\partial^2}{\partial x \partial y})\right] \tag{3.8}$$

where $\nu = e^{-\alpha}(n + 3/2)$, m is the mass of the Dirac field, and (x,y) are pairs of fermionic variables such as (M_{np}, S_{np}) or (T_{np}, R_{np}). Because of the anti-commuting nature of x and y, H_n can be regarded as acting on a 4-dimensional vector space spanned by the elements 1, x, y and xy. Within this space, one can diagonalize $N^{-1}H_n$ to find a "ground state," a "1-particle state," a "1-antiparticle state" and a "particle-antiparticle state." An inner product and creation and annihilation operators can be found such that this 4-state space can be described in familiar Fock space language [4]. By putting together states for different values of the indices n, p, one can form a complete Fock space for the fermions, thus making contact between our wave-functional description and a more familiar one. Note, however, that our definition of "particle" and "antiparticle" states, introduced for convenience by diagonalizing the instantaneous Hamiltonian, will not correspond to measurements made by observers moving (say) on timelike geodesics (see later).

With this choice of non-local variables, the Hartle-Hawking state corresponding to a particular radius e^{α} of the 3-sphere, a particular homogeneous scalar field ϕ, and specified fermionic data $M_{np}, R_{np}, S_{np}, T_{np}$ on the 3-sphere is

$$\Psi(\alpha, \phi; M_{np}, R_{np}, S_{np}, T_{np}) = \int_{\substack{compact \\ Euclidean}} \exp(-\hat{I}) \mathcal{D} e_{\mu}^{AA'} \mathcal{D}\phi \mathcal{D}\phi_A \mathcal{D}\tilde{\phi}_{A'} \mathcal{D}\chi^A \mathcal{D}\tilde{\chi}^{A'} \tag{3.9}$$

where the integration is over all compact Riemannian 4-geometries bounded by the 3-sphere (Fig. 2), with regular scalar and Dirac fields which match the prescribed boundary data.

It is in the case of massless fermions that our non-local choice of fermionic data is more or less forced upon us. To see this, note that one expects to be able to approximate Ψ semi-classically as $\Psi \sim A \exp(-\hat{I}_{classical})$, where $\hat{I}_{classical}$ is the Euclidean action of a smooth classical solution (if such exists) of the coupled Einstein-scalar-Dirac equations on a compact geometry subject to the prescribed data on the bounding 3-sphere, and A is one-loop prefactor. If the fermionic data included a part corresponding to an eigenvalue of the wrong spin on the 3-sphere, then the classical solution for $m = 0$ would be expected to blow up in the interior. For example, in solving the first-order massless Weyl equation $\nabla_{AA'}\phi^A = 0$ in a region of flat Euclidan 4-space bounded by a 3-sphere centered on the origin, subject to boundary data with eigenvalue of the wrong sign, one finds that the solution is proportional to a negative power of the radius and diverges at the origin. Only when the data $(M_{np}, R_{np}, S_{np}, T_{np})$ are chosen will there be a regular classical solution for $m = 0$. Correspondingly one expects the path integral (3.9) to give a sensible quantum expression in this case. Similarly, when generalizing the prescription (3.9) to allow for data on an arbitrary compact 3-surface, one expects again to specify only that half of the fermionic data with the correct sign of the eigenvalue of the 3-dimensional Dirac operator. [The boundary-value problem for the Weyl equation on a general compact manifold is discussed by Atiyah, Patodi and Singer[1].] In the massive Dirac case, with $m > 0$, there is more freedom in specifying fermionic data. For example one could still specify local data ϕ_A and χ_A, and then in a semi-classical expansion find regular solutions to the elliptic second-order versions of the Dirac equation for ϕ_A and χ_A, then finally construct the remainder of the classical solution by defining a regular $\tilde{\chi}_{A'}$ and $\tilde{\phi}_{A'}$ through the Dirac equations $\sqrt{2}e^{\mu}_{AA'}D_\mu\phi^A = im\tilde{\chi}_{A'}$, $\sqrt{2}e^{\mu}_{AA'}D_\mu\chi^A = im\tilde{\phi}_{A'}$.

We can briefly summarize the conclusions of [4] for the Hartle-Hawking state (3.9) of this model, which are similar to those for the bosonic perturbations described in section 2.

1. The Wheeler-DeWitt equation (2.4) again leads to a Schrödinger equation (2.11) for the evolution of each fermionic mode along a Lorentzian classical trajectory of the background model.

2. The Wheeler-DeWitt equation also leads to an equation (2.8) for the wave function of the background geometry and scalar field, in which the potential $V(\alpha, \phi)$ is modified by the energy of the fluctuations of the fermion fields. After regularization, this modification has only a small effect on the quantum evolution of the background.

3. The wave functions for each fermionic mode again start in their ground state at early times in a Lorentzian background. They then evolve by the Schrödinger equation to give creation of "particle-antiparticle" pairs, where the notion of "particle" is that specified previously by diagonalization of the instantaneous Hamiltonian.

4. However, a particle detector moving on a geodesic at early times during the inflationary de Sitter epoch will measure a thermal spectrum of spin-$\frac{1}{2}$ particles at the de Sitter temperature, which here depends on the value of the scalar field ϕ. In fact the whole fermionic quantum state is approximately de Sitter-invariant at early times–a property closely connected with thermality [7].

Various questions are raised by this work, particularly in connection with the divergences associated with quantum field theory. For example, a somewhat arbitrary choice was made in the factor ordering both of the background Hamiltonian operators in (2.8) and in the fermionic Hamiltonian (3.8). Further, the back-reaction of the bosonic and fermionic perturbations on the background is given by a divergent mode sum of energies, which must be regularized to give a finite contribution to $V(\alpha, \phi)$. It is possible that these unsatisfactory features may be avoided by working with supergravity theories [5], which may yield a unique factor ordering [3] and which have improved ultraviolet properties due to cancellations between bosonic and fermionic modes.

BIBLIOGRAPHY

1. M.F. Atiyah, V.K. Patodi and I.M. Singer, "Spectral asymmetry and Riemannian geometry. I", Math. Proc. Camb. Phil. Soc. 77 (1975), 43-69.
2. F.A. Berezin, The method of second quantization, Academic Press, New York, 1966.
3. P.D. D'Eath, "Canonical quantization of supergravity", Phys. Rev. D29 (1984), 2199-2219.
4. P.D. D'Eath and J.J. Halliwell, "Fermions in quantum cosmology", to appear in Phys. Rev. D.
5. P.D. D'Eath, and D.I. Hughes, in preparation.
6. L.D. Faddeev and A.A. Slavnov, Gauge fields, Benjamin/Cummings, Reading, Mass., 1980.
7. G.W. Gibbons and S.W. Hawking, "Cosmological event horizons, thermodynamics and particle creation", Phys. Rev. D15 (1977), 2738-2751.
8. G.W. Gibbons and S.W. Hawking, "Action integrals and partition functions in quantum gravity", Phys. Rev. D15 (1977), 2752-2756.
9. J.J. Halliwell and S.W. Hawking, "Origin of structure in the universe", Phys. Rev. D31 (1985), 1777-1791.
10. J.B. Hartle, "Quantum cosmology", in High energy physics: Proceedings of the Yale advanced study institute, edited by M.J. Bowick and F. Gürsey, World Scientific, Singapore, 1985.
11. J.B. Hartle and S.W. Hawking, "Wave function of the universe", Phys. Rev. D28 (1983), 2960-2975.
12. S.W. Hawking, "The quantum state of the universe", Nucl. Phys. B239 (1984), 257-276.
13. S.W. Hawking, "Quantum cosmology", in Relativity, groups and topology II, edited by B.S. DeWitt and R. Stora, North-Holland, Amsterdam, 1984.
14. J.E. Nelson and C. Teitelboim, "Hamiltonian formulation of the theory of interacting gravitational and electron fields", Ann. Phys. (NY) 116 (1978), 86-104.

P.D. D'Eath DEPARTMENT OF APPLIED MATHEMATICS AND
THEORETICAL PHYSICS
SILVER STREET
CAMBRIDGE CB3 9EW, ENGLAND

Current Address
Department of Physics
University of California
Santa Barbara, CA 93106, U.S.A.

J.J. Halliwell DEPARTMENT OF APPLIED MATHEMATICS AND
THEORETICAL PHYSICS
SILVER STREET
CAMBRIDGE CB3 9EW, ENGLAND

Contemporary Mathematics
Volume **71**, 1988

GAUGE SYMMETRIES OF STRING FIELD THEORY

Michael E. Peskin

ABSTRACT

I review the covariant and gauge-invariant formulation of the field theory for the simple open bosonic string. I first display the Virasoro operators L_n and show that the action of these operators on string functionals $\Phi\,[x\,(\sigma)]$ generates the gauge motions of their component local fields. I then develop the formulation of a gauge-invariant action principle. This requires extending Φ to a functional of coordinates and ghost variables: $\Phi\,[x(\sigma), c(\sigma), b(\sigma)]$. The ghost variables b, c lead to a nilpotent operator Q_{BRST}, and this in turn generates a set of natural differential operators d, δ satisfying $d^2 = \delta^2 = 0$, and $d\,\delta + \delta\,d = 2\,K \downarrow$, where $K = p^2 + M^2$ for the string. I finally assemble these objects into the gauge-invariant action: $S = < \Phi \mid Q_{BRST} \mid \Phi >$.

For further discussion, see the following two references:

1. T. Banks and M.E. Peskin, "Gauge Invariance of String Fields", Nucl.Phys. B 264, 513 (1986).

2. T. Banks, M.E. Peskin, C.R. Preitschopf, and E. Martinec, "All Free String Theories are Theories and Forms", Nucl. Phys. B 274, 71 (1986).

Michael E. Peskin

SLAC

P.O. Box 4349

Stanford, CA 94305

Contemporary Mathematics
Volume **71**, 1988

TWISTORS AND STRINGS

William T. Shaw[1]

ABSTRACT. Classical string theory is examined in terms of a null curve decomposition. Null curves have a natural twistor description in certain dimensions. Open and closed strings are examined from this point of view and some simple examples are given. A classical mode expansion may be given in terms of a twistor Fourier series. The usual twistor description of null lines in Minkowski space corresponds to the excitation of one mode while massive strings correspond to the excitation of two or more modes. The usual constraint equations for the string are eliminated in twistor space, but other constraints may emerge as reality conditions if one seeks solutions corresponding to real strings. These constraints may be solved explicitly in dimensions 3,4,6 and 10, giving a procedure for constructing real, general and covariant solutions to the bosonic string equations.

1. INTRODUCTION. The purpose of this note is to describe the status of the "twistor transform" of string theory. The notion of a "twistor transform" refers to a device for transforming partial differential equations and associated real manifolds into holomorphic objects and certain associated complex manifolds. The key feature of the transform is that the differential equations are eliminated, leaving free holomorphic data. A string could refer to a classical or quantum minimal surface, timelike or spacelike, open, closed (or neither), bosonic or supersymmetric in a flat space-time of any dimension and signature in which it makes sense to talk about the object.

The main reason for considering the connections between the two types of theory concerns the manner in which string theory is quantized. At the classical level the equations describing a string (in the "conformal gauge") consist of some free wave equations and some constraints. It is trivial to solve the wave equations. One is then left with a choice about whether to quantize before or after imposing the constraints. If one quantizes before imposing the constraints one must check that the quantum constraints eliminate the unphysical

1980 Mathematics Subject Classification (1985 Revision): 32/53A/81E.

[1]Supported in part by NSF Grant DMS 8603523.

states. The other approach involves "solving" the constraints before quantization. The conventional procedure for doing so involves the "light-cone gauge", where one breaks Lorentz invariance by choosing a special coordinate system in which one can formally solve the constraints. After quantization one must check that the theory is Lorentz invariant. (These issues have been discussed at length by many authors-see e.g. [1] and the references contained therein.) Both these procedures have been thoroughly investigated. However, the light-cone gauge approach is unsatisfactory not only in that it breaks covariance, but also because it does not explicitly solve the constraints-one must always resort to quadrature to determine one of the coordinates of the world-sheet (usually referred to as X^-). This would be a minor point but for the hope that one might be able to do better-indeed, the complete solution of the constraints, without residual quadrature, was given (in the complex setting) for three dimensions by Weierstrass in 1866 [2] and for four dimensions by Montcheuil [3] (see also [4]).

Generalizations of these results to higher dimensions, also maintaining explicit covariance, would be interesting for the construction and classification of classical solutions, and might provide a way of quantizing string theory in a way which is both covariant and free of constraints. The main results of this paper are best described as series of tricks for constructing the general solutions for a real classical bosonic string in dimensions 3,4,6 and 10. The results arose out of an attempt to solve the reality constraints on a twistor description of strings in dimension 4 [5,6]. In dimension 4 the twistor description of a minimal surface is a holomorphic curve in complex projective 3-space $\mathbb{P}^3$ and gives the geometric interpretation of the construction in [4]. The way one solves these constraints generalizes to any dimension d in which it is possible to characterize a null vector as a product of unconstrained spinor variables. The first few even dimensions in which this is possible are 4,6 and 10. The case d = 8 is specifically excluded because of certain purity constraints. The case d = 3 is discussed because it is very similar to d = 10 and is important historically. The twistor interpretation of the d = 6 case is discussed by Hughston and Shaw [7].

The plan of this work is as follows. In section two the role of null curves in describing a string will be reviewed. A characterization of the null curves needed to describe both open and closed strings is obtained. From here on the focus is on open strings, which are characterized by one null curve subject to a periodicity condition (closed strings are described by two such curves). In section three the twistor translation of a real rigidly rotating string (d = 4) is obtained. This serves as a model for more complicated real surfaces. In section four general open strings are considered in twistor terms.

These are represented in twistor space by certain quasi-periodic (more descriptively-helical) curves on $\mathbb{P}^3$ subject to three reality conditions. By expanding the twistor curves in a Fourier series these constraints can be expressed as constraints on the twistor Fourier modes. The solution corresponding to the excitation of just one mode describes a null geodesic, while two modes describe the rigidly rotating string. A procedure for finding the general solution to the constraints is obtained in a very simple way. One consequence of the analysis is that for some numbers of twistor modes, no non-trivial solutions are possible (e.g. there are no non-trivial 3-twistor strings). The free variables in the general $d = 4$ solution are described as a list of 2-component spinors (which represent the Fourier expansion of a loop in spin-space) and an arbitrary real 4-vector describing the "centre" of the null curve. The null curve itself is obtained by integrating the null vector field defined as the Hermitian square of the spinor field. The integration can be done explicitly. In section five the corresponding procedures for $d = 3$ and $d = 6$ Minkowski space-times are described. In these dimensions a null vector can be expressed as a product of spinors unconstrained apart from reality conditions. This fails when $d = 8$ because the spinor description of a null vector must contain a spinor which is itself pure in the sense of Cartan [8] (see also [9]). When $d = 10$ the purity conditions are circumvented because of certain identities obeyed by the Γ-matrices. After a brief review of relevant facts about $d = 10$ spinors, the solution for an open string in $d = 10$ is described.

An interesting feature of this approach is the fact that the solutions obtained by writing down a loop in spin space contain a degree of redundancy. More precisely, if one describes a null vector as a product of spinors then many spinors can give the same null vector. In dimension $d(d = 3,4,6,10)$ then the spinors which describe the same null vector lie on a quadric (sphere) in $\mathbb{R}^{d-2}-(0)$. This redundancy has some interesting consequences in relation to the introduction of supersymmetry, and, when $d = 10$, to the notion of purity. Some comments on this are made in section six.

2. NULL CURVE DECOMPOSITIONS. In attempting to find a twistor description of a system described by differential equations in a space-time, it is helpful to follow two guidelines. Firstly, since twistor theory is most naturally applied to conformally invariant theories, one seeks some conformally invariant system of equations to transform. Secondly, since twistor theory also refers naturally to complex space-times, one attempts to complexify the system. The equations governing a string do <u>not</u> possess the kind of conformal invariance we need! The action of a string is just the area of the world-sheet, which is manifestly not conformally invariant. (The conformal invariance in question

should not be confused with the invariance of the equations under reparameterizations of the world-sheet coordinates.) Fortunately a slight shift in viewpoint is all that is needed to deal with this problem.

Consider the equations of motions of a string in the "conformal gauge". One introduces coordinates τ and σ for the world-sheet, chosen so that the metric is of the form

$$ds^2 = f(\tau,\sigma)[d\tau^2 - d\sigma^2]. \tag{1}$$

If we give flat d-dimensional Minkowski space-time $\mathbb{M}_d$ coordinates X^a, $a = 0,1,\ldots,d-1$, then the world-sheet $X^a(\tau,\sigma)$ must satisfy the equations

$$X^a{}_{,\tau\tau} - X^a{}_{,\sigma\sigma} = 0, \tag{2}$$

$$X_{,\sigma}.X_{,\tau} = 0 \ ; \ X_{,\tau}.X_{,\tau} = -X_{,\sigma}.X_{,\sigma} \ , \tag{3}$$

where $U.V = U^aV^b\eta_{ab}$, and η_{ab} is the flat metric on $\mathbb{M}_d$. These can be solved directly to give

$$X^a(\tau,\sigma) = A^a(\tau - \sigma) + B^a(\tau + \sigma) \tag{4}$$

where the curves $A^a(s)$ and $B^a(s)$ are null:

$$A'(s).A'(s) = 0 = B'(s).B'(s). \tag{5}$$

Thus the problem of determining solutions to the equations of motion for a string reduce to the problem of finding the null curves in $\mathbb{M}_d$. This is now a problem in conformal geometry. At this stage one can make a trivial complexification by allowing the curves $A(s)$ and $B(s)$ to be complex-valued, whilst keeping s real. This is appropriate for the description of real timelike surfaces. For spacelike strings s is itself a complex variable.

Assuming that one has a knowledge of the null curves, one can proceed to examine the properties of the theory. The first question concerns the initial/boundary conditions for (2). One assumes that the string is defined for all τ, with initial conditions given at some time τ_0, subject to the constraints (3). Provided one gives appropriate boundary conditions, the motion of the string is determined for $\tau > \tau_0$ and $\tau < \tau_0$. Usually one of two types of boundary condition are specified. The string is said to be closed if periodic boundary conditions are applied:

$$X^a(\sigma + 2\pi,\tau) = X^a(\sigma,\tau). \tag{6}$$

The periodicity can be chosen to be 2π since the equations are reparameterization invariant. The string is said to be open if the boundary conditions

(7) $$X^a{}_{,\sigma}(0,\tau) = X^a{}_{,\sigma}(\pi,\tau) = 0$$

are applied, where the parameter range is chosen to be $[0,\pi]$. The boundary conditions constrain the null curves as follows. If the string is open then (7) implies

(8) $$B'_a(\tau) = A'_a(\tau),$$
(9) $$B'_a(\tau + \pi) = A'_a(\tau - \pi).$$

The first condition shows that one can arrange that $A^a(s) = B^a(s)$. The second condition then implies that

(10) $$A^a(\tau + \pi) = A^a(\tau - \pi) + C^a,$$

where C^a is a constant vector. An open string is therefore determined by one such null curve $A^a(s)$, in the form

(11) $$X^a(\tau,\sigma) = A^a(\tau - \sigma) + A^a(\tau + \sigma).$$

For a closed string $A^a(s)$ and $B^a(s)$ may be distinct but must satisfy

(12) $$A^a(s + 2\pi) - A^a(s) = C^a,$$
(13) $$B^a(s + 2\pi) - B^a(s) = C^a,$$

for some constant vector C^a. Note that whether the string is open or closed,

(14) $$X^a(\tau + 2\pi,\sigma) - X^a(\tau,\sigma) = 2C^a.$$

For a real string to propagate to the future it is necessary to demand that C^a is real, timelike and future-pointing.

The basic entity from which one can construct strings of either type is therefore a null curve $\phi^a(s)$ subject to

(15) $$\phi^a(s + 2\pi) - \phi^a(s) = C^a$$

for some constant future-pointing timelike vector C^a. Note that the null vector field $\phi'^a(s)$ is periodic with period 2π. Such a curve will be called a helical null curve. One's next task in this approach is to find all the null curves in $\mathbb{M}_d$ and select those that are helical.

At this point the specialization to d = 4 will be made, in order to use standard twistor techniques. However, certain features of these techniques certainly generalize to all dimensions, although the details depend on the dimension and are only understood in any detail for d = 3,4,6 and 10. When d = 4 one defines a curve in twistor space by the conditions

$$\text{(16)} \qquad \omega^A(s) = i\phi^{AA'}(s)\pi_{A'}(s), \quad \phi'^{AA'}(s)\pi_{A'}(s) = 0.$$

Twistor space, complex projective 3-space $\mathbb{P}^3$, is given homogeneous coordinates $[\omega^A, \pi_{A'}]$. For higher dimensions ω and π become pure spinors for the group O(d), the pair $[\omega, \pi]$ forms a pure spinor for the group O(d+2), and relations similar to (16) are imposed. For details of the case d = 6 see [7]. Given a curve $[\omega^A(s), \pi_{A'}(s)]$ in twistor space, the curve $\phi^a(s)$ is defined as a solution of the relations

$$\text{(17)} \qquad \omega^A(s) = i\phi^{AA'}(s)\pi_{A'}(s) \qquad \omega'^A(s) = i\phi^{AA'}(s)\pi'_{A'}(s)$$

and is null by construction. See [5,6] for detailed discussions of the geometry associated with this system. For the null curve to be helical $\phi'(s)$ must be periodic with period 2π. The $\pi_{A'}$-curve in twistor space must therefore define a periodic curve on $\mathbb{P}^1$, and with an appropriate scaling $\pi_{A'}$ is itself periodic. Then the ω^A-curve must satisfy the constraint

$$\text{(18)} \qquad \omega^A(s+2\pi) = \omega^A(s) + iC^{AA'}\pi_{A'}(s) .$$

Geometrically, considering $\mathbb{P}^3$ as a bundle over $\mathbb{P}^1$, the curve must project to a loop on $\mathbb{P}^1$ and satisfy (18). Note that for the string to propagate to the future one is forced to consider curves in $\mathbb{P}^3$ which are not themselves loops. This will also hold in any dimension when the spaces are suitably generalized. The constant C^a is closely related to the total momentum of the string. If one defines the conserved quantity

$$\text{(19)} \qquad p^a = \int d\sigma\, X^a{}_{,\tau}$$

then $P^a = C^a$ for an open string and $P^a = 2C^a$ for a closed string.

3. SOME SIMPLE EXAMPLES. To illustrate the ideas of the last section it is useful to consider some simple examples. Let C^a be a fixed timelike vector and let $C^a = 2\pi v t^a$ where $t.t = 1$. Let U^a be another constant real vector. A simple example of a curve in twistor space satisfying (18) is obtained as follows. Let $\pi_{A'}(s)$ have period 2π and let ω^A be "linear" in both $\pi_{A'}$ and s:

(20) $$\omega^A(s) = i[svt^{AA'} + iU^{AA'}]\pi_{A'}(s).$$

The resulting space-time curve is easily found to be

(21) $$\phi^a(s) = svt^a + iU^a - v(\pi^{A'}t^{AB'}\pi_{B'})/(\pi_{C'}\pi^{'C'}).$$

This is certainly null but is not necessarily real. It is real if and only if

(22) $$U^a = v\ \mathrm{Im}\{(\pi^{A'}t^{AB'}\pi_{B'})/(\pi_{C'}\pi^{'C'})\}.$$

This implies that $U.t = 0$, and also that

(23) $$U^{AA'}\bar{\pi}_A\pi_{A'} = 0,$$

(24) $$2\,U^{AA'}\dot{\bar{\pi}}_A\pi_{A'} + ivt^{AA'}\bar{\pi}_A\pi_{A'} = 0\ ,$$

(25) $$2\,U^{AA'}\dot{\bar{\pi}}_A\dot{\pi}_{A'} + ivt^{AA'}(\bar{\pi}_A\dot{\pi}_{A'} - \dot{\bar{\pi}}_A\pi_{A'}) = 0.$$

Thus U^a is necessarily spacelike and partially determines $\pi_{A'}$. These constraints can be solved for $\pi_{A'}$ for any spacelike U^a. To fix ideas, let y^a be a unit spacelike vector and wrote $U^a = \psi y^a$. Spinor coordinates can be chosen so that (conventions as in [5]).

(26) $$\sqrt{2}\,t^{AA'} = \begin{bmatrix} 1 & 0 \\ 0 & 1 \end{bmatrix} \quad \sqrt{2}\,z^{AA'} = \begin{bmatrix} 1 & 0 \\ 0 & -1 \end{bmatrix}$$

(27) $$\sqrt{2}\,y^{AA'} = \begin{bmatrix} 0 & -i \\ i & 0 \end{bmatrix} \quad \sqrt{2}\,x^{AA'} = \begin{bmatrix} 0 & 1 \\ 1 & 0 \end{bmatrix} \ .$$

Also write $\pi_{A'} = (\pi_{0'}, \pi_{1'})$, and define $\zeta = \pi_{0'}/\pi_{1'}$. The constraints above then reduce to demanding that ζ is real and satisfies

(28) $$2\psi\zeta' = v(1 + \zeta^2).$$

Writing $\zeta = \tan(\theta)$ we have $2\psi d\theta = v ds$. The curve $\zeta(s)$ has a period 2π if $\theta = ns/2$, so that $v = n\psi$ for some non-zero integer n. The associated space-time curve is then

$$\phi^a(s) = \psi[nst^a + iy^a - n(\pi^{A'}t^{AB'}\pi_{B'})/(\pi_{C'}\pi^{'C'})] \tag{29}$$

in terms of standard (t,x,y,z) coordinates the locus of the curve is given by, with $\psi = 1/n$,

$$t = s;\ y = 0;\ x = n^{-1}\cos(ns);\ z = n^{-1}\sin(ns) \tag{30}$$

which is indeed a null helix in the $y = 0$ plane.

The above construction assumes that the string world-sheet is real. If one relaxes this requirement and considers strings as maps into complex Minkowski space-time $\mathbb{CM}_4$, a much larger class of solutions can be obtained in this way, since the constraints (23)-(25) no longer apply. Staying in the real setting for the present, it is easy to see the form of the resulting string. The open strings obtained by combining advanced and retarded forms of (30) are realised as spinning straight line segments, with angular velocity n/2. The twistor curves of the form (20) describe elementary classical states of the open string. The possibilities for the closed strings are more complex, as the advanced and retarded modes can have different n-values and lie in different Minkowskian 3-spaces. For each choice of n and U^a one obtains a null helix $\phi^a[n,U](s)$ satisfying $U\cdot\phi = 0$. The closed strings are therefore of the form

$$X^a(\tau,\sigma) = \phi^a[n,U_1](\tau-\sigma) + \phi^a[m,U_2](\tau+\sigma) \tag{31}$$

for integers m,n and spacelike vectors U_1,U_2. The world-sheet is realised as the locus of a spinning loop with zero linear momentum. The loop typically has many cusps and/or self-intersections. Note that the strings obtained in this way do not truly represent the classical ground state. The special case when C^a is null plays this role. In this case, up to a phase $\pi_{A'}$ is constant and one can see that C^a is proportional to $\bar{\pi}^A\pi^{A'}$. The string world-sheet, if real, then degenerates to a null geodesic and is represented by a point in twistor space. The null helical curve is the first excited state beyond the null geodesic. This is part of a pattern that will become apparent when one considers a general mode expansion for the null curves.

4. CLASSICAL MODE EXPANSIONS IN TWISTOR SPACE. To generalize the examples given in the last section requires a deeper understanding of the structure of the curves in Twistor space. The goal is to find all real null curves $\phi^a(s)$ in $\mathbb{M}_4$ satisfying the condition (15). It is both convenient and appropriate to include the limiting case where C^a is null. Suppose one assumes that $\phi'_a(s)$ is null, future-pointing and real. Then one may write

(32) $$\phi'_a(s) = \bar{\pi}_A(s)\pi_{A'}(s)$$

for some spinor field $\pi_{A'}(s)$. Note that $\pi_{A'}(s)$ is only defined up to a phase. Also, since $\phi'_a(s)$ is periodic one may expand $\pi_{A'}(s)$ as a Fourier series:

(33) $$\pi_{A'}(s) = \sum_{n=-\infty}^{\infty} \pi^n{}_{A'}\exp(ins) \quad .$$

Now define

(34) $$\Pi^a = \sum_{n=-\infty}^{\infty} \bar{\pi}^n{}_A \pi^n{}_{A'} \cdot$$

Clearly one has $C^a = 2\pi\Pi^a (= P^a$ for an open string). One wants to make a similar expansion for $\omega^A(s)$ also, but since $\omega^A(s)$ is not periodic this is not possible directly. However, one can use (34) to define a field satisfying (18) and then add in a Fourier series to obtain the most general possibility. So write

(35) $$\omega^A(s) = is\Pi^{AA'}\pi_{A'}(s) + \Omega^A(s)$$

(36) $$\Omega^A(s) = \sum_{n=-\infty}^{\infty} \omega_n{}^A\exp(ins)$$

Writing $Z(s) = [\omega^A(s), \pi_{A'}(s)]$ defines a curve in (non-projective) twistor space satisfying (18). Given this curve one can define a space-time curve using (17) and this is then a null curve in $\mathbb{CM}_4$, but for arbitrary $\Omega^A(s)$ it is not real and its derivative does not satisfy (32), but only

(37) $$\phi'_a(s) = \lambda_A(s)\pi_{A'}(s)$$

for a certain (computable) $\lambda_A(s)$ which forms part of the dual twistor description of the curve. Since one wants to find real strings it is necessary to impose extra condtions. The reality conditions are [6]:

$$(38)\qquad Z.\bar{Z} = 0 = Z.\dot{\bar{Z}} = \dot{Z}.\dot{\bar{Z}}$$

for all s. (Here and below the "dot" between twistors dentoes contraction of the twistor indices: $Z.W = Z^{\alpha}W_{\alpha}$). These follow directly from the requirement that $\phi_a(s)$ is real for all s.

The reality conditions can be translated into algebraic constraints on the Fourier coefficients $Z_n = \{\omega_n{}^A(s), \pi^n{}_{A'}(s)\}$. Define, for any integer N, the following quantities:

$$(39)\qquad A_N = \sum_{n=-\infty}^{\infty} Z_n.\bar{Z}_{(n-N)}$$

$$(40)\qquad B_N = \sum_{n=-\infty}^{\infty} \{nZ_n.\bar{Z}_{(n-N)} + \Pi^{AA'}\pi^n{}_{A'}\bar{\pi}^{(n-N)}{}_A\}$$

$$(41)\qquad C_N = \sum_{n=-\infty}^{\infty} \{n^2 Z_n.\bar{Z}_{(n-N)} + 2n\Pi^{AA'}\pi^n{}_{A'}\bar{\pi}^{(n-N)}{}_A\}.$$

Then the reality conditions are equivalent to demanding that

$$(42)\qquad A_N = 0;\ B_N = 0;\ C_N = 0$$

for all N. Note however that these constraints satisfy the identities

$$(43)\qquad (A_N)^* = A_{-N}$$

$$(44)\qquad (B_N)^* = B_{-N} + NA_{-N}$$

$$(45)\qquad (C_N)^* = C_{-N} + 2NB_{-N} + N^2 A_{-N}$$

so it is sufficient to require that (42) holds for $N \geq 0$.

In general it is a fairly involved procedure to solve these algebraic conditions. Consider first what happens when only one mode is excited. That is, $Z_n^{\alpha} = 0$ unless $n = N_0$. Furthermore, since the twistor curve is only defined up to a phase one may set $N_0 = 0$. Thus $\pi_{A'}(s) = \pi^0{}_{A'}$ and $\Pi_a = \pi^0{}_{A'}\bar{\pi}^0{}_A$ is <u>null</u>. Then

$$(46)\qquad \omega^A(s) = is\,\Pi^{AA'}\pi^0{}_{A'} + \omega_0{}^A = \omega_0^A$$

The twistor curve degenerates to a single point. Examining the constraints one notes that $C_N = 0 = B_N$ for all N, and $A_N = 0$ for all $N \neq 0$. Furthermore,

(47) $$A_0 = \bar{Z}_0 . Z_0$$

so the only constraint is that Z_0^α is a null twistor. The corresponding space-time curve is just the real null geodesic

(48) $$\phi_a(s) = \phi_a(0) + s\bar{\pi}^0{}_A \pi^0{}_{A'}$$

where $\phi_a(0)$ is any solution of

(49) $$\omega_0{}^A = i\phi^{AA'}(0)\pi^0{}_{A'}.$$

Evidently when just one twistor mode is excited one recovers the standard twistor description of a real null geodesic as a point in twistor space subject to the vanishing of (47).

Consider now the case of two excited modes. By making an appropriate phase change one may assume that these are Z_0^α and Z_p^α, where $p > 0$. Thus

(50) $$\pi_{A'}(s) = \pi^0{}_{A'} + \pi^p{}_{A'}\exp(ips)$$

and

(51) $$\Pi^a = \bar{\pi}^0{}_A \pi^0{}_{A'} + \bar{\pi}^p{}_A \pi^p{}_{A'}$$

is now future-pointing and timelike (unless $\pi^0{}_{A'}$ and $\pi^p{}_{A'}$ are proportional). Given the formula for Π^a the only independent constraints which do not vanish identically are A_0, A_p and B_0. That is,

(52) $$Z_0.\bar{Z}_0 = -Z_p . \bar{Z}_p = p^{-1}\Pi^a\Pi_a ,$$

(53) $$Z_0.\bar{Z}_p = 0.$$

Note that the mass-squared of the system (with an open string) is given here by

(54) $$m^2 = P^a P_a = 4\pi^2\Pi^a\Pi_a = -4\pi^2 p Z_p.\bar{Z}_p$$

and in general the constraint B_0 gives

(55) $$m^2 = -4\pi^2 \sum_{n=-\infty}^{\infty} n Z_n.\bar{Z}_n .$$

In the present case, if $p > 0$ one must arrange that $Z_0 . \bar{Z}_0 > 0$, so that $Z_p . \bar{Z}_p < 0$ and $m^2 > 0$. In terms of the spinor components the constraints are

$$(56) \qquad 0 < \Pi^a \Pi_a = 2|\pi^0{}_{A'} \pi^{pA'}|^2 = p[\omega_0{}^{A'} \bar{\pi}^0{}_A + \bar{\omega}_0{}^{A'} \pi^0{}_{A'}]$$
$$= -p[\omega_p{}^A \bar{\pi}^p{}_A + \bar{\omega}_p{}^{A'} \pi^p{}_{A'}]$$

$$(57) \qquad \omega_0{}^A \bar{\pi}^p{}_A + \bar{\omega}_p{}^{A'} \pi^0{}_{A'} = 0.$$

It is straightforward to show (by solving for the spinors component by component, for example) that the general solution to these constraints yields a null curve which is essentially the null helix described in section three (unless $\pi^0{}_{A'} \propto \pi^p{}_{A'}$). All the solutions are obtained from (30) by applying an orthochronous Poincaré transformation and a constant rescaling. Thus the first excited state, a two twistor mode, describes the rigidly rotating open string. The particular null curve (30) is obtained by choosing $p = n$, and setting

$$(58) \qquad \pi^0{}_{A'} = 2^{-3/4}(+i,1)$$

$$(59) \qquad \pi^p{}_{A'} = 2^{-3/4}(-i,1)$$

$$(60) \qquad \omega_0{}^A = -U^{AA'} \pi^0{}_{A'}$$

$$(61) \qquad \omega_p{}^A = -U^{AA'} \pi^p{}_{A'}$$

where $U^a = p^{-1} y^a$ and $y^{AA'}$ is as given by (27). When $\pi^0{}_{A'} \propto \pi^p{}_{A'}$ the string degenerates to a null geodesic with a bizarre parameterization, and if $\pi^p{}_{A'} = \chi \pi^0{}_{A'}$ then $Z^\alpha_p = \chi Z^\alpha_0$.

To solve the constraints in this way when there are three or more twistors is a fairly involved procedure. Fortunately there is a more straightforward alternative. Consider again the expressions (32) and (33), giving an expansion of the curve's derivative in terms of spinors:

$$(62) \qquad \phi'_a(s) = \bar{\pi}_A(s) \pi_{A'}(s) \ ,$$

$$(63) \qquad \pi_{A'}(s) = \sum_{n=-\infty}^{\infty} \pi^n{}_{A'} \exp(ins) \ ,$$

so one can write $\phi'_a(s)$ as

$$(64) \qquad \phi'_a(s) = \sum_{p=-\infty}^{\infty} \phi^p{}_a \exp(ips).$$

The coefficients $\phi^p{}_a$ are given by

$$(65) \qquad \phi^p{}_a = \sum_{n=-\infty}^{\infty} \bar{\pi}^{n-p}{}_A \, \pi^n{}_{A'} .$$

It is trivial to integrate (64), giving

$$(66) \qquad \phi_a(s) = \gamma_a + s\phi^0{}_a + \sum_{p\neq 0} \phi^p{}_a (ip)^{-1} \exp(ips),$$

where γ_a(real) is the average of $\phi_a(s)$ on $[-\pi,\pi]$. The expression (66) gives the <u>general</u> null curve satisfying (15), where $C_a = 2\pi\phi^0{}_a$. The corresponding ω-field is found as $\omega^A = i\phi^{AA'}\pi_{A'}$. The curves correspond to the set of <u>unconstrained</u> variables $\{\gamma_a, \pi^n{}_{A'}; -\infty<n<\infty\}$. [Modulo the overall phase invariance: e.g. $\{\pi^n{}_{A'}\}$ and $\{\pi^{n+k}{}_{A'}\}$ describe the same null curve if k is a fixed integer.] Note that this approach relies on the characterization of a general null vector as a product of two spinors. For the vector to be real one chooses one spinor to be the complex conjugate of the other. With appropriate modifications to deal with the action of complex conjugation in higher dimensions, this procedure can be made to work in any dimension where null vectors are characterized as being a product of <u>unconstrained</u> spinor variables. If the spinors themselves are constrained, there is little point in trying to use them to solve the nullity conditions, which may be simpler than the spin-space constraints! Whether or not there are constraints depends on the space-time dimension. In the next section some of the possibilities will be described.

To conclude this section some of the strings arising from null curves corresponding to several twistors will be described. If one tries to solve (39)-(41) directly with three non-vanishing Z^α_n one finds inconsistencies. Except for some highly degenerate cases, 3-twistor states are not allowed. To see why, consider specifying three π-fields: $\pi^0{}_{A'}, \pi^p{}_{A'}$ and $\pi^q{}_{A'}$ $(0 < p < q)$. If one computes $\phi^a(s)$ from (66) and the $\omega_n{}^A$ from (35) and (36) one finds that in general <u>six</u> $\omega_n{}^A$ are non-zero! Specifically, if $q \neq 2p$ one finds that (generically)$\omega_n{}^A \neq 0$ for $n = 0, p, q, p+q, p-q, q-p$. In the special case $q = 2p$ five ω_n^A are non-zero, for $n = 0, p, 2p, -p, 3p$. Evidently the next open string state above the rigid rotator is a 5-twistor state. An example of such a string is obtained by writing (c.f. 58,59)

(67) $\pi^{0}{}_{A'} = 2^{-3/4}(i,0)$

(68) $\pi^{p}{}_{A'} = 2^{3/4}(-i,1)$

(69) $\pi^{2p}{}_{A'} = 2^{-3/4}(0,1)$

which generates a null curve $\phi^a(s)$ with coordinates

(70) $t = s, \qquad x = (4p)^{-1}\cos(2ps)$

(71) $y = -s/2 + (4p)^{-1}\sin(2ps),\ z = p^{-1}\sin(ps)$

Another interesting curve is obtained by regarding (67-69) as a degenerate case of the four-π set:

(72) $\pi^{0}{}_{A'} = 2^{-3/4}(i,0)$

(73) $\pi^{p}{}_{A'} = 2^{-3/4}(-i,0)$

(74) $\pi^{r}{}_{A'} = 2^{-3/4}(0,1)$

(75) $\pi^{r+p}{}_{A'} = 2^{-3/4}(0,1)$

To recover (67-69) one sets $r = p$, and the set (58-59) is obtained by setting $r = 0$ and $p = n$. The corresponding null curve is found by integrating

(76) $$\sqrt{2}\phi'^{AA'}(s) = \begin{bmatrix} 1 + \cos(ps) & -e^{-irs}\sin(ps) \\ -e^{irs}\sin(ps) & 1-\cos(ps) \end{bmatrix}$$

and if $r \neq p$, with $\gamma^a = 0$ one finds

(77) $$t = s, \qquad z = p^{-1}\sin(ps),$$
$$x = 1/2(p+r)^{-1}\cos[(p+r)s] + 1/2(p-r)^{-1}\cos[(p-r)s],$$
$$y = 1/2(p+r)^{-1}\sin[(p+r)s] - 1/2(p-r)^{-1}\sin[(p-r)s].$$

5. SOLUTIONS IN OTHER DIMENSIONS. When one considers space-time dimensions other than four one must work with more complicated spin-spaces. One must also

decide whether one is seeking a full "twistor transform" of the system or merely looking for a simple way of solving the equations. A knowledge of the former may be relied on to give the latter but may be much harder to obtain. Here the emphasis will be on just solving the equations. The twistorial aspects for low dimensions have been considered elsewhere. (For d = 3 see [10,5]; d = 4; [5,6] d = 6: [7]).

The first interesting case is actually d = 3. This is important both historically and because it serves as a model for the d = 10 case. The properties of spinors in a 3-dimensional Minkowski space have been discussed elsewhere (see e.g. [11]). There is one kind of 2-component spinor λ^A, carrying a representation of SU(1,1). Complex conjugation is represented by an involutory operation†. There is a full set (i.e., a set whose complex span is all spinors) of real spinors λ^A satisfying

(78) $$\lambda^{\dagger A} = \lambda^A .$$

The 3-metric h_{ab}, signature (+--) is represented by the spinor

(79) $$h_{AKBL} = h_{(AK)(BL)} = -1/2(\varepsilon_{AB}\varepsilon_{KL} + \varepsilon_{KB}\varepsilon_{AL}).$$

The important property of this spinor is that $h_{(AKB)L} = 0$. A real vector V^a is null if and only if its spinor equivalent $V^{AK} = V^{(AK)}$ can be written as

(80) $$V^{AK} = \lambda^A \lambda^K$$

for some real λ^A. Real strings in d=3 can therefore be obtained from null curves $\phi^a(s)$ satisfying

(81) $$\phi'^{AK}(s) = \lambda^A(s)\lambda^K(s)$$

where $\lambda^A(s)$ is periodic and real;

(82) $$\lambda^A(s) = \sum_{n=-\infty}^{\infty} \lambda^A{}_n \exp(ins); \quad \lambda^A{}_{-n} = \lambda^{\dagger A}{}_n.$$

The null curve is therefore specified by its average γ^a and the set of spinors: $\{\lambda^A{}_n | n \geq 0,\ \lambda^A{}_0 = \lambda^{\dagger A}{}_0\}$.

When d=6 the reduced (Weyl) spinors are 4-component objects Z^α and W_α (the usual twistors for complex 4-space). Vectors X^α are represented by skew 2-index spinors $X^{\alpha\beta}$ and the metric is represented by the totally skew quantity $\varepsilon_{\alpha\beta\gamma\delta}$:

$$X^a X_a = X^{\alpha\beta}X_{\alpha\beta} = 1/2\varepsilon_{\alpha\beta\gamma\delta}X^{\alpha\beta}X^{\gamma\delta}. \tag{83}$$

When the signature is (+-----) complex conjugation on spinors is represented by a non-involutory operation * such that ** = -1. In this dimension a vector V^a is null if and only if its spinor form is simple. If V^a is real also one must have

$$V^{\alpha\beta} = i\psi^{[\alpha}\psi^{*\beta]} \tag{84}$$

for some complex ψ^α. Thus the null curves are obtained by integrating

$$\phi'^{\alpha\beta}(s) = i\psi^{[\alpha}(s)\psi^{*\beta]}(s), \tag{85}$$

where

$$\psi^\alpha(s) = \sum_{n=-\infty}^{\infty} \psi_n^\alpha \exp(ins). \tag{86}$$

The free variables are then the average and the set $\{\psi_n^\alpha \; -\infty<n<\infty\}$. This construction does have an interesting twistor interpretation when the relevant twistors are the pure spinors for 8-space. (See [7].)

When d=8 this approach seems to fail, because of problems caused by the triality principle associated with 8-dimensional spinors. If one expresses a null vector as a product of spinors, then at least one of the spinors must be pure (see [9]). The condition that a spinor be pure is a condition equivalent to the original space-time nullity condition, so it seems that there is nothing to gain by working in spin space.

When d=10 one can actually avoid the purity conditions, although the pure spinors have an intimate relation with the null-cone structure. The simplifications occurs because the spinor equivalent of the space-time metric has a certain symmetry similar to that in 3 dimensions, and one can write down a simple expression for null vectors in terms of real spinors. To describe this one needs a few facts about ten-dimensional spinors, and it is convenient to review some of their properties.

The remarks made here are based on the approaches and results of Cartan [8] Kugo and Townsend [12] and Penrose and Rindler [13]. Remarks by Nillson [14] and Witten [15] were also useful in clarifying various points. The space-time metric g_{ab} has signature (+---------). The Γ-matrices are 32×32 matrices

$\Gamma^{a}{}_{\alpha}{}^{\beta}$ satisfying

(87) $$\Gamma^{a}{}_{\alpha}{}^{\beta}\Gamma^{b}{}_{\beta}{}^{\gamma} + \Gamma^{b}{}_{\alpha}{}^{\beta}\Gamma^{a}{}_{\beta}{}^{\gamma} = 2g^{ab}\delta_{\alpha}{}^{\gamma}$$

and act on 32-component Dirac spinors ψ_{α}. The space of Dirac spinors splits into two inequivalent 16-dimensional spaces;

(88) $$\psi_{\alpha} = \begin{pmatrix} \psi_A \\ \phi^A \end{pmatrix} .$$

The two reduced spin spaces $\mathbb{S}_A$ and $\mathbb{S}^A$ are naturally dual to one another, but there is no way to raise or lower an A-index. However, 32-component indices may be raised and lowered with the matrix $C^{\alpha\beta}$;

(89) $$C^{\alpha\beta} = \begin{bmatrix} 0 & \delta^A_B \\ \delta^A_B & 0 \end{bmatrix} \qquad C_{\beta\alpha} = \begin{bmatrix} 0 & \delta^A_B \\ \delta^A_B & 0 \end{bmatrix}$$

where $\hat{\psi}^{\alpha} = C^{\alpha\beta}\psi_{\beta} = (\Phi^A, \Psi_A)$ and $\psi_{\alpha} = \psi^{\beta}C_{\beta\alpha}$. The matrices $C\Gamma^a$ are symmetric. One can decompose the Γ-matrices as

(90) $$\Gamma^{a}{}_{\alpha}{}^{\beta} = \begin{bmatrix} 0 & \gamma^{a}{}_{AB} \\ \gamma_{aAB} & 0 \end{bmatrix}$$

and the $\gamma^{a}{}_{AB}$ and γ^{aAB} are symmetric, satisfying

(91) $$\gamma^{a}{}_{AB}\gamma^{bBC} + \gamma^{b}{}_{AB}\gamma^{aBC} = 2g^{ab}\delta_{A}{}^{C}.$$

In the case of 10-dimensional Minkowski space-time, it is straightforward to define the action of complex conjugation on spinors. The complex conjugate of a spinor is a spinor of the same type:

(92) $$(\psi_A)^* \in \mathbb{S}_A; \ (\psi^A)^* \in \mathbb{S}^A$$

and the operation is involutory; $** = +1$. It can be defined to be component by component complex conjugation:

(93) $$(\Psi_A)^* = \Psi^*{}_A.$$

A real (Majorana) spinor is one for which

(94) $$\Psi^*{}_A = \Psi_A$$

and similarly for ψ^A. With this choice $\gamma^a{}_{AB}$ and γ^{aAB} are real also. A consequence of the choice of (charge conjugation) matrix (89) and the conventions on complex conjugation is that for some choice of time direction t^a the matrices $\gamma^0{}_{AB} = t_a\gamma^a{}_{AB}$ and γ^{0AB} are the 16 × 16 identity matrices. More generally, in 10 dimensions, the Majorana condition on a Dirac spinor Ψ is stated as $\Psi^* = C\Gamma^0\Psi$ [12], with the matrices $C\Gamma^a$ symmetric. On real spinors, the quadratic form

(95) $$\psi^A \longrightarrow \psi^A\gamma^0_{AB}\,\psi^B$$

is then positive-definite.

By taking products of successive γ's and removing symmetrized pairs using (91) one obtains a sequence of spinor forms. Specifically,

(96) $$\gamma^{ab}{}_A{}^B = \gamma^{[a}{}_{AC}\gamma^{b]CB}$$

(97) $$\gamma^{abc}{}_{AB} = \gamma^{abc}{}_{[AB]} = \gamma^{[a}{}_{AE}\gamma^{b|EC|}\gamma^{c]}{}_{CB}$$

(98) $$\gamma^{abcd}{}_A{}^B = \gamma^{[a}{}_{AE}\gamma^{b|EC|}\gamma^c{}_{CF}\gamma^{d]FB}$$

(99) $$\gamma^{abcde}{}_{AB} = \gamma^{abcde}{}_{(AB)} = \gamma^{[a}{}_{AE}\gamma^{b|EF|}\gamma^c{}_{FG}\gamma^{d|GH|}\gamma^{e]}{}_{HB}$$

(and similarly for pairs of indices upstairs). Then [8] any <u>symmetric</u> 2-index spinor F_{AB} can be expanded as

(100) $$F_{AB} = \gamma^a{}_{AB}F_a + \gamma^{abcde}{}_{AB}F_{abcde}\,,$$

any skew 2-index spinor corresponds to a 3-form:

(101) $$H_{[AB]} = \gamma^{abc}{}_{AB}H_{abc}\,,$$

while any $F_A{}^B$ can be expanded as

(102) $$F_A{}^B = F\delta_A{}^B + F_{ab}\,\gamma^{ab}{}_A{}^B + F_{abcd}\gamma^{abcd}{}_A{}^B.$$

These forms satisfy various identities. In particular, direct application of the Clifford algebra shows that

(103) $$\gamma_m{}^{AB}\gamma^{abcde}{}_{BC}\gamma^{mCD} = 0$$

and

$$(104) \qquad \gamma^{AB}{}_{f}\gamma^{abcde}{}_{AB} = 0$$

which are useful in projecting out the vector and 5-form component of a symmetric 2-index spinor.

Given a vector V^a, one may define its spinor forms as

$$(105) \qquad V_{AB} = 1/4\gamma_{aAB}V^a$$

$$(106) \qquad V^{AB} = 1/4\gamma_a{}^{AB}V^a \ .$$

Then $V^a = 1/4\gamma^{aAB}V_{AB} = 1/4\gamma^a{}_{AB}V^{AB}$. The spinor translations of the space-time metric are

$$(107) \qquad G^{ABCD} = 1/16\gamma^{aAB}\gamma_a{}^{CD}$$

$$(108) \qquad G_{ABCD} = 1/16\gamma^a{}_{AB}\gamma_{aCD}$$

$$(109) \qquad G^{AB}{}_{CD} = 1/16\gamma^{aAB}\gamma_{aCD}$$

and these satisfy some obvious identities

$$(110) \qquad G^{ABCD} = G^{(AB)(CD)} = G^{CDAB}$$

$$(111) \qquad G^{AC}{}_{BC} = 5/8\delta^A{}_B \ ; \ G^{AB}{}_{AB} = 10$$

$$(112) \qquad G^{ABCD}G_{CDEF} = G^{AB}{}_{EF}$$

$$(113) \qquad G^{ABCD}G^{EF}{}_{CD} = G^{ABEF}$$

as well as the not so obvious symmetry conditions:

$$(114) \qquad G^{(ABC)D} = 0 = G_{(ABC)D}$$

To prove (114), let F_{CD} be an arbitrary symmetric spinor, and decompose it into a vector and a 5-form as

$$(115) \qquad F_{CD} = 1/4\gamma^a{}_{CD}V_a + \gamma^{abcde}{}_{CD}F_{abcde}.$$

Clearly,

(116) $$G^{CDPQ}F_{CD} = V^{PQ}.$$

Also,

(117) $$\gamma_g{}^{PC}F_{CD}\gamma^{gDQ} = 1/4\gamma_g{}^{PC}\gamma^a{}_{CD}\gamma^{gDQ}V_a + 0$$

so

(118) $$16G^{PCDQ}F_{CD} = -8V^{PQ}$$

and

(119) $$G^{PCDQ}F_{CD} = -1/2V^{PQ} = G^{CPDQ}F_{CD}.$$

Adding (116) to each half of (119), using the symmetries of G and F,

(120) $$[G^{CDPQ} + G^{PCDQ} + G^{DPCQ}]F_{CD} = 0,$$

which proves (114).

It is now easy to prove the following elementary result: A real vector V^a is null and future-pointing if and only if it can be written as

(121) $$V^a = 1/4\gamma^a{}_{AB}\lambda^A\lambda^B$$

where λ^A is real. Clearly, if V^a has the given form then

(122) $$g_{ab}V^aV^b = G_{ABCD}\lambda^A\lambda^B\lambda^C\lambda^D = G_{(ABC)D}\lambda^A\lambda^B\lambda^C\lambda^D = 0.$$

Conversely, if V^a is null the matrix V_{AB} is real and has rank 8 (this follows from the Clifford algebra by elementary arguments). So there is an 8 (real) dimensional family of spinors ε^A such that

(123) $$V_{AB}\varepsilon^B = 0 .$$

For any of these ε^A, let $\varepsilon^a = 1/4\gamma^a{}_{AB}\varepsilon^A\varepsilon^B$. Then ε^a is real and null and satisfies $\varepsilon_aV^a = 0$, so that $\varepsilon^a \propto V^a$. Note that $\gamma^0{}_{AB}\varepsilon^A\varepsilon^B > 0$ so ε^a is future-pointing. One can now rescale ε^A by a real scale factor μ so that $\lambda^A = \mu\varepsilon^A$ satisfies (121).

Given this characterization of a null vector one can let $\lambda^A(s)$ be a loop in spin-space and make a Fourier expansion, obtaining a null curve just as in lower dimensions. The d=10 case is seen to be very similar to the d=3 case; the metric spinor has the same crucial symmetry and the reality conditions take the same form.

As the dimension of the space-time increases the redundancy in the spinor description grows. With d=3, there is a 1-2 correspondence between null vectors and spinors. Specifically, there is a $\mathbb{Z}_2$ invariance group, since λ^A and $-\lambda^A$ define the same null vector. When d=4 there is a U(1) or S^1 invariance group because the null vector is invariant under

(124) $$\pi_A \longrightarrow \exp(i\theta)\pi_A.$$

When d=6 the invariance is U(2) or S^3: $\phi^{[\alpha}\phi^{*\beta]}$ and $\psi^{[\alpha}\psi^{*\beta]}$ define the same null vector iff

(125) $$\phi^\alpha = a\psi^\alpha + b\psi^{*\alpha}$$

where a and b are complex and $|a|^2 + |b|^2 = 1$. When d=10 the invariance is an S^7, represented by the spinors in the 8-dimensional kernel of V_{AB}, normalized by $\gamma^0{}_{AB}$.

The pattern of the redundancy is that in dimension d(d=3,4,6,10) the spinors defining the given non-zero null vector lie on a positive definite quadric in $\mathbb{R}^{d-2}-\{0\}$. If one relaxes the normalization, for example, by allowing reparameterizations of the null curves ϕ^a, which rescales the null derivative $\phi'^a(s)$, then the spinors defining the given null *direction* form a real space of dimension d-2. The geometry and interpretation of this redundancy will be considered in the next section.

6. REDUNDANCY, PURIFICATION AND SUPERSYMMETRIC SYSTEMS. The redundancy in the spinor solutions emerges because a null vector can be represented as a product of spinors in several different ways. The dimension of the redundancy increases linearly with the dimension of the space-time. In ten dimensions there is an S^7's worth of spinors λ^A generating the same null vector via the formula

(126) $$V^a = 1/4\gamma^a{}_{AB}\lambda^A\lambda^B$$

This relation is intimately related to the notion of purity in ten dimensions. A pure spinor $\pi^A \in S^A$ is one for which

(127) $$\gamma^a{}_{AB}\pi^A\pi^B = 0$$

or equivalently,

$$G_{ABCD}\pi^C\pi^D = 0. \tag{128}$$

Such a spinor defines a self-dual 5-plane which is both simple and null. The corresponding 5-form is computed from (104) and (115), the purity condition being equivalent to the 2-spinor $\pi^A\pi^B$ having no vector part. A non-zero pure spinor is essentially complex (c.f. [14]) since (127) implies that neither $Re(\pi^A)$ nor $Im(\pi^A)$ can vanish unless both are zero. Suppose one writes a non-zero pure spinor π^A as

$$\pi^A = \psi^A + i\phi^A \tag{129}$$

where ψ^A and ϕ^A are real. The real and imaginary parts of (127) give

$$\gamma^a{}_{AB}\psi^A\psi^B = \gamma^a{}_{AB}\phi^A\phi^B \tag{130}$$

$$\gamma^a{}_{AB}\psi^A\phi^B = 0 . \tag{131}$$

It follows that

$$\gamma^a{}_{AB}\pi^A\bar{\pi}^B = \gamma^a{}_{AB}\psi^A\psi^B + \gamma^a{}_{AB}\phi^A\phi^B . \tag{132}$$

Thus the real and imaginary parts of a pure spinor define the same null vector, and this vector can be expressed as $1/8\gamma^a{}_{AB}\pi^A\bar{\pi}^B$.

Now suppose that one is given a real spinor λ^A. How many different ways can this spinor be "purified"? One looks for real spinors μ^A such that the spinor $\lambda^A + i\mu^A$ is pure. Firstly, given λ^A (real) set $\Lambda^a{}_B = \gamma^a{}_{AB}\lambda^A$ and consider the equation

$$\Lambda^a{}_B\mu^B = 0 \tag{133}$$

for real μ^B. Superficially this is 10 equations in 16 unknowns. However, at most 9 of the equations are independent because of the identity

$$\lambda^C\lambda^D\gamma_{aCD}\Lambda^a{}_B = 0. \tag{134}$$

Hence the nullity of Λ is at least 7. Suppose it is $n + 1 \geq 7$ and that μ^A sovles (133). Now contract the identity $G_{A(BCD)} = 0$ with $\lambda^A\lambda^B\mu^C\mu^D$. One finds that

(135) $$G_{ABCD}\lambda^A\lambda^B\mu^C\mu^D = -2G_{ACBD}\lambda^A\mu^C\lambda^B\mu^D$$

$$= -1/8\lambda^A\mu^C\gamma_{aAC}\gamma^a{}_{BD}\lambda^B\mu^D$$

$$= 0.$$

Hence the real null vectors $\gamma^a{}_{AB}\lambda^A\lambda^B$ and $\gamma^a{}_{AB}\mu^A\mu^B$ are orthogonal and hence proportional. Since they are both future-pointing one can rescale μ^A by a real constant to make them equal. The normalization condition could be taken to be

(136) $$\gamma^0{}_{AB}\lambda^A\lambda^B = \gamma^0{}_{AB}\mu^A\mu^B.$$

Thus, for a given real λ^A, there is a family of real μ^A such that $\pi^A = \lambda^A = i\mu^A$ is pure, and this family lies on a quadric Q_n (136) of dimension $n \geq 6$ contained in the kernel of Λ, dimension $n + 1 \geq 7$. This construction realizes the space of non-zero pure spinors as $\{\mathbb{R}^{16} - \{0\}\} \times Q_n$: 16 real dimensions of real spinors, each with its own n-quadric. Since the space of pure spinors has complex dimension 11. [13] in fact n = 6. So for each real λ^A, there is a 6-(real) dimensional family of μ^A such that $\pi^A = \lambda^A + i\mu^A$ is pure. This is one short of the dimensionality of the space of spinors defining a given null vector. However, the null vector $1/8\gamma^a{}_{AB}\pi^A\bar{\pi}^B$ is also invariant under $\pi^A \longrightarrow \exp(i\theta)\pi^A$, or

(137) $$\lambda^A \longrightarrow \cos(\theta)\lambda^A - \sin(\theta)\mu^A,$$

which recovers the extra degree of freedom.

Instead of regarding the redundancy in the spinor solution as a nuisance, one might consider that the loop in spin space characterizes the dynamics of a larger system, consisting of the null curve and some extra spinor variables. Let $P^a(s)$ be the null vector field and $\lambda^A(s)$ an associated curve in spin space:

(138) $$P^a(s) = 1/4\lambda^a{}_{CD}\lambda^C(s)\lambda^D(s).$$

Now let $\theta^A(s)$ be any curve in spin space satisfying ($P_{AB} = 1/4\gamma_{aAB}P^a$)

(139) $$P_{AB}\dot{\theta}^B = 0$$

Clearly $\dot{\theta}^B \propto \lambda^B$ is a possibility. Also, for any $\varepsilon_C(s)$, if θ^B satisfies (139), so does

(140) $$\hat{\dot\theta}^B = \dot\theta^B + P^{BC}\varepsilon_C.$$

Since the nullity of P^{BC} is 8, (140) realizes the 8-dimensional family of solutions to (139). With a minor change in perspective, this 8-dimensional thickening of the null curve in spin-space is precisely its supersymmetric extension.

Given $P^a(s)$ and $\theta^B(s)$ satisfying (139), consider a space-time curve $X^a(s)$ satisfying

(141) $$\frac{dx^a}{ds} = P^a - \gamma^a{}_{AB}\theta^A\dot\theta^B \quad .$$

Writing (140) as

(142) $$\delta\dot\theta^B(s) = P^{BC}\varepsilon_C$$

induces, via (141)

(143) $$\delta\dot x^a(s) = -\gamma^a{}_{AB}\theta^A\delta\dot\theta^B.$$

If one imagines that $\theta^A(s)$ is really a fermionic curve, (142) and (143) realize, in the tangent space to the curve $\{x^a(s),\ \theta^A(s)\}$, the local supersymmetry associated with the world-line of a massless superparticle ([16]; see also [15]). The adjoinment of the variable $\theta^A(s)$ defines a supersymmetric extension of the null curve. Apart from an increase in the spinorial degrees of freedom , this is entirely analogous to the d = 4 extension discussed in [6].

Now consider how these ideas relate to twistor theory proper in ten dimensions. At one level, a twistor in any dimension can be thought of as a pair of Dirac spinors (ω,π) satisfying the conformal Killing spinor equation.

(144) $$\nabla^a\omega(x) = \Gamma^a\pi(x).$$

Many of the properties of this equation in general dimension have been discussed in [17]. In dimension ten the reduced spinor parts of (144) are

(145) $$\nabla^{CD}\omega^A(x) = G^{CDAB}\pi_B(x),$$

(146) $$\nabla_{CD}\omega_A(x) = G_{CDAB}\pi^B(x).$$

Suppose we work with (145) and define a twistor as a pair $\{\omega^A,\pi_A\}$ satisfying (145). Equivalently,

(147) $$\nabla^{(CD}\omega^{A)}(x) = 0,$$

because of the symmetries of G^{CDAB}. Application of the Ricci identity implies $\nabla_a\pi_B = 0$, and hence

(148) $$\omega^A(x) = \omega^A + X_{CD}G^{CDAB}\pi_B = \omega^A + X^{AB}\pi_B$$

where ω^A is also constant. Incidence of a space-time point with a twistor is defined, as in lower dimension, by the zeroes of $\omega^A(x)$:

(149) $$\omega^A + X^{AB}\pi_B = 0 .$$

If now X^{AB} is a null curve $X^{AB}(s)$, then $\pi_B(s)$ is to be chosen so that

(150) $$\dot{X}^{AB}(s)\pi_B(s) = 0,$$

so that $X^{AB}(s)$ is also incident with $\{\dot{\omega}^A,\dot{\pi}_B\}$:

(151) $$\dot{\omega}^A + X^{AB}\dot{\pi}_B = 0.$$

These relations are similar to those imposed when d = 4 or 6.

In ten dimensions, at this level one has a degree of choice about what type of spinors $\{\omega^A,\pi_B\}$ should be. WIth X real, one could consider choosing each of ω^A and π_B to be real, in which case (150) is realized by choosing π_B so that

(152) $$\dot{X}^{AB} = G^{ABCD}\pi_C\pi_D.$$

Alternatively, one could demand that π_B is pure, in which case (150) is realized as

(153) $$\dot{X}^{AB} = G^{ABCD}\pi_C\bar{\pi}_D.$$

Note that if π_B is pure, and ω^A satisfies (149), then the symmetries of G^{ABCD} require that ω^A is pure and that

(154) $$\omega^A\pi_A = 0.$$

Also, the choice of π pure "contains" the choice π real, by taking the real and imaginary parts of (149). Also, making π pure recovers the geometric notion of a twistor as a null simple self-dual 5-plane, and is also necessary for the existence of contour integral formulae for solutions of massless ten-dimensional wave equations.

Using the geometric notion of a twistor implies that the full twistor description of a null curve is highly complex. The curve $\{\omega^A(s), \pi_A(s)\}$ has to satisfy the purity condition, as does its derivative. The general (complex) solution to these constraints is rather more elusive than in lower dimensions. In d = 4 there are no constraints and in d = 6 the purity condition is a single scalar constraint like (154). In d = 6 this constraint is soluble for quite general null curves [7]. In dimension ten the construction given in section five gives a partial solution for a restricted type of curve (i.e., those of interest in string theory): the purification process applied to the real loop $\lambda^A(s)$. It would be interesting to obtain a more general solution in ten dimensions.

ACKNOWLEDGEMENTS. This work has profited from discussions with several people. I especially wish to thank V.W. Guillemin and L.P. Hughston.

ADDITION. After this work was completed the author learned of related work by Fairlie and Manogue [18], in which parameterizations of null curves are described in terms of properties of the Division Algebras. The dimensions of the Division Algebras are such their approach works in precisely the same dimensions as discussed here.

BIBLIOGRAPHY

1. Horowitz, G.T. Introduction to String Theories, University of California at Santa Barbara preprint, (1985). To appear in the proceedings of the School on "Topological Properties and Global Structure of Spacetime", Ettore Majorana Center for Scientific Culture.

2. Weierstrass, K. Monatsberichte der Berliner Akademie, (1866), 612-625.

3. Montcheuil, M., "Résolution de L'équation $ds^2 = dx^2 + dy^2 + dz^2$." Bulletin de la Société Mathématique de France, 33(1905), 170-171.

4. Eisenhart. L.P., "A fundamental parametric representation of space curves," Annals of Math. (Series II), XIII, (1911), 17-35.

5. Shaw, W.T., "Twistors, minimal surfaces and strings," Class. Quant. Grav. 2(1985), L113-L119.

6. Shaw, W.T., "An ambitwistor description of bosonic or supersymmetric minimal surfaces and strings in four dimensions," Class. Quant. Grav. 3 (1986), to appear.

7. Hughston, L.P. and Shaw, W.T., "Minimal curves in six dimensions," Oxford/M.I.T. preprint, (1986), Also in Twistor Newsletter no. 21. Oxford preprint, February 1986.

8. Cartan, E., "The Theory of Spinors," (1937): 1966, Dover, New York.

9. Hughston, L.P., "Applications of SO(8) Spinors," In Gravitation and Geometry, a volume in honour of I. Robinson, (ed. Rindler, W. and Trautman, A.) Bibliopolis, Naples (1986). Also in Twistor Newsletter no. 17, Oxford preprint, January 1984.

10. Hitchin, N.J., "Monopoles and Geodesics," Comm. Math. Phys., 83, (1982), 579-602.

11. Shaw, W.T., "Twistor theory and the energy-momentum and angular momentum of the gravitational field at spatial infinity," Proc. R. Soc. Lond. A390 (1983), 191-215.

12. Kugo, T., and Townsend, P. "Supersymmetry and the Division Algebras," Nucl. Phys. B221 (1983), 357-380.

13. Penrose, R. and Rindler, W., "Spinor and space-time Vol. 2," 1986. Cambridge, University Press.

14. Nillson, B.E.W., "Pure spinors as auxiliary fields in the ten-dimensional supersymmetric Yang-Mills theory," Class. Quant. Grav. 3(1986), L41-L45.

15. Witten, E. "A Twistor-like transform in Ten Dimensions," Princeton preprint, 1985. To appear in Nucl. Phys. B.

16. Brink, L. and Schwarz, J.H., "Quantum Superspace," Phys. Lett. 100B (1981), 310-312.

17. Awada, M.A., Gibbons, G.W. and Shaw, W.T., "Twistors, Conformal Supergravity and the Super BMS Group," Annals of Physics 170 (1986), 1.

18. Fairlie, D.B. and Manogue, C.A., "Lorentz invariance and the Composite String," Princeton preprint, (1986).

MATHEMATICS DEPARTMENT,
MASSACHUSETTS INSTITUTE OF TECHNOLOGY
CAMBRIDGE, MA 02139.

LIST OF PARTICIPANTS

Dean E. Allison
Department of Mathematics
Southern Illinois University at Carbondale
Carbondale, Illinois 62901

Judith M. Arms
Department of Mathematics
University of Washington
Seattle, Washington 98195

Abhay V. Ashtekar
Physics Department
Syracuse University
Syracuse, New York 13244-1130

David D. W. Bao
Department of Mathematics
University of Houston
University Park
Houston, Texas 77004

Charles P. Boyer
Department of Mathematics
Clarkson University
Potsdam, New York 13676

David E. Campbell
Department of Mathematics
University of Texas
Austin, Texas 78714

Demetrios Christodoulou
Department of Mathematics
Syracuse University
Syracuse, New York 13244-1130

Peter David D'Eath
D.A.M.T.P.
University of Cambridge
Cambridge, England

Arthur Fischer
Department of Mathematics
University of California
Santa Cruz, California 95064

Frank Flaherty
Department of Mathematics
Oregon State University
Corvallis, Oregon 97331

John Friedman
Department of Physics
University of Wisconsin at Milwaukee
Milwaukee, Wisconsin 53201

Gregory Galloway
Department of Mathematics
University of Miami
Coral Gables, Florida 33124

David Garfinkle
Department of Physics
Washington University
St. Louis, Missouri 63130

Richard S. Hamilton
Department of Mathematics
University of California
San Diego, California 92109

Steven Harris
Department of Mathematics
Oregon State University
Corvallis, Oregon 97331

William A. Hiscock
Department of Physics
Montana State University
Bozeman, Montana 59717

Darryl D. Holm
T-7, MS B 284 Mathematical
Modeling
Los Alamos National Lab.
Los Alamos, New Mexico 87545

Gary T. Horowitz
Department of Physics
University of California
Santa Barbara, California 93106

James A. Isenberg
Department of Mathematics
University of Oregon
Eugene, Oregon 97403

Theodore A. Jacobson
Department of Physics
University of California
Santa Barbara, California 93106

Ronald J. Knill
Department of Mathematics
Tulane University
New Orleans, Louisiana 70118

Sergiu Klainerman
Department of Mathematics
New York University
New York, New York 10012

Marek B. Kossowski
Department of Mathematics
Rice University
Houston, Texas 77001

Rabinder K. Koul
Department of Physics
Syracuse University
Syracuse, New York 13244-1130

Karel V. Kuchar
Department of Physics
University of Utah
Salt Lake City, Utah 84112

Demir N. Kupeli
Department of Mathematics
University of Alabama at
Birmingham
Birmingham, Alabama 35294

Lee A. Lindblom
Department of Physics
Montana State University
Bozeman, Montana 59715

Anne M. Magnon
Department of Physics
Syracuse University
Syracuse, New York 13210

Jerrold E. Marsden
Department of Mathematics
University of California
Berkeley, California 94720

Jonathan W. Morrow-Jones
Department of Physics
University of California
Santa Barbara, California
93106

Robert J. McKellar
Department of Mathematics
and Statistics
University of New Brunswick
Fredericton, New Brunswick
Canada E3B 5A3

Phillip E. Parker
Department of Mathematics
Wichita State University
Wichita, Kansas 67208

Roger Penrose
Mathematical Institute
Oxford University
Oxford, OX1-3LB
United Kingdom

Jung S. Rno
Department of Physics
Univerity of Cincinnati
Cincinnati, Ohio 45236

Kristin Schleich
Department of Physics
University of California
Santa Barbara, California 93106

Richard M. Schoen
Department of Mathematics
University of California at
San Diego
La Jolla, California 92093

William T. Shaw
Department of Mathematics
Massachusetts Institute of
Technology
Cambridge, Massachusetts 02139

Bonnie J. Shulman
Department of Mathematics
and Physics
University of Colorado
Boulder, Colorado 86303

Lee Smolin
Department of Physics
Yale University
New Haven, Connecticut 06511

Rafael D. Sorkin
Department of Physics
Syracuse University
Syracuse, New York 13210

Victor Szczyrba
Department of Physics
Yale University
New Haven, Connecticut 06511

Jennie Traschen
EFI
University of Chicago
Chicago, Illinois 60637

Albert L. Vitter III
Department of Mathematics
Tulane University
New Orleans, Louisiana 70118

Robert M. Wald
Enrico Fermi Institute
University of Chicago
Chicago, Illinois 60637

Don. C. Wilbour
Department of Mathematics
University of Washington
Seattle, Washington 98105

Donald M. Witt
Department of Physics
University of Wisconsin
at Milwaukee
Milwaukee, Wisconsin 53201

Philip B. Yasskin
Department of Mathematics
Texas A&M University
College Station, Texas 77843

S. T. Yau
Department of Mathematics
University of California
at San Diego
La Jolla, California 92093

Ulvi H. Yurtsever
Department of Theoretical
Physics
California Institute of
Technology
Pasadena, California 91125

ABCDEFGHIJ — 898